BAYESIAN INFERENCE

BAYESIAN
INFERENCE
WITH ECOLOGICAL
APPLICATIONS

———

WILLIAM A. LINK

USGS Patuxent Wildlife Research Center, USA

RICHARD J. BARKER

*Department of Mathematics and Statistics, University of Otago,
New Zealand*

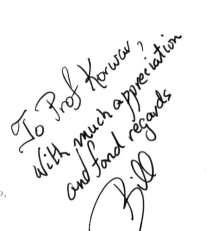

To Prof Korwar?
With much appreciation
and fond regards
Bill

ELSEVIER

AMSTERDAM • BOSTON • HEIDELBERG • LONDON
NEW YORK • OXFORD • PARIS • SAN DIEGO
SAN FRANCISCO • SINGAPORE • SYDNEY • TOKYO

Academic Press is an imprint of Elsevier

Academic Press is an imprint of Elsevier
32 Jamestown Road, London, NW1 7BY, UK
30 Corporate Drive, Suite 400, Burlington, MA 01803, USA
525 B Street, Suite 1900, San Diego, California 92101-4495, USA

First edition 2010

Library of Congress Cataloging-in-Publication Data
A catalog record for this book is available from the Library of Congress

British Library Cataloguing-in-Publication Data
A catalogue record for this book is available from the British Library

ISBN: 978-0-12-374854-6

For information on all Academic Press publications
visit our website at www.elsevierdirect.com

Typeset by: diacriTech, India

Printed and bound in China
10 11 12 13   10 9 8 7 6 5 4 3 2 1

Short Contents

PART I
PROBABILITY AND INFERENCE

1. Introduction to Bayesian Inference   3
2. Probability   13
3. Statistical Inference   23
4. Calculating Posterior Distributions   47

PART II
THE BAYESIAN MĀRAMATANGA

5. Bayesian Prediction   77
6. Priors   109
7. Multimodel Inference   127

PART III
APPLICATIONS

8. Hidden Data Models   163
9. Closed-Population Mark-Recapture Models   201
10. Latent Multinomial Models   225
11. Open Population Models   239
12. Individual Fitness   271
13. Autoregressive Smoothing   287

PART IV
APPENDICES

A. Probability Rules   301
B. Probability Distributions   307

Full Contents

Preface xi
Acknowledgments xiii

PART I
PROBABILITY AND INFERENCE

1. Introduction to Bayesian Inference 3

1.1 Introduction 3
1.2 Thomas Bayes 4
1.3 The Bayesian/Frequentist Difference 7
1.4 Understanding the Basis of
     Thomas Bayes' System of Inference 8

2. Probability 13

2.1 What is Probability? 14
2.2 Basic Probability Concepts for
     Bayesian Inference 16

3. Statistical Inference 23

3.1 Likelihood 25
3.2 Confidence Intervals 29
3.3 Bayesian Interval Estimation 36
3.4 Summary and Comparison of
     Inferential Systems 42

4. Calculating Posterior Distributions 47

4.1 Conjugacy 49
4.2 Monte Carlo Methods 54
4.3 Markov Chain Monte Carlo 59

PART II
THE BAYESIAN MĀRAMATANGA

5. Bayesian Prediction 77

5.1 The Bayesian Māramatanga 78
5.2 Estimating Mean Residual Lifetime 81
5.3 Derived Parameters in Summary
     Analyses: BBS Case Study 85
5.4 Derived Parameters and Out of Sample
     Inference in a Dose–Response Study 93
5.5 Product Binomial Representation
     of the CJS Model 98
5.6 Posterior Predictive Model
     Checking 104

6. Priors 109

6.1 An Example where Priors Matter 110
6.2 Objective Bayesian Inference 115
6.3 Afterword 123

7. Multimodel Inference   127

7.1   The BMI Model   129
7.2   Bayes Factors   134
7.3   Multimodel Computation   139
7.4   Indices of Model Acceptability:
      AIC and DIC   149
7.5   Afterword   157

PART III

APPLICATIONS

8. Hidden Data Models   163

8.1   Complete Data Likelihood   164
8.2   Randomized Response Data   165
8.3   Occupancy Models as
      Hierarchical Logistic Regressions   176
8.4   Distance Sampling   181
8.5   Finite Population Sampling   186
8.6   Afterword   199

9. Closed-Population Mark-Recapture Models   201

9.1   Introduction   201
9.2   Mark-Recapture Models and
      Missing Data   202
9.3   Time and Behavior Models   204
9.4   Individual Heterogeneity Models   209
9.5   Example: Koalas   218
9.6   Afterword   223

10. Latent Multinomial Models   225

10.1   Model M_t 226
10.2   Model $M_{t,\alpha}$ 226
10.3   Gibbs Sampling for Model $M_{t,\alpha}$ 232
10.4   An Implementation of Model $M_{t,\alpha}$ 234
10.5   Extensions   236

11. Open Population Models   239

11.1   Continuous-Time Survival
       Models   240
11.2   Open Population Mark-Recapture –
       Band-Recovery Models   244
11.3   Open Population Mark-Recapture –
       The CJS Model   247
11.4   Full Open Population
       Modeling Formalities   252
11.5   CMSA Model and Extensions   255
11.6   Multiple Groups   260
11.7   Robust Design   261
11.8   Multistate Models and Other
       Extensions of the CJS Model   268
11.9   Afterword   269

12. Individual Fitness   271

12.1   Population Fitness   272
12.2   Individual Fitness   273
12.3   Realized Individual Fitness   274
12.4   Individual Fitness in Group Context   275
12.5   Analysis of Individual Fitness:
       An Example   276
12.6   Population Summaries of Fitness   281
12.7   Afterword   286

13. Autoregressive Smoothing   287

13.1   Dove Data and Preliminary Analyses   288
13.2   Modeling Differences
       in Parameter Values   291
13.3   Results for Dove Analysis   292
13.4   Higher Order Differences   293
13.5   Afterword   295

PART IV

APPENDICES

A. Probability Rules   301

A.1   Properties of Moments   301
A.2   Conditional Expectations   303
A.3   Independence and Correlation   304

B. Probability Distributions   307

B.1   Uniform Distribution   307
B.2   Discrete Uniform Distribution   308

B.3   Normal Distribution   308
B.4   Multivariate Normal Distribution   309
B.5   Bernoulli Trials and the
       Binomial Distribution   311
B.6   Poisson Distribution   312
B.7   Multinomial Distribution   313
B.8   Exponential Distribution   315
B.9   Gamma Distribution   317

B.10   Beta Distribution   319
B.11 t-Distribution   320
B.12   Negative Binomial Distribution   321

Bibliography   323

Index of Examples   331

Index   333

Preface

This book is written for scientists involved in the analysis and interpretation of data. It is written for a fairly general audience, though some knowledge of calculus is assumed, as also some familiarity with basic statistical methods. There are occasional references to linear algebra, which we hope will be useful for some while not off-putting to others.

The book is organized into 3 sections. The first section includes reviews of basic concepts of probability, comparisons of Bayesian and frequentist approaches to inference, and an introduction to Bayesian computation. We encourage those that might be inclined to skip over this section to spend a moment with Sections 1.1-1.3, the introductory material in Chapter 2, and Section 3.4; these are our sales pitch, briefly summarizing the philosophical background and appeal of Bayesian inference.

Both authors were initially attracted to Bayesian statistics because of its usefulness, and later became convinced of its simplicity and beauty; we hope that this will be the experience of our readers. Section 2 highlights the ease with which posterior distributions are used for inference, emphasizing the unified approach to inference for all unknown quantities, be they parameters, summaries of collections of parameters, missing values, future observables, hypothetical replicates, or even mathematical models.

Section 3 has a sampling of applications, primarily based on our experience working with ecologists and wildlife biologists. This section emphasizes the importance of the 'complete data likelihood', describing 'the data we wish we had' as the starting point of hierarchical modeling. We regard the introductory material in Chapter 8 and Sections 8.1-8.2 as especially important in providing a simple example of general interest to illustrate the principles applied in this section.

The book includes several dozen 'panels' of *WinBUGS* code. These and related data files are available for download from our website (www.maramatanga.com) along with a list of errata. We have checked all of the code repeatedly, and are optimistic that there aren't too many mistakes. We appreciate your input and comments; these can be sent to wlink@usgs.gov or rbarker@maths .otago.ac.nz.

Use of trade, product, or firm names does not imply endorsement by the U.S. Government.

W.A. Link
R.J. Barker
15 May 2009

Acknowledgments

We are indebted to many colleagues who have both encouraged and supported our endeavor in writing this book. Deserving of special mention is Jim Nichols, mentor and friend to a generation of quantitative biologists and ecological statisticians. For their enthusiastic contributions to debates on statistical inference, we thank Andy Royle and Matt Schofield. Reviews of earlier drafts of this material were also provided by Emmanuelle Cam, Mike Conroy, Evan Cooch, Chris Fonnesbeck, and John Shanks.

For their forbearance and love, not to mention the countless cups of coffee which lubricate progress, we dedicate this book to:

Carol and Lynnette

PROBABILITY AND INFERENCE

CHAPTER

1

Introduction to Bayesian Inference

OUTLINE

1.1 Introduction   3
1.2 Thomas Bayes   4
    1.2.1 The Doctrine of Chances   5

1.3 The Bayesian/Frequentist
    Difference   7

1.4 Understanding the Basis of
    Thomas Bayes' System of
    Inference   8
    1.4.1 Conditional Probability   8
    1.4.2 Bayes' Theorem   9

1.1 INTRODUCTION

This book is about the Bayesian approach to statistics. We wrote it having worked for many years with a cadre of topnotch ecologists and wildlife biologists. Our experience has been that the most successful have been those that take a lively interest in the modeling and analysis of their data. We wrote this book with such folks in mind. Although most of the examples we present will be of special interest to biologists working in the fields of ecology, wildlife management, and environmental studies, we are confident that the book's theme will be of interest to other scientists, researchers with the same enthusiasm for doing the best possible job of analyzing their data.

The training of such researchers has traditionally included a course or two in statistics, presenting the basic ideas of estimation and hypothesis testing under what is commonly known as the "classical" or frequentist paradigm. There is much to be learned: estimation, with its assortment of optimality criteria; hypothesis testing, with all of the complications of multiple comparisons and model selection; linear models in their various incarnations (regression, anova, ancova, etc.), nonparametric methods, experimental design, asymptotic theory. It is natural that biologists, having acquired a large toolbox of frequentist statistical methods, should tend toward a skeptical view of Bayesian statistics. Doesn't embracing the "new" philosophy mean discarding the old and familiar tools, painstakingly acquired, and which (after all) work well enough? Why bother with Bayesianism? Isn't the approach philosophically flawed? Isn't it true that "Bayesian methods constitute a radically different way of doing science ... [and that] Bayesians categorically reject various tenets of statistics and the scientific method that are currently widely accepted in ecology and other sciences" (Dennis, 1996). Isn't the field of Bayesian inference an esoteric quagmire?

"No one after drinking old wine desires new; for he says, 'The old is good.'"
(Luke 5:39, RSV)

Our answer is that the Bayesian approach is attractive because it is useful. Its usefulness derives in large measure from its simplicity. Its simplicity allows the investigation of far more complex models than can be handled by the tools in the classical toolbox.

There are interesting philosophical issues to discuss in comparing frequentist and Bayesian approaches; some of these require more than a superficial understanding of mathematics and probability theory. It is unfortunate that the philosophical discussion is often pushed to the fore, its complexity scaring the classically trained researcher back into the fold, without perhaps a fair exposure to Bayesian methods. The issues are worth addressing and are addressed in this text. But our priority in writing this text is utilitarian. The authors were first attracted to Bayesian inference because of its usefulness, and gradually became convinced of its beauty, as well; we hope that this will be the experience of our readers.

1.2 THOMAS BAYES

Little is known of the man who first described the mathematical form of reasoning eponymically known as "Bayesian" inference.[1] Thomas Bayes was born in 1701 (or 1702) in London, England (or perhaps, Hertfordshire). He was admitted to the University of Edinburgh in 1719, where he studied logic and theology. Like his father, he was a nonconformist minister (i.e., Protestant but not Anglican); he is included in a 1727 listing of "Approved Ministers of the Presbyterian Denomination."

Bayes was elected a Fellow of the Royal Society in 1742, having been recommended as "a Gentleman of known merit, well skilled in Geometry and all parts of Mathematical and Philosophical Learning." It is thought that this nomination came in consequence of his "*An Introduction to the Doctrine of Fluxions, and a Defence of the Mathematicians Against the Objections of the Author of the Analyst*" (1736), in which he defended Isaac Newton against the Bishop Berkeley, who had referred to Newton as an "infidel Mathematician."

Apart from his defense of Newton and a well-received contribution to a debate on the purpose of God in creation ("*Divine Benevolence, or an Attempt to Prove That the Principal End of the Divine Providence and Government is the Happiness of His Creatures*", 1731) no other works of Bayes' are known to have been published during his lifetime.

1. "Eponymous" is a good word, like supercalifragilisticexpialidocious – if you say it loud enough, you'll always sound precocious. Bayesian writers seem unable to resist it.

1.2.1 The Doctrine of Chances

The work which preserved Bayes' place in history is *"An Essay towards solving a Problem in the Doctrine of Chances"* published posthumously (1763) through the good offices of Bayes' friend Richard Price. The essay begins as follows:

> *"Given* the number of times in which an unknown event has happened and failed: *Required* the chance that the probability of its happening in a single trial lies somewhere between any two degrees of probability that can be named."

Bayes' "unknown event" is what we would today call a Bernoulli trial: a random event with two possible outcomes, labeled "success" and "failure." The flip of a coin, an attempt to guess the correct answer on a multiple choice exam, whether a newly hatched chick survives to fledging; each is a Bernoulli trial.

The "probability of the event happening in a single trial" is referred to as the success parameter of the Bernoulli trial, and denoted by p. Sometimes, the nature of the trial leads to a specification of p. A coin is deemed "fair" if "Heads" is as likely as "Tails", so that the probability of getting "Heads" is $p = 1/2$. On a multiple choice exam with five options per question, we might assume that the probability of guessing correctly is $p = 1/3$ if we could rule out two of the options and were forced to guess among the remaining three. In both cases, the nature of the event leads to a specification for p.

More frequently, however, there is no basis for knowing p. Indeed, even if we had a basis for an educated guess, we might wish to put aside whatever prejudices we have, and allow the data to speak for themselves. Suppose that having observed 10 newly hatched chicks, we note that only four survive to fledging. Bayes' essay seeks means to translate such data into conclusions like "The probability that p is between 0.23 and 0.59 is 0.80." Given "two degrees of probability," say a and b, the problem is to determine $\Pr(a \leq p \leq b)$, i.e., the probability that p is in the interval from a to b.

For those familiar with such things, it may sound as though Bayes were trying to construct a confidence interval for p. Doubtless, Bayes' goal was the same as that of the frequentist statistician who computes a confidence interval: to quantify uncertainty about an unknown quantity, using data and probabilities. As we shall see, however, Bayes' way of accounting for uncertainty about p is quite different from the frequentist approach, which developed later. The difference lies in the Bayes' use of the term "probability." We illustrate with a simple example.

The authors of this text have a deformed New Zealand 50 cent piece; it is convex, rather than flat, with the Queen's image on the side that bulges outward (see Fig. 1.1). Consider the event that the coin lands, when flipped, with the Queen's image facing upward ("Heads"); the unknown quantity of interest is $p = \Pr(\text{Heads})$.[2] The convexity of the coin raises doubts about whether $p = \frac{1}{2}$.

---

2. There is an implicit *ceteris paribus* in defining the event, as also in Bayes' essay: individual flips of the coin are replicates, in the sense that conditions are assumed to be relatively constant with regard to the outcome. For instance, the authors have observed that p appears to depend on the hardness of the surface on which the coin lands. Speaking entirely in terms of the coin, and not in terms of a mathematical model, it seems impossible to explicitly define what we mean by "replicates," and whatever ambiguity remains in that definition is carried over into our definition of the probability p.

FIGURE 1.1   New Zealand coins including a deformed 50¢ piece.

What would the reader guess? Will "Heads" prevail ($p > 1/2$), will the coin tend to land, saucer-like, with the Queen's image down ($p < 1/2$)? or is the coin fair, despite being bent? We asked a class of 28 students: 15 said $p < 1/2$, 6 said $p = 1/2$, and 7 said $p > 1/2$. Guesses ranged from 0.10 to 0.75, with mean of 0.46 and standard deviation of 0.16.

We flipped the coin 20 times, and observed 17 Heads. Given these data, we are inclined to believe that $p > 0.50$, probably even greater than 0.70. The methods outlined in Bayes' essay lead us to quantify our uncertainty about p by the following statements:

"the odds are 30:1 in favor of p being at least 0.65"

or

"the probability that p is less than 0.5 is less than 0.001"

or

"there is a 95% chance that $0.660 \leq p \leq 0.959$."

Our impression is that most people feel comfortable with statements like these; their sense is evident enough, even if one does not know the mathematics behind their computation.

Indeed, their flavor is so consistent with colloquial usage, that it takes some thought to understand why they are rejected as nonsense under the frequentist paradigm. The frequentist view is that p is a fixed quantity, and that consequently it makes no sense to make probability statements about it. The value of p is either in the interval (0.660, 0.959), or it is not. It makes no more sense, from the frequentist perspective, to assign a probability to the statement $0.660 < p < 0.959$, than it would for a primitive mathematician to say that $\Pr(3.0 < \pi < 3.2) = 0.95$, where

π is the ratio of the circumference to the diameter of a circle. The unknown probability p, like π, Avogadro's number and the speed of light, is a fixed quantity: it is a fixed characteristic of the bent coin; there is nothing random about it, and thus there is no sense in assigning probabilities to whether or not it lies in some particular interval.

1.3 THE BAYESIAN/FREQUENTIST DIFFERENCE

The essential difference between Bayesian and frequentist philosophies is in the use of the term "probability." Frequentists restrict the application of the term probability to summaries of hypothetical replicate data sets, whereas Bayesians use probability to describe all unknown quantities. We will discuss the distinction in more detail subsequently, noting for now that there is no "right" or "wrong" in the choice between the two paradigms: the distinction is purely a matter of definitions. If there is any special virtue in Bayes' usage of the term "probability," it is that it appears to be more consonant with informal use of the concept than the restrictive sense required by frequentists, with the consequence that Bayesian statistics may be more easily intelligible than corresponding frequentist statistics. For example, before presenting the Bayesian summaries of our 20 coin flips, we wrote that the data indicate that p is "probably even greater than 0.70" – a statement which makes no sense, for frequentists, but is completely consistent with informal usage, and formalized in the Bayesian setting.

Bayes' goal was to present methods for quantifying uncertainty about p. It is worth noting that while a frequentist might not be satisfied with Bayes' statement of the problem, or the method of solution he proposed, the frequentist's solution results in numbers quite similar to those produced using Bayes' approach. Without stopping now to describe its development, the frequentist "exact 95% confidence interval" for p based on our data is [0.621, 0.968]. This interval is not created to address Bayes' question, but as a statement of reasonable bounds on the unknown value p. It is the product of statistical machinery that is 95% effective in similar circumstances. By this we mean that if our 20 flips of the bent coin were replicated by a large number of statisticians, and each computed an exact 95% confidence interval, then at least 95% of the intervals produced would include the true value of p. As for the particular interval [0.621, 0.968], we cannot know whether p is included or not. Our "confidence" in the interval arises from the typical performance of the machinery that produced it.

Bayes' method yields the interval [0.660, 0.959]; frequentist methods yield the interval [0.621, 0.968]. Given the similarity of these intervals, wrangling over the differences of interpretation might seem a tempest in a teapot. Indeed, for most simple problems, given adequate data, Bayesians and frequentists reach similar conclusions. So why bother with Bayes?

Here's why:

1. **Simplicity**: The approach Bayes developed extends in a straightforward fashion to analysis of far more complex data and models, including cases where no frequentist method exists. It is particularly appropriate for hierarchical models.
2. **Exactness**: Bayes' approach provides mathematically sensible methods of analysis without the need for asymptotic approximations, and a precise inferential system even when dealing with small sample sizes and limited data.

3. **Coherency**: Bayesian inference is a self-consistent and qualitatively simple system of reasoning. All unknown quantities – be they parameters, latent variables, or predictions – are treated as random variables. Existing knowledge about unknown quantities is summarized and explicit mathematical expression by means of probability distributions is given. Consequently, Bayesian inference provides a formal mechanism for incorporating and updating prior knowledge, and a proper accounting for all sources of uncertainty.

1.4 UNDERSTANDING THE BASIS OF THOMAS BAYES' SYSTEM OF INFERENCE

Bayes' essay on the *Problem in the Doctrine of Chances* is not the easiest reading for the modern reader. Many of the refinements of modern mathematical notation and terminology were not available prior to his death in 1761. In addition, Bayes put considerable effort into computational aspects of the problem, a major obstacle in his time, but a triviality in the era of fast computers. The most important part of his essay, however, is its contribution to the problem of reasoning from data to causes, and here the concepts are readily intelligible.

The basic tool of Bayes' inference is the theorem that bears his name. Bayes' theorem is a simple statement of facts about probability, so simple that one wonders how many contemporary statisticians harbor a vague sense of having been cheated by fate: "if I'd only been born several centuries earlier, it could have been 'Smith's Theorem.' " The theorem is not remarkable in itself. The remarkable (and controversial) aspect of Bayes' work was his application of the theorem to what is now known as statistical inference.[3]

Thus, study of Bayesian inference begins with consideration of conditional probability and Bayes' Theorem, which we now describe.

1.4.1 Conditional Probability

Table 1.1 summarizes data on 238 fishing vessels trawling primarily for squid and shrimp off the coast of New Zealand during the 1987/88 through 1995/96 fishing seasons. New Zealand government officials present on these vessels reported accidental bycatch of 45 rare and endangered New Zealand sea lions (*Phocarctos hookeri*), with 41 mortalities. Table 1.1 is a classification of the vessels by nation of origin and the incidence of bycatch.[4]

Suppose that we have a listing of the 238 vessels, and that a record has been sampled at random. The probability that the selected record corresponds to a Russian vessel is $0.517 = 123/238$. Fair odds for a wager would be in the ratio 123:115, that is, 1.07:1. If you were betting on the vessel being Russian, fairness would require that you wager $1.07 against my dollar.

---

3. Bayes might not have been the first to propose this application of the theorem. Stigler (1983) points out that David Hartley's *Observations on Man* (1749) includes a statement of the same problem as Bayes considered, with the same solution, attributed to "an ingenious friend." Stigler's entertaining paper suggests that the ingenious friend was *not* Thomas Bayes.
4. Vessels' nation of origin were associated with a variety of potential explanatory variables (see Manly *et al.*, 1997, for a detailed analysis).

TABLE 1.1 Cross-classification of vessels fishing off the coast of New Zealand, 1987/88 through 1995/96 seasons, by country of origin and accidental sea lion bycatch.

	Japan	NZ	Russia	Total
No Bycatch	18	90	100	208
Bycatch	1	6	23	30
Total	19	96	123	238
Pr(Nation)	0.080	0.403	0.517	
Pr(Bycatch|Nation)	0.053	0.063	0.187	
Pr(Nation|Bycatch)	0.033	0.200	0.767	
Odds Ratio	0.397	0.370	3.072	

Now suppose that as the bet is being arranged, we are informed that the sampled record was for a vessel that had accidental bycatch. We know that roughly 5% (1/19) of the Japanese and 6% (6/96) of the New Zealand vessels had accidental bycatch, but that nearly 19% (23/123) of the Russian vessels had accidental bycatch. The additional information about bycatch makes the wager more appealing to you, enhancing the chance that the selected vessel is Russian. Bayes' theorem describes the way in which this additional information modifies the probabilities for our wager.

The national bycatch rates are summarized in Table 1.1 by the row labeled "Pr(Bycatch| Nation)." We write $Pr(A|B)$ as shorthand for "the probability of A given B." These numbers are *conditional probabilities*, reflecting that they have been calculated subject to specified restrictions. Thus, $Pr(Bycatch|Russian) = 0.187$ was calculated by restricting our attention entirely to the 123 Russian vessels, and paying no regard to the remaining 115 records.

Fairness dictates that the stakes of our wager be determined in light of the information that the sampled vessel had accidental bycatch. The probability that the sampled vessel is Russian, given that it had bycatch, is $Pr(Russian|Bycatch) = 0.767$ (23/30); fair odds are in the proportion 23:7, or 3.29:1; you should wager $3.29 against my dollar. Note that $Pr(Bycatch| Russian)$ is an entirely different quantity than $Pr(Russian|Bycatch)$.

The effect of the additional information (Bycatch) was to change the probability that the vessel is Russian from 0.517 to 0.767, and the odds from 123/115 to 23/7. The effect is summarized by an *odds ratio* $(23/7)/(123/115) = 3.07$. This is the factor by which you must increase your wager to maintain a fair bet, in light of the additional knowledge. Were you wagering on the vessel being of Japanese rather than Russian origin, the odds ratio would be less than 1 (see Table 1.1 , final row); fairness would dictate that your wager be reduced by 60%, because the additional information augurs against you.

1.4.2 Bayes' Theorem

A *joint probability* is the probability that a list of events occurs. We write $Pr(B, A)$ for the joint probability of events B and A.

The relation between joint probabilities $\Pr(B,A)$ and conditional probabilities $\Pr(B|A)$ is the basis of Bayes' theorem. Consulting Table 1.1, we see that the probability a randomly selected vessel is Russian *and* has bycatch is $23/238$, while the probability the selected vessel is Russian *given that* it has bycatch is $23/30$. These stand in the ratio

$$\frac{\Pr(\text{Russian, Bycatch})}{\Pr(\text{Russian}|\text{Bycatch})} = \frac{\frac{23}{238}}{\frac{23}{30}} = \frac{30}{238};$$

not coincidentally, the ratio is the probability a randomly selected vessel has bycatch.

Rearranging terms, we have

$$\Pr(\text{Russian, Bycatch}) = \Pr(\text{Russian}|\text{Bycatch})\,\Pr(\text{Bycatch}).$$

This illustrates the basic relation

$$\Pr(B,A) = \Pr(B|A)\,\Pr(A). \tag{1.1}$$

Bayes' theorem is easily established by fiddling around with Eq. (1.1). First, by simply switching the roles of A and B, we have

$$\Pr(A,B) = \Pr(A|B)\,\Pr(B). \tag{1.2}$$

Then, since $\Pr(B,A)$ is the same as $\Pr(A,B)$, we may substitute the right-hand side of Eq. (1.1) in the left-hand side of Eq. (1.2), obtaining

$$\Pr(B|A)\,\Pr(A) = \Pr(A|B)\,\Pr(B);$$

finally, we divide both sides by $\Pr(A)$, obtaining

$$\Pr(B|A) = \frac{\Pr(A|B)\,\Pr(B)}{\Pr(A)}. \tag{1.3}$$

Now suppose that instead of a single event B, we have a set $B_j, j = 1, 2, \ldots, k$ of *mutually exclusive* and *exhaustive* events: only one of them can happen at a time, and one or another of them must happen. For the sea lion bycatch example, $k = 3$ and $B_j = $ Japan, New Zealand, or Russia, as $j = 1, 2,$ or 3. The events B_j being mutually exclusive and exhaustive, it follows from the basic notions of probability and (1.2) that

$$\Pr(A) = \sum_j \Pr(A, B_j) = \sum_j \Pr(A|B_j)\,\Pr(B_j). \tag{1.4}$$

Bayes' theorem is established by substituting a specific event B_i for B in (1.3), and then substituting (1.4) in the denominator of the right-hand side of (1.3). Thus Bayes' theorem is that

$$\Pr(B_i|A) = \frac{\Pr(A|B_i)\,\Pr(B_i)}{\sum_j \Pr(A|B_j)\,\Pr(B_j)}. \tag{1.5}$$

Bayes' Theorem as a Formal Mechanism for Inference

It is worthwhile taking a step back from this imposing looking formula, having a good squint at it, and considering how it can be used to formalize the process of inference.

The "mutually exclusive and exhaustive" set of events $B_j, j = 1, 2, \ldots, k$ can be thought of as states of nature, alternative explanations of how things are. A is an event, the probability of which depends on the state of nature, quantified by $\Pr(A|B_j)$. The probability of state B_j, assessed without knowledge of the event A is $\Pr(B_j)$.

Bayes' theorem gives a recipe for calculating the probability of a particular state of nature, B_i, in light of the observation A. One simply calculates the product

$$\Pr(A|B_i)\Pr(B_i)$$

and divides this product by the sum of corresponding products for all possible states of nature. Consequently, Eq. (1.5) is often expressed as

$$\Pr(B_i|A) \propto \Pr(A|B_i)\Pr(B_i),$$

the symbol "$\propto$" being read as "is proportional to."

Consider once again the bent coin previously described. Suppose that somehow we knew that $p = \Pr(\text{Heads})$ was either 0.55 or 0.90. Imagine, for instance, that we had two coins manufactured to physical specifications guaranteeing these success rates, that one had been received, without labeling, and to determine which coin we had received, we flipped the coin 20 times, observing the event $A = $ "heads on exactly 17 flips."

Using standard probability calculations (discussed in more detail in Chapter 2), we find $\Pr(A|p=0.55) = 0.0040$, and $\Pr(A|p=0.90) = 0.1901$, corresponding to the two possible states of nature. Following are two applications of Bayes' theorem for inferring which of the states of nature prevails.

Scenario 1   On receipt of the coin, our uncertainty as to which we had received might well be expressed by supposing that $\Pr(p=0.55) = \frac{1}{2}$. There being no reason to suppose one coin or the other was shipped earlier, or was transported more rapidly, we model the processes determining which arrived first as equivalent to the flip of a fair coin. We then calculate the values

$$\Pr(A|p=0.55)\Pr(p=0.55) = 0.0040 \times \frac{1}{2} = 0.0020,$$

and

$$\Pr(A|p=0.90)\Pr(p=0.90) = 0.1901 \times \frac{1}{2} = 0.0951;$$

and conclude by means of Eq. (1.5) that

$$\Pr(p=0.55|A) = \frac{0.0020}{0.0020+0.0951} = 0.0206.$$

Thus we could say that, given our observation of 17 heads in 20 flips, the odds were greater than 47:1 in favor of this being the coin with $p=0.90$.

Scenario 2   Bayes' theorem can be useful to evaluate perceptions of the probable state of nature, whether we endorse them or not. Suppose that we contact the shipping clerk at the company which manufactured our bent coin, who says "I'm 95% sure I sent the coin with $p=0.55$." Interpreting this as an assertion that $\Pr(p=0.55) = 0.95$, we calculate

$$\Pr(p=0.55|A) = \frac{0.0040(0.95)}{0.1901(0.05)+0.0040(0.95)} = 0.2859,$$

and respond, "Even so, given our observation, the odds would be better than 2.5:1 against this being the coin with $p = 0.55$."

These calculations for the bent coin and the sea lion example illustrate the use of Bayes' theorem as a tool for inference. We begin with a probabilistic statement of uncertainty about states of nature (the nation of origin of a randomly selected fishing vessel, the success parameter $p = \Pr(\text{Heads})$ for the bent coin). Next, we describe the probabilities of outcomes (bycatch, number of "heads" in n tosses) as determined by the various states of nature. Bayes' theorem is employed to revise our statement of uncertainty about the state of nature, as informed by an observed outcome.

The examples given so far all involve the use of Bayes' theorem in an inferential setting, but only begin to give the flavor of what we would call a fully Bayesian analysis. To give the full sense of what Bayesian analysis is, we need first to review the concepts of probability distribution and likelihood. We do this in Chapter 2, establishing basic notions and notation.

CHAPTER

2

Probability

OUTLINE

2.1	What is Probability?	14	2.2.2 Percentiles, Moments, and Independence	18
2.2	Basic Probability Concepts for Bayesian Inference	16	2.2.3 Common Probability Distributions	20
	2.2.1 Joint, Marginal, and Conditional Distributions	17	2.2.4 Transforming Variables	20

The field of statistics has its foundations in probability. Both endeavors have to do with uncertainty: probability, with uncertain events; statistics, with uncertain mechanisms. Probability addresses questions like "given these conditions, what sort of outcomes are to be anticipated?" Statistics works the opposite direction: "what sort of conditions existed in order to produce these outcomes?" Thus, statistical inference was originally and aptly designated "inverse probability."[1]

The foundations of statistics in probability are explicitly evident in Bayesian applications, arguably more so than in frequentist applications. Frequentism uses probability to develop tools for statistical inference; Bayesianism uses probability itself as the tool for statistical inference. Indeed, Bayesian inference can be succinctly described as the process of assigning and refining probability statements about unknown quantities.

Perhaps the greatest challenge to learning Bayesian methods is in learning to use probability as a model of uncertainty for *all* unknown quantities. We feel comfortable enough thinking about our data as realizations of random processes, perhaps consisting of stochastic ecological

---

1. Unfortunately, to use the phrase "inverse probability" for "statistics" nowadays is to risk confusion, because the meaning of the phrase has evolved, being attached to specific and various statistical methods, rather than to the field generally. Historical review (Dale, 1991) shows that it was applied in a general sense originally, then specifically to Bayes' methods, both by protagonists and antagonists, then later (by R.A. Fisher) of what are now known as "likelihood" methods.

signals contaminated by random noise related to sampling processes and measurement errors. We feel comfortable describing our data using probabilities, even though in reality there is nothing random about them at all: they are simply columns of numbers, as fixed as can be.

But it can be difficult to learn to use probability to describe other quantities, such as parameter values. These we have been taught to think of as fixed and unknown, period. To illustrate: we assume that most readers do not know the millionth digit of π; call it X. There is nothing random about X; it does not change from day to day, nor does it vary from north to south, east to west or along any other gradient. Thus, many would be uncomfortable with saying "there's a 10% chance that $X = 5$." As a practical description of your uncertainty, however, the probability statement is perfectly reasonable. Suppose that the correct value were to be announced for the first time tomorrow, and that a friend with no inside knowledge offered a bet on whether or not the digit were a 5. You would probably be willing to accept odds greater than 9:1 against $X = 5$, but unwilling to accept odds less than 9:1. Your decision would be guided by comparisons of the situation to probability models involving a random event with one chance in ten of success.[2]

The mathematical theory of probability provides a model for uncertainty. The model is a rigorously developed mathematical entity, based on axioms reflecting commonsense descriptions of uncertainty. Most practicing statisticians never need to make reference to the details of sets, σ-algebras, and measures briefly described in Section 2.1. Our purpose in presenting this material is to highlight the notion of probability as a mathematical model for uncertainty. Bayesian thinking embraces this notion, using probability to describe any and all uncertainties, thus attaining a pleasing philosophical simplicity.

Building from its axiomatic foundations, probability theory produces a few simple concepts and tools which are needed for Bayesian inference. We briefly review these in Section 2.2, and establish notational conventions to be used throughout the text. Anticipating that many readers will already have some familiarity with these, we relegate further details to appendices: probability concepts to Appendix A, probability distributions to Appendix B.

2.1 WHAT IS PROBABILITY?

We know what we mean when we say that "there's a 70% chance of rain tomorrow" but it is fairly difficult to provide a concrete definition. We need a mathematical model to point to, one for which we can say "my uncertainty about whether or not it will rain tomorrow is represented by that model."

What do we mean when we say "there's a 50% chance that the coin I am about to flip will show 'Heads' "? This question may seem easier than the first: we may say that there are two equally likely outcomes, one of which is "Heads," ergo, the probability of "Heads" is 0.50. But this begs the question: what do we mean by "likely"? We might respond with reference to long experience of flipping this and similar coins, but be interrupted by an annoying interlocutor who refocuses our attention on the question, saying "I'm not asking about the coin's past

---

2. As it turns out, the millionth digit of π is 1; this, at least, on the authority of several web pages. Would we say $\Pr(X = 1) = 100\%$? Or might we allow that there could have been some error in calculation that the various websites have shared, and operate as though $\Pr(X = 1)$ is slightly less than one?

history, but about the flip that is about to take place." Again, what is needed is a mathematical model, an abstraction representing our understanding and uncertainty about the process under consideration.

Probability theory is the mathematical abstraction of uncertain events. By "uncertain events," we mean events which can occur in multiple, distinct ways; events which might not have an inevitable outcome. Thus we distinguish events and outcomes. Tomorrow's weather is an event, rain and snow are outcomes. In probability theory, *outcomes* are formalized as elements ω in a universal set Ω, and *events E* (we drop the word "uncertain") as sets of outcomes.

Probability is formally defined as a function P on a collection F of subsets of Ω. The collection of subsets F must be a σ-*algebra*: it must be nonempty and closed under complementation and countable unions. The function P is a probability measure, meaning that it has three basic characteristics corresponding naturally with intuitive descriptions of chance. These are, that for any set $E \in F$,

$$0 \leq P(E) \leq 1, \tag{2.1}$$

and

$$P(E^C) = 1 - P(E), \tag{2.2}$$

where $E^C = \Omega \backslash E$ is the complement of E (i.e., that event E did not occur); also, that if $E_1, E_2, E_3 \ldots$ are disjoint sets, and E_* is their union, then

$$P(E_*) = \sum_i P(E_i). \tag{2.3}$$

Simply put, these require Eq. (2.1) that chances are between zero and one, Eq. (2.2) that the chance that an event *does not* occur is 1 minus the chance that it does occur, and Eq. (2.3) that chances of mutually exclusive events can be added, to determine the probability that one of them occurs. The universal set Ω, the σ-algebra F, and the probability measure P make up a *probability space*. A (real-valued) random variable is a function X on Ω with the property that $\{\omega : X(\omega) \leq t\}$ is in the set F, for all values t.

Example: A Probability Space and a Random Variable

Consider the following collection of subsets of $\Omega = [0, 1]$: $F = \{E_1 = [0, 0.50], E_2 = (0.50, 1], E_3 = [0, 1], E_4 = \emptyset\}$, where $\emptyset$ is the empty set, with no elements. We can define a probability function P on F as satisfying

$$P(E_1) = 0.5, P(E_2) = 0.5, P(\Omega) = 1, \text{ and } P(\emptyset) = 0.$$

Define

$$X(\omega) = \begin{cases} 0, & \text{if } \omega \leq 0.5 \\ 1, & \text{otherwise} \end{cases}.$$

The collection of sets F is a σ-algebra, and P is a probability measure satisfying (2.1)–(2.3), so $\{\Omega, F, P\}$ is a probability space. The function X is a random variable taking values 0 and 1, each with probability $1/2$.

Suppose that I am about to flip a coin, one which I perceive to be fair. I can describe my uncertainty about the outcome by analogy to the probability space $\{\Omega, F, P\}$ just defined, with E_1 corresponding to "Heads," E_2 to "Tails," E_3 to "Either Heads or Tails," and E_4 to "Neither Heads nor Tails" (e.g., the coin lands on its edge, or bursts into flame and disappears in mid-air). Alternatively, we might associate "Heads" and "Tails" with the two values of the random variable X.

The probability space $\{\Omega, F, P\}$ might seem an unnecessary abstraction, and we might be inclined to think of probabilities, as frequentists do, as relating to long-term frequencies. A frequentist explains the statement "Pr(Heads) = $^1\!/_2$" as meaning that if we were to flip the coin an arbitrarily large number of times, the observed relative frequency of "Heads" will converge to 0.50. Two comments are worth making: first, that if we define the statement "Pr(Heads) = $^1\!/_2$" by reference to an abstract probability space, the convergence of the long-term relative frequency to 0.5 is a mathematical consequence.[3] Thus, what is taken by some as a *definition* of probability, is a *feature* of probability, if defined in terms of mathematical models.

The second comment is that the long-term frequency definition limits the application of the term "probability." It is difficult to conceive, for instance, of a large number of replicate tomorrows, so as to define the statement that "there is a 70% chance of rain tomorrow." It is impossible to conceive of replicate values of π, each having a different millionth digit. This limitation explains frequentist reluctance to use probability as a measure of uncertainty about states of nature, or as a description of personal uncertainty.

One way or another, we seldom need to give too much thought to what we mean by probability. It would be a shame if, having planted a tree, we had to dig it up every day to examine its roots. We will have no need to refer to probability spaces, σ-algebras, set functions or anything of the sort subsequently. However, it will be useful to embrace the broader definition of probability allowed by these abstractions: that probability is a mathematical model of uncertainty, a model which obeys three simple rules (Eqs. (2.1)–(2.3)) corresponding to our intuition about how uncertainty works. This step beyond the long-term frequency definition of probability provides the basis of Bayesian inference, enabling us to use probability to describe any sort of uncertainty.

2.2 BASIC PROBABILITY CONCEPTS FOR BAYESIAN INFERENCE

To conduct Bayesian inference, we need some familiarity with a few basic concepts of probability. We need to know about probability distributions, and about moments and percentiles as descriptions of them. We need to know about joint distributions for collections of random variables, about conditional distributions, about marginal distributions. Bayesian modeling consists of the specification of a joint distribution for data and unknown quantities; Bayesian inference is based on conditional distributions of unknowns, given data.

Here, we give a brief overview of such concepts. Our purpose is primarily to introduce notation to be used subsequently; further details are given in Appendix A.

---

3. This is the weak law of large numbers, proved by Jakob Bernoulli (1654–1705); for details, see Appendix A.

2.2.1 Joint, Marginal, and Conditional Distributions

Distribution Functions

Suppose that X is some random quantity, possibly vector-valued. The *probability distribution function* (*pdf* or simply *distribution function*) $f(x)$ is that function which, when integrated or summed, as appropriate, yields $\Pr(X \in R)$ for regions R in the range of X. The *support* of X is the set of values x such that $f(x) > 0$.

For example, the function

$$f_{Y,Z}(y,z) = \frac{y}{5z}, \quad \begin{cases} 0 < y < z \\ z = 1,2,3,4 \end{cases} \tag{2.4}$$

is the distribution of a vector $X = (Y, Z)'$. The random variable Z takes values 1, 2, 3, or 4; Y is continuous, with range limited by the corresponding value of Z. If we wish to find the probability that X is in the region R defined by $\{Y \le 1.5, Z \le 2\}$, we compute

$$\sum_{z=1}^{2} \int_{0}^{\min\{1.5,z\}} f_{Y,Z}(y,z)\, dy$$

$$= \int_{0}^{1} f_{Y,Z}(y,1)\, dy + \int_{0}^{1.5} f_{Y,Z}(y,2)\, dy = \frac{17}{80}.$$

A distribution function for a vector-valued random variable is often referred to as the joint distribution of its components. The distributions of individual components of the vector are called marginal distributions; the name derives from tables with cross-classifications of joint probabilities with probabilities for individual components along the margins, obtained by summing along rows or columns.

Marginal Distribution Functions

For our sample joint distribution Eq. (2.4), suppose we wish to calculate the probability that $Z = z$. This is the marginal distribution

$$f_Z(z) = \Pr(Z = z) = \Pr(0 < Y < z, Z = z)$$

$$= \int_{0}^{z} f_{Y,Z}(y,z)\, dy = \int_{0}^{z} \frac{y}{5z}\, dy = \frac{z}{10}, \tag{2.5}$$

for $z = 1, 2, 3, 4$. The process of obtaining this marginal distribution is often described as "integrating y out of the joint distribution" or "marginalizing" over y.

Conditional Distribution Functions

The conditional distribution of Y given Z is obtained by dividing the joint distribution by the marginal for Z, viz.,

$$f_{Y|Z}(y|z) = \frac{f_{Y,Z}(y,z)}{f_Z(z)} = \frac{2y}{z^2}, \tag{2.6}$$

for $0 < y < z$.

Note that the conditional distribution $f_{Y|Z}(y|z)$ is regarded as a function of y, for fixed z. If we regard the joint distribution $f_{Y,Z}(y,z)=y/(5z)$ in the same way, we note that the two are proportional: each has a single y in it, in the numerator. The only difference is the fixed quantity which multiplies it, $1/(5z)$ for the joint distribution, $2/z^2$ for the conditional distribution. Thus given the joint distribution we could immediately identify $f_{Y|Z}(y|z) \propto g(y)=y$. To fully specify the conditional distribution, we need only to scale $g(y)$ so that the result integrates to 1. The integral of $g(y)$ for $0<y<z$ is $z^2/2$, so the conditional distribution is $y/(z^2/2)=2y/z^2$.

Bracket Notation

In Eqs. (2.4)–(2.6), we have used fairly standard, but somewhat elaborate notation. The subscripts (Y,Z), Z, and $(Y|Z)$ are included because otherwise we would be using the same f in three different ways. One solution, of course, would be to use different letters for different functions. But an even easier solution is to drop the letters altogether, and to use what is known as *bracket notation*. Thus we write $[Y,Z]$ for the joint distribution, $[Y|Z]$ for the conditional distribution of Y given Z, and $[Z]$ for the marginal distribution of Z. Equation (2.6) is then simply

$$[Y|Z] = \frac{[Y,Z]}{[Z]},$$

and our comment about the proportionality of the joint and conditional distribution becomes $[Y|Z] \propto [Y,Z]$, it being understood that we are treating Z as a fixed value.

We almost exclusively use bracket notation in this book. Our experience is that the notation brings clarity to modeling, especially when we get to complex hierarchical models. Suppose that X represents data, and that θ represents unknown quantities, including parameters. Bayesian inference is based on the conditional distribution of distribution of θ, given the data, that is

$$[\theta|X] \propto [X,\theta] = [X|\theta][\theta]. \qquad (2.7)$$

Not wanting to get too far ahead of ourselves at this point, we mention that the left-hand side of Eq. (2.7) is called the posterior distribution of θ, and the two terms on the right-hand side are (as functions of θ) the likelihood and the prior distribution. The Bayesian difference, outlined in Section 2.1, is in using probability to describe all unknown quantities. Thus there are distribution functions associated with θ; all Bayesian inference is based on posterior distributions obtained from likelihood and prior. Philosophically, it's all quite simple.

2.2.2 Percentiles, Moments, and Independence

Given that all of Bayesian inference is based on probability distributions, it is important that we have some vocabulary to describe their features. The basic descriptors are *percentiles* and *moments*. We briefly describe these, with more details in Appendix A.

Percentiles

Let p be a number between 0 and 1. If X is a continuous random variable, then the $100p$th percentile of $[X]$ is that number x_p such that $\Pr(X \le x_p) = p$ and $\Pr(X \ge x_p) = 1 - p$. The 50th percentile is called the *median*; the 25th and 75th percentiles are called the first and third *quartiles*.

This definition of percentiles does not work so well for discrete random variables. For example, the random variable Z with distribution function (2.5) has $\Pr(Z \le z) = 0.10, 0.30, 0.60$, and 1.00 for $z = 1, 2, 3$, and 4. There is no number $z_{0.50}$ such that $\Pr(Z \le z_{0.50}) = 0.50$. A more flexible definition of percentiles is required:

A $100p$th percentile of $[X]$ is any number x such that
$$\Pr(X < x) \le p \text{ and } \Pr(X > x) \ge 1 - p.$$

We say "a" $100p$th percentile rather than "the" $100p$th percentile, because the number might not be uniquely defined. For instance, any number between 2 and 3 is a 50th percentile of $[Z]$, given by Eq. (2.5). To make life a bit easier, noting that 2 and 3 are the 30th and 60th percentiles, and that 50 is 2/3 of the way from 30 to 60, we might choose to go 2/3 of the way from 2 to 3, and report 2.67 as "the" median. But the best solution is to simply say that the median is between 2 and 3.

Moments

Probability distributions are often summarized by typical values of quantities sampled from them. Suppose X is a random variable, possibly vector-valued, with density function $f_X(x)$, and that $g(x)$ is a function. Then the expected value of $g(X)$ is defined as

$$E(g(X)) = \int g(x) f_X(x) \, dx. \tag{2.8}$$

Equation (2.8) uses somewhat informal notation which will be used throughout this book. More formally, we would require multiple integrals for the continuous components of X, and summations for discrete components; the ranges of all the variables would have to be indicated as well. But the basic idea should be clear enough with this informal notation. As Mark Twain put it "Plain clarity is better than ornate obscurity."

For a univariate random variable X, the expected value of $g(X) = X^k$ is referred to as the kth moment of X. A number of useful summaries of $[X]$ are based on moments. For univariate random variables, we have

The mean: $E(X)$

The variance: $\text{Var}(X) = \sigma^2(X) = E(X^2) - E(X)^2$

The standard deviation: $\text{SD}(X) = \sigma(X) = \sqrt{\text{Var}(X)}$.

For pairs of random variables (X, Y), we have

The covariance: $\text{Cov}(X, Y) = E(XY) - E(X)E(Y)$

The correlation: $\text{Corr}(X, Y) = \rho(X, Y) = \dfrac{\text{Cov}(X, Y)}{\sigma(X)\sigma(Y)}$.

Details on the variance as a measure of spread, and the correlation as a measure of linear association, are deferred to Appendix A.

The reader might wish to verify that for the distribution (2.4), the mean values of Y and Z are 2 and 3, that $\text{Var}(X)=\text{Var}(Y)=1$, and that $\rho(X,Y)=2/3$.

Independence

A pair of random variables (X,Y) are *independent* if their joint distribution factors as $[X,Y]=[X][Y]$; a set of n random variables are *mutually independent* if $[X_1,\ldots,X_n]=[X_1]\cdots[X_n]$. Mutual independence implies pairwise independence, but not the other way round; it is possible to have pairwise independence but not mutual independence.

In Bayesian modeling we will make frequent use of *conditional independence*: X and Y are conditionally independent of Z if $[X,Y|Z]=[X|Z][Y|Z]$. Random samples from a common distribution with unknown parameter θ are often modeled as mutually independent, conditional on the unknown parameter: $[X_1,\ldots,X_n|\theta]=[X_1|\theta]\cdots[X_n|\theta]$. Such random variables are often described as "independent and identically distributed" (iid), but it is important to remember that the independence is conditional on the unknown parameter. Conditional independence is, for practical purposes, the same as "exchangeability" a term which is often used in Bayesian applications, and which we will use in this text.[4]

2.2.3 Common Probability Distributions

Entire texts have been written cataloguing families of probability distributions (Johnson *et al.*, 2005; Kotz *et al.*, 1994, 1995, 1997, 2000) and vast literatures exist on characterizing[5] specific families of distributions, especially the normal and exponential distributions (e.g., Galambos and Kotz, 1978).

Fortunately, the vast majority of ecological models are based on only a few probability distributions. The most important for ecological examples, by our reckoning, are those given in Table 2.1; details on these and several others are given in Appendix B. Of special importance are the multinomial distribution (for categorical data, including mark-recapture data), the Poisson distribution (which serves as the basis for many analyses of count data), and the exponential distribution (a starting point for continuous time survival analysis). The beta and gamma distributions are primarily used as convenient prior distributions for unknown parameters.

We draw the reader's attention to the second column of the table, as providing convenient notation that will be used throughout the book. For instance $X \sim P(\lambda)$ will indicate that X has a Poisson distribution with parameter λ.

2.2.4 Transforming Variables

A common problem in statistics is one where given a random variable X, with *pdf* $f_X(x)$, we need to find the distribution of Y, where $Y=g(X)$ for an invertible function g. Since g is

4. A sequence $\{X_i\}, i=1,2,\ldots$ is *exchangeable* if its joint distribution is not changed under permutation of any finite number of its indices.
5. A characterization is an "if and only if" statement. For instance, suppose that $X_1, X_2,\ldots,X_n$ are iid. Then if their distribution is exponential, $\Pr(\min\{X_1,\ldots,X_n\}\leq t)=\Pr(X_1/n\leq t)$, for all $t>0$. But what is more, the only distributions with this property are exponential; this condition characterizes exponential distributions.

TABLE 2.1   Important probability distributions for ecological modeling.

Name	Notation	Parameters	Mean	Variance
Normal	$N(\mu,\sigma^2)$	$-\infty < \mu < \infty, \sigma > 0$	μ	σ^2
Binomial	$B(N,p)$	Integer $N > 0, 0 \leq p \leq 1$	Np	$Np(1-p)$
Multinomial	$M_k(N,\boldsymbol{\pi})$	Integers $N, k > 0, \boldsymbol{\pi} = (\pi_1,\ldots,\pi_k)', \pi_i \geq 0, \sum \pi_i = 1$	$N\boldsymbol{\pi}$	$N\{\mathrm{diag}(\boldsymbol{\pi}) - \boldsymbol{\pi}\boldsymbol{\pi}'\}$
Poisson	$P(\lambda)$	$\lambda > 0$	λ	λ
Exponential	$E(\lambda)$	$\lambda > 0$	$1/\lambda$	$1/\lambda^2$
Gamma	$Ga(\alpha,\beta)$	$\alpha, \beta > 0$	α/β	α/β^2
Uniform	$U(a,b)$	$-\infty < a < b < \infty$	$\dfrac{a+b}{2}$	$\dfrac{(b-a)^2}{12}$
Beta	$Be(a,b)$	$a,b > 0$	$\dfrac{a}{a+b}$	$\dfrac{ab}{(a+b)^2(a+b+1)}$

invertible, we can write $X = g^{-1}(Y)$. If X is a discrete random variable, then Y is also discrete and has as its *pdf*

$$\Pr(Y = y) = \Pr(X = g^{-1}(y))$$

with the sample space of y, denoted by $\mathcal{S}_y$, equal to the set of values $\{g(x_1), g(x_2), \ldots\}$ computed for each $x \in \mathcal{S}_x$ where $\mathcal{S}_x$ is the sample space of x.

We might think that for continuous variables we do exactly the same operation on the *pdf* for X. Not quite; if X is continuous, and provided $g(X)$ is also differentiable, then the *pdf* of Y is given by

$$f_Y(y) = f_X(g^{-1}(y)) \times \left| \frac{dg^{-1}(y)}{dy} \right|,$$

where $dg^{-1}(y)/dy$ is the derivative of the inverse transformation $X = g^{-1}(Y)$. The support of Y is given by the range of the transformation $g(X)$. This result is often referred to as the *change of variables theorem*.

To illustrate, we derive the lognormal distribution, corresponding to a random variable $Y = e^X$, where $X \sim N(\mu, \sigma^2)$. The inverse transformation is $g^{-1}(y) = \ln(y)$ with derivative $1/y$, and so

$$f_Y(y) = \frac{1}{\sqrt{2\pi}\sigma} e^{-\frac{1}{2\sigma^2}(\ln(y)-\mu)^2} \frac{1}{y},$$

with support $y > 0$.

The change of variables theorem has a multivariate extension. Suppose we have a continuous random vector $\boldsymbol{X} = (X_1, \ldots, X_p)$ and we are interested in the transformation to $\boldsymbol{Y} = \boldsymbol{g}(\boldsymbol{X})$. If $\boldsymbol{g}$ is invertible and differentiable, then

$$f_Y(\boldsymbol{y}) = f_X(\boldsymbol{g}^{-1}(\boldsymbol{y})) |J_{g^{-1}}(\boldsymbol{y})|,$$

where $J_{g^{-1}}(\boldsymbol{y})$ is the Jacobian determinant of the inverse transformation (i.e., the determinant of the matrix of partial derivatives).

Statistical Inference

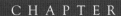
OUTLINE

3.1	**Likelihood**	25		3.3.2 *Bayesian Inference for*	
3.2	**Confidence Intervals**	29		*Binomial Success Rate*	37
	3.2.1 *Approximate CI's for*			3.3.3 *Incorporating Prior Knowledge*	
	Binomial Success Rate	30		*in Interval Estimate for*	
	3.2.2 *Exact CI for Binomial*			*Binomial Success Parameter*	41
	Success Parameter	32	**3.4**	**Summary and Comparison of**	
	3.2.3 *Confidence Intervals –*			**Inferential Systems**	**42**
	Summary	35		3.4.1 *Point Estimation*	43
3.3	**Bayesian Interval Estimation**	**36**		3.4.2 *Conclusion (peroration)*	45
	3.3.1 *Basics of Bayesian Inference*	36			

A coin-flip, an attempt to guess the correct answer on a multiple choice exam, whether a newly hatched chick survives to fledging: each is a binary event; each results in one of two mutually exclusive outcomes. Such random binary events are generically termed as *Bernoulli trials* with outcomes labeled "success" or "failure." As simple as they are, Bernoulli trials are of fundamental importance to mathematical modeling of ecological processes. They are also completely satisfactory as a basis for comparing alternative approaches to statistical inference: we take an observation about the rare and endangered roseate tern (*Sterna dougallii*) as our text for this chapter. Shealer and Spendelow (2002) write that

> "During 748 hours of observation from 1995 to 1998, we identified ten roseate terns that we considered to be habitual kleptoparasites.[1] Of these ten kleptoparasites, eight were females and two were males. The probability

---

1. Food thieves; "habitual" ones, at that. Some of the females would even – but no, the shameful depths to which these avian sociopaths lower themselves cannot be mentioned without indelicacy or included in a statistics text which, we hope, will be read around the family hearth.

of eight or more kleptoparasites out of ten being of the same sex by random chance alone is low (binomial test, P = 0.055)."

We will use these data to compare the fundamentals of frequentist and Bayesian approaches to inference. The question at hand is whether habitual kleptoparasitism (K) is more associated with the female roseate tern than with the male roseate tern. That is, we wish to decide whether

$$\Pr(K|F) > \Pr(K|M), \tag{3.1}$$

where F and M denote female and male, respectively.

We address the question by imagining the genders of the 10 habitual parasites to be independent Bernoulli trials, with "female" labeled as "success," and unknown success parameter p. This model serves as a starting point for analysis. Its use, regardless of statistical paradigm, requires thought about possible violations of the independence assumption and about the interpretation of the parameter p.

The first thing to note is that p is not $\Pr(K|F)$ but $\Pr(F|K)$.[2] It is fairly easy to show that Eq. (3.1) is equivalent to $\Pr(F|K) > \Pr(F)$.[3] Thus if $\Pr(F)$ were known, inference about Eq. (3.1) could be based on the observations of gender among habitual kleptoparasites. Shealer and Spendelow (2002) appear to assume that $\Pr(F) = 0.50$ and to examine whether the data are consistent with the assumption $\Pr(F|K) = \Pr(F)$.

Another issue, perhaps even more fundamental, is whether p is, in fact, equal to $\Pr(F|K)$. To be more precise, $p = \Pr(F|K,O)$, where O denotes "bird is observed." If male habitual kleptoparasites are better at hiding their larceny than their female counterparts, observers will tend to see disproportionately many females. For p to be the same as $\Pr(F|K)$, we must assume that O and F are conditionally independent, provided K is given. Then

$$p = \Pr(F|K,O) = \frac{\Pr(F,O|K)}{\Pr(O|K)}$$

$$= \frac{\Pr(F|K)\,\Pr(O|K)}{\Pr(O|K)} = \Pr(F|K).$$

The requisite assumption is that for K birds, gender and observation are independent. The issue of the potential confounding of observation processes with the response of interest commonly arises in ecology. It is a theme that we explore in the examples that we consider in detail later in this book.

Consideration of these assumptions is a necessary part of the inferential process, whether we choose a frequentist or Bayesian approach. We will put such considerations behind us for now, and, assuming that we have an observation $X = 8$ of a random variable $X \sim B(10, p)$, consider the process of statistical inference for p. We are content to assume that the number X of females among the 10 kleptoparasitic terns was a random event that could have

---

2. Confusing $\Pr(A|B)$ and $\Pr(B|A)$ is quite common in inferential settings; the mistake is similar to the logical fallacy of *affirming the consequent*: "I know that A implies B; B has occurred, therefore A must have occurred."

3. Give it a try! The proof follows from the definition of conditional probability and the observations that $\Pr(M) = 1 - \Pr(F)$ and $\Pr(K) = \Pr(K,M) + \Pr(K,F)$.

resulted in $X = x$, for any integer $x \in \{0, 1, \ldots, 10\}$, and that the probabilities of these various outcomes were

$$\Pr(X = x) = B(x; 10, p) = \binom{10}{x} p^x (1 - p)^{10-x}, \tag{3.2}$$

determined by an unknown quantity p. Our goal is to learn about p, based on the observation $X = 8$.

3.1 LIKELIHOOD

The first use of the binomial distribution $B(x; n, p)$ is for probability calculations. These address the probability of specified events, given known values of parameters. Statistical inference turns the problem around: given observed events, what can we say about the unknown values of parameters governing them? Consider the kleptoparasitism data. The process of inference begins with inspection of Table 3.1, which displays the probabilities of outcomes $X = x$, for a binomial random variable with $n = 10$, for a range of values of p, calculated using Eq. (3.2).

Our interest is drawn to the column for $x = 8$, corresponding to our observation. We note that the probability of observing $x = 8$ is quite small for small values of p, then increases to a maximum value of 0.302, for $p = 0.8$, then decreases to 0 once again as p approaches 1.0. On this basis, it seems reasonable to estimate the unknown value of p by 0.80. We write $\hat{p} = 0.80$, the circumflex over the p indicating that this value is an estimate, rather than the true value of the parameter.

It is worth looking closely at the process used in choosing this value for $\hat{p}$. Probability calculations use *rows* of Table 3.1 to find probabilities of outcomes x based on known values

TABLE 3.1   Binomial probabilities $B(x; n, p)$.

p	x										
	0	1	2	3	4	5	6	7	8	9	10
0	1.000	0.000	0.000	0.000	0.000	0.000	0.000	0.000	**0.000**	0.000	0.000
0.1	0.349	0.387	0.194	0.057	0.011	0.001	0.000	0.000	**0.000**	0.000	0.000
0.2	0.107	0.268	0.302	0.201	0.088	0.026	0.006	0.001	**0.000**	0.000	0.000
0.3	0.028	0.121	0.233	0.267	0.200	0.103	0.037	0.009	**0.001**	0.000	0.000
0.4	0.006	0.040	0.121	0.215	0.251	0.201	0.111	0.042	**0.011**	0.002	0.000
0.5	0.001	0.010	0.044	0.117	0.205	0.246	0.205	0.117	**0.044**	0.010	0.001
0.6	0.000	0.002	0.011	0.042	0.111	0.201	0.251	0.215	**0.121**	0.040	0.006
0.7	0.000	0.000	0.001	0.009	0.037	0.103	0.200	0.267	**0.233**	0.121	0.028
0.8	0.000	0.000	0.000	0.001	0.006	0.026	0.088	0.201	**0.302**	0.268	0.107
0.9	0.000	0.000	0.000	0.000	0.000	0.001	0.011	0.057	**0.194**	0.387	0.349
1.0	0.000	0.000	0.000	0.000	0.000	0.000	0.000	0.000	**0.000**	0.000	1.000

of p.[4] For the purpose of statistical inference, we have used a *column* of Table 3.1. Our knowledge of an outcome is used to make educated guesses at the value of the unknown value of p.

In considering a column, rather than a row of Table 3.1, we are regarding $B(x;10,p)$ as a function of p for fixed x, rather than as a function of x for fixed p. For probability calculations, $B(x;10,p)$ is used as a function of x alone, with fixed p. For statistical inference, we are considering $B(x;10,p)$ as a function of p alone, with the outcome x fixed.

When we fix a value of x, and let p vary, $B(x;10,p)$ is no longer a distribution function. The numbers in the column for $x=8$ do not add up to one. Their only meaning, in this inferential context, is relative to one another. For instance, consulting Table 3.1 one finds that the probability of observing $X=8$ is four times larger for $p=0.5$ than it is for $p=0.4$ (0.044 vs. 0.011). The ratio of the probabilities, rather than the difference between them, is the basis of comparison; one concludes that the data provide four times the support for $p=0.5$ than for $p=0.4$.

If we were to carry out the calculation longhand instead of consulting Table 3.1, we would write

$$\frac{B(8;10,0.5)}{B(8;10,0.4)} = \frac{\binom{10}{8}0.5^8(1-0.5)^{10-8}}{\binom{10}{8}0.4^8(1-0.4)^{10-8}}$$

$$= \frac{0.5^8(1-0.5)^{10-8}}{0.4^8(1-0.4)^{10-8}} = 4.14,$$

and note along the way that it is unnecessary to calculate the combinatorial term, which cancels from numerator and denominator, and depends only on the fixed value of x in $B(x;10,p)$. Thus, statisticians routinely omit multiplicative components of probability distributions that are independent of the parameter of interest and define the *likelihood function*[5] as

$$L(p) = p^x(1-p)^{n-x}. \tag{3.3}$$

We can verify analytically what we observe from Table 3.1 that with $X=8$ and $n=10$, the maximum value of the likelihood occurs when $p=0.8$. First, we note that it is easier to maximize the logarithm of the likelihood than the likelihood itself. Since the extrema of $\ln(L(p))$ and $L(p)$ coincide, it suffices to maximize

$$l(p) = \log(L(p)) = x\log(p) + (n-x)\log(1-p)$$

by setting

$$l'(p) = \frac{dl(p)}{dp} = \frac{x}{p} - \frac{n-x}{1-p} = 0,$$

and solving for p. The solution is easily found and denoted by $\hat{p}=x/n$. We note again that the circumflex over the p indicates that this value is an estimate of the true but unknown value p; this particular estimate is known as the *maximum likelihood estimator* (MLE).

---

4. Thus, Shealer and Spendelow's "binomial test, P=0.055" is seen to be $\Pr(X \geq 8)$ for $X \sim B(10,0.50)$.

5. More precisely, we might say that Eq. (3.3) is "a version of the likelihood." Statisticians typically substitute the symbol "$\propto$" meaning "proportional to" for the equals sign in Eq. (3.3), or write $L(p) = C\, p^x(1-p)^{n-x}$, where C is an arbitrary constant. The important thing is to understand that the likelihood values have no meaning *per se* but only as they stand in relation to other values as ratios.

Interval Estimates Needed

How well does the MLE work? How much trust should we put in our estimate that 80% of kleptoparasitic roseate terns are female?

Consulting Table 3.1, we see that if p were 0.80, the most probable outcome would be $X=8$, but that outcomes $X=7$ or $X=9$ would not be unexpected. There is probabilistic uncertainty in the outcome; various outcomes in the row labeled $p=0.80$ are consistent with the single, fixed parameter.

Similarly, focusing attention on a single column of Table 3.1, that headed with $x=8$, we note that p is the most *likely* value (i.e., maximizes the likelihood) but that there are other values with reasonably high likelihood. The observation $X=8$ is not inconsistent with $p=0.70$, or even with $p=0.60$. There is statistical uncertainty about the parameter p.

Thus, it is unwise to limit our inference about p to the reporting of a maximum likelihood estimator. Instead, we report a collection of p's, an *interval estimate*, and all of the values we deem consistent with the observed data. The method we use to decide on which collection of p's to report is determined by whether we follow the frequentist or Bayesian philosophy of statistical inference. Before presenting these, we take a moment to describe interval estimation based solely on the likelihood function.

Likelihood Intervals

Define the *scaled likelihood* as the likelihood divided by its maximum value, denoted by

$$L^*(p) = \frac{L(p)}{L(\hat{p})}.$$

Note that the scaled likelihood takes values ranging from zero to one and that the scaled likelihood of the MLE is one (Fig. 3.1). With $n=10$ and $X=8$, $L^*(p) \geq 0.25$ if and only if $0.549 \leq p \leq 0.950$ (Fig. 3.1); values of p in the interval are at least 25% as likely as the MLE,

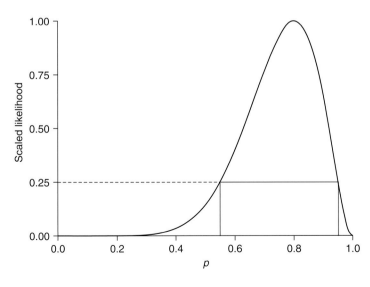

FIGURE 3.1 Scaled likelihood for observation $X=8$ based on $X \sim B(10,p)$. The scaled likelihood exceeds 0.25 for $0.549 \leq p \leq 0.950$.

points outside the interval are less well supported. The interval consists of values p with scaled likelihood within 75% of the maximum. The set (0.549, 0.950) is an interval estimate of p, used rather than simply reporting a *point estimate* (such as the MLE) to quantify statistical uncertainty. To distinguish this interval estimate from others developed subsequently, we designate $\{p \in [0,1] | L^*(p) \geq \alpha\}$ as a $100(1-\alpha)\%$ *scaled likelihood interval*. The quantity α measures the uncertainty associated with the interval; the smaller the value of α, the longer the interval.

It is interesting to note that if $n = 100$ and $X = 80$, the MLE is still 0.80, but the corresponding interval estimate shrinks to (0.728, 0.861) (see Fig. 3.2). Many values of p that had reasonable support with $n = 10$ are not well supported with the larger sample size. For example, with $n = 10$ and $\hat{p} = 0.80$, the scaled likelihood at $p = 0.65$ is 0.582, whereas with $n = 100$ and $\hat{p} = 0.80$, the scaled likelihood at $p = 0.65$ is 0.004. The larger sample size has had the predictable effect of reducing statistical uncertainty. The second interval is 33.1% as long as the first, a reduction in length by a factor of approximately $\sqrt{10/100}$; the uncertainty associated with our estimate has dropped by a factor approximately equal to the square root of the ratio of the sample sizes. The inferential value of a data set often increases at a rate roughly proportional to the square root of the sample size.

The interval estimates described so far have two deficiencies: first, they are by no means easily calculated, and second, there was something arbitrary in the choice of the factor 0.25, that is, in our choice of how well supported a value p must be to be included in the interval. Some arbitrariness might be inevitable: a more conservative scientist might want to include p's with lower support than the minimum scaled likelihood of 0.25. More importantly, however, the meaning of the phrase "values of p that are at least one-fourth as likely as the most likely value" is somewhat vague, and if interpreted in terms of the likelihood function, tautologous.

How then should we construct interval estimates? We begin by considering the frequentist approach, continuing with the example of inference for the binomial proportion p.

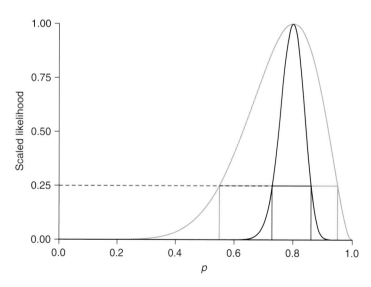

FIGURE 3.2 Scaled likelihood for observation $X = 80$ based on $X \sim B(100, p)$. The scaled likelihood exceeds 0.25 for $0.728 \leq p \leq 0.861$. Corresponding graphic for $X = 8$, $n = 10$ in background, for comparison.

3.2 CONFIDENCE INTERVALS

The frequentist interval estimate is called a confidence interval (CI): it is an interval of the form (l_X, u_X) with endpoints determined by the observation X, defined with certain criteria in mind. Before describing these criteria, we stress that the CI is itself a random variable, just like the data X.

The concept of data as realizations of random variables is crucial to statistical inference. There is a fixed set of stochastic processes by which data are produced; there is a fixed set of parameters governing these processes. There is also an observed data set, a single realization of these random processes. So, for instance, we flip a coin and observe the outcome "Heads." We distinguish a random process X, the coin flip, and a single outcome, "Heads." For simplicity, we may write $X =$ "Heads," but we maintain two concepts in our understanding: that of a random process X, and that of a single realization, the data.

Again, suppose that we observe eight successes in 10 Bernoulli trials. We say that $X \sim B(10, p)$ and $X = 8$. There is nothing random about the number eight, yet we are comfortable regarding X as a random variable because we carry along a conception of a stochastic mechanism, which could have produced $X = 5$ or $X = 9$, and so forth, but in the current case, happened to have produced $X = 8$.

These observations carry over to our understanding of CI's. Given that we have observed $X = 8$, the CI will be (l_8, u_8), a fixed interval. This particular interval is a single outcome of a stochastic process. The same random mechanisms could have resulted in $X = 5$ and a "realized value" of the CI of (l_5, u_5). The definition of CI is based on properties of the random interval (l_X, u_X), rather than those of particular realizations.

The concept of a random interval is just unfamiliar enough that we might take a moment to consider a simple example. Suppose that we choose two numbers at random on the unit interval (0, 1) and consider the interval extending from the smaller to the larger of the two values. What can we say about this random interval? Its features are random: its length will vary from 0 (if the two variables happen to coincide) to 1 (if the values are as far apart as possible); on the average, the length of the interval will be 1/3. We might ask whether the interval will include specific values, such as 0.6. The answer: sometimes it will, sometimes it will not. If both endpoints are less than 0.6 (probability $= 0.6^2 = 0.36$) or if both endpoints are above 0.6 (probability $= 0.4^2 = 0.16$) the interval will not include 0.6, but otherwise, it will. Thus, the chance that the interval will include 0.6 is $0.48 = 1 - 0.36 - 0.16$.

Imagine a computer program that generates random intervals, as just described. In the long run, 48% of the intervals will include the value 0.6. Suppose the program has been written, tested, is known to function properly, and is about to be used to create one final interval. We can say in advance that we have 48% confidence that the interval will cover the value 0.6.

Let us change the scenario somewhat. Suppose that there is a constant p, unknown to us, and a program that produces random intervals, which include p with probability 0.48. The program's writer knows p and has verified that the program works correctly. In the long run, 48% of the intervals will include the value p. The program runs, and from across the room, the sound of a laser printer warming up and paper being fed through rollers is heard. A single sheet of paper appears in the print tray, containing two numbers, the lower and upper endpoints of an interval. We have 48% confidence that p will be in the interval produced. We read the numbers: 0.328 and 0.832. It is tempting to say that there is a 48% chance that p is between 0.328 and 0.832, but from a frequentist perspective the statement would be incorrect: p is a fixed quantity, and it either *is* in the interval or *is not* in the interval; there is no probability involved with p itself. The only probabilities are those associated with generating the interval. The interval (0.328, 0.832) is described as a 48% CI for p; the "confidence" is a description of the mechanisms producing the interval rather than of the interval itself.

Here, then, is the definition of a CI for the binomial success rate, which we will use for inference about the gender of kleptoparasitic terns.

CI for Binomial Success Rate

Suppose that $X \sim B(n,p)$, an interval C_X is said to be a $100(1-\alpha)\%$ CI for p if $\Pr(p \in C_X | p) \geq 1 - \alpha$ for all p.

Setting $\alpha = 0.05$ (the most commonly used value), the definition says that a 95% CI is a random interval, defined in such a way as to ensure that no matter what the value of p may be, the observed data X will lead to the creation of an interval C_X, which includes p with probability of at least 95%.[6] As before, the quantity α measures the uncertainty associated with the interval; the smaller the value of α, the longer the interval.

We describe techniques for creating such intervals in the following sections. For the moment, we stress that the confidence one may have in a CI is with regard to the technique and not in the particular outcome. Using techniques described below, a 90% CI for p based on eight successes in 10 Bernoulli trials is (0.541, 0.931). It is not correct to say "there is a 90% chance that p is between 0.541 and 0.931"; rather, one may say that the technique used works in 90% of similar circumstances.

A final comment before moving on to particulars. Note that the definition of a CI involves the conditional probability $\Pr(p \in C_X | p)$. We have conditioned on p, that is, treated it as a fixed quantity. The probability statement is about random variation associated with the data X alone. We ask the reader's indulgence if we belabor the point: our reason is that this most elementary topic in statistics, this fundamental concept in the frequentist paradigm, is very commonly misunderstood. Since other aspects of the frequentist paradigm (e.g., hypothesis testing, estimator evaluation) have the same epistemological foundations, it is not surprising that other aspects of this prevailing paradigm are also commonly and similarly misunderstood.

3.2.1 Approximate CI's for Binomial Success Rate

We now describe the most common method of constructing a CI for the binomial success parameter.

Using the normal approximation to the binomial distribution (see B.5), it follows that the MLE $\hat{p} = X/n$ has a distribution that is approximately the same as that of a normal random variable, with mean equal to the true value of p and standard deviation $\sigma(\hat{p}) = \sqrt{p(1-p)/n}$. Consequently, letting $z_{\alpha/2}$ denote the upper $(1-\alpha/2)$ quartile of the standard normal

---

6. The reader might question why the definition requires probability of "at least 95%" rather than "exactly 95%." The answer is that it may be impossible to create an interval with exact $100(1-\alpha)\%$ probability, especially when dealing with discrete data. The distinction is important and will be discussed subsequently.

distribution, it follows that

$$1-\alpha \approx \Pr\left(-z_{\alpha/2} \le \frac{\hat{p}-p}{\sigma(\hat{p})} \le z_{\alpha/2}\right),$$ (3.4)

which after simple rearrangement becomes

$$1-\alpha \approx \Pr\left(\hat{p}-z_{\alpha/2}\,\sigma(\hat{p}) \le p \le \hat{p}+z_{\alpha/2}\,\sigma(\hat{p})\right).$$ (3.5)

For simplicity, the interval defined on the right-hand side of Eq. (3.5) is typically written as $\hat{p} \pm z_{\alpha/2}\,\sigma(\hat{p})$. Because $\sigma(\hat{p})$ depends on the unknown value of p, we substitute $\hat{p}$ in the formula for $\sigma(\hat{p})$, obtaining an estimator

$$\hat{\sigma}(\hat{p}) = \sqrt{\hat{p}(1-\hat{p})/n}.$$

Substituting this estimator in Eq. (3.5), we obtain an approximate $100(1-\alpha)\%$ CI for p based on the observation X, namely

$$I_X = \hat{p} \pm z_{\alpha/2}\,\hat{\sigma}(\hat{p}).$$

The chance is approximately $100(1-\alpha)\%$ that the interval I_X will include p. The approximation is good if n is reasonably large and p is neither too close to 0 nor too close to 1. Note that the length of the interval is inversely proportional to $\sqrt{n}$, so that, as we observed with the likelihood interval, the length of the interval (an expression of statistical uncertainty) decreases at a rate proportional to the square root of the sample size.

The interval I_X is an approximate CI. It is natural to ask how close the approximation is to reality. The definition of a CI requires that for each value of p, the probability of the set of X's which lead to interval estimates including p should be at least $(1-\alpha)$. Coverage rates of the approximate 90% CI with $n=100$ are given in Fig. 3.3; note that for $0.10 \le p \le 0.90$, the coverage rate is reasonably close to 90%, although typically slightly too low.

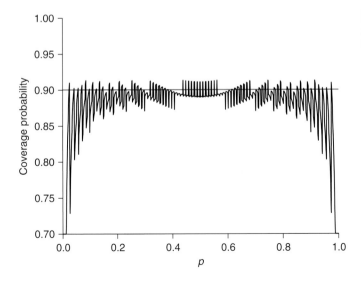

FIGURE 3.3 Coverage rates for approximate 90% confidence interval I_X, $n=100$.

There is something decidedly odd about using an approximation which works pretty well for *many* p's, when the definition of CI requires its satisfactory performance for *all* p's.

An alternative approximate CI with better coverage properties than those of I_X can be obtained by a more complicated rearrangement of Eq. (3.4), which avoids the need to substitute $\hat{\sigma}(\hat{p})$ for $\sigma(\hat{p})$.[7] The resulting CI for p is

$$I_X^* = \frac{2n\hat{p}+z_{\alpha/2}^2}{2(n+z_{\alpha/2}^2)} \pm \frac{z_{\alpha/2}}{2(n+z_{\alpha/2}^2)} \sqrt{4n\hat{p}(1-\hat{p})+z_{\alpha/2}^2}.$$

A comparison of coverage rates for I_X and I_X^* is given in Fig. 3.4. The CI I_X^*, although having considerably better coverage properties than the simpler approximation I_X, is seldom used, probably because of its imposing looking formula. It should be noted that both methods are approximate, and both fail to satisfy the defining criterion of CI's, that the coverage probability should be *at least* $(1-\alpha)$ for every value of p. Ideally, one would routinely use an exact CI, such as we now describe.

3.2.2 Exact CI for Binomial Success Parameter

Let

$$F_U(p) = \sum_{k=x}^{n} B(k;n,p) \quad \text{and} \quad F_L(p) = \sum_{k=0}^{x} B(k;n,p);$$

these are the *upper tail probability*, $\Pr(X \geq x)$, and the *lower tail probability*, $\Pr(X \leq x)$, respectively, for $X \sim B(n,p)$. These are illustrated in Fig. 3.5 for the roseate tern data ($n=10, x=8$).

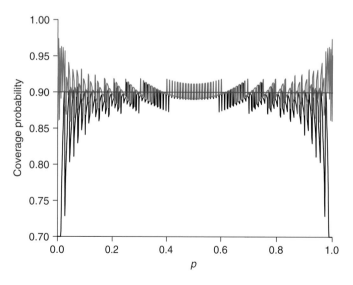

FIGURE 3.4 Coverage rates for nominal 90% confidence intervals, I_X^*(blue) and I_X(black), $n=100$.

7. We must solve for the set of p's satisfying $\left|\hat{p}-p\right|/\sqrt{p(1-p)/n} \leq z_{\alpha/2}$. This is done by squaring both sides of the inequality, rearranging terms, and applying the quadratic formula.

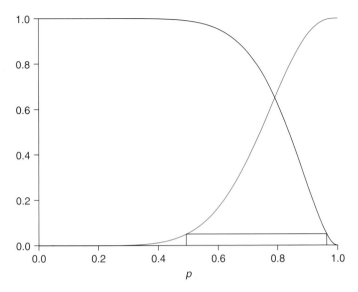

FIGURE 3.5 $F_U(p) = \Pr(X \geq 8 | p)$ (red) and $F_L(p) = \Pr(X \leq 8 | p)$ (blue) for $X \sim B(10, p)$. Figure shows construction of exact 90% confidence interval for p. Height of box (black line) is 0.05.

We can use the curves $F_U(p)$ and $F_L(p)$ to identify values of p that are inconsistent with the observed data. Note that $F_U(p)$ (red curve) is an increasing function of p; thus, since $F_U(0.493) = 0.05$, $\Pr(X \geq 8) \leq 0.05$ for any $p \leq 0.493$. The chance of having eight or more females among 10 randomly selected individuals is no more than 5%, if $p \leq 0.493$. Similarly, $F_L(p)$ (blue curve) is a decreasing function of p, and because $F_L(0.963) = 0.05$, $\Pr(X \leq 8) \leq 0.05$ for any $p \geq 0.963$. For these large values of p, the chance of observing eight or fewer females in a random sample of 10 birds is also no more than 5%.

A 90% exact CI for p is obtained by combining these two facts: the interval $(0.493, 0.963)$ consists of all the values remaining after ruling out those too large to account for an observation of eight or less and those too small to account for an observation of eight or more. The formal definition of such intervals follow the exact CI for binomial success rate.

Exact CI for Binomial Success Rate

Suppose that $X \sim B(n, p)$. Set $p_L(0, \alpha/2) = 0$, and for $x = 1, 2, \ldots, n$ define

$$p_L(x, \alpha/2) = \max_p \{ p : F_U(p) \leq \alpha/2 \}.$$

Set $p_U(n, \alpha/2) = 1$, and for $x = 0, 1, \ldots, n-1$ define

$$p_U(x, \alpha/2) = \min_p \{ p : F_L(p) \leq \alpha/2 \}.$$

Then the interval $J_X = \left(p_L(X, \alpha/2), p_U(X, \alpha/2) \right)$ is an exact $100(1 - \alpha)\%$ CI for p.

Although we have labeled this interval an "exact confidence interval," it remains to be shown that it has the requisite property that $\Pr(p \in J_X | p) \geq 1 - \alpha$ for all p. This requirement, combined with the discreteness of the binomial distribution leads to the unpleasant consequence that the coverage probability is substantially greater than $1 - \alpha$ for many values of p. Consider, for instance, the performance of the exact 70% CI for p based on sample size $n=25$. There are 26 possible outcomes for X, hence 26 possible CI's as summarized in Table 3.2.

For any given value of p, it is a straightforward matter to calculate the coverage probability of the exact CI. For instance, if $p=0.3595$, we see that the interval will include p if and only if $7 \le X \le 11$. If $p=0.3595$, $\Pr(7 \le X \le 11)=0.703$, close to the desired level. However, if we shift our attention slightly, to $p=0.3591$, we see that the interval will include p if and only if $6 \le X \le 11$. For $p=0.3591$, $\Pr(6 \le X \le 11)=0.792$, considerably larger than the nominal 70% coverage rate of the interval.

Figure 3.6 displays the true coverage rates as a function of p, for $n=25$, $\alpha=0.30$.

While it is gratifying to see that the coverage rate is always at least as large as it ought to be, it is somewhat disconcerting to note that for many values of p it is considerably larger than it need be. This phenomenon is not merely associated with the case $n=25, \alpha=0.30$ considered

TABLE 3.2 Exact 70% CI's for p based on x successes in $n=25$ Bernoulli trials.

x	Jx	x	Jx	x	Jx
0	[0.0000, 0.0731]	9	[0.2502, 0.4835]	18	[0.5985, 0.8196]
1	[0.0065, 0.1287]	10	[0.2863, 0.5235]	19	[0.6407, 0.8532]
2	[0.0275, 0.1788]	11	[0.3230, 0.5629]	20	[0.6839, 0.8857]
3	[0.0540, 0.2263]	12	[0.3604, 0.6015]	21	[0.7281, 0.9168]
4	[0.0832, 0.2719]	13	[0.3985, 0.6396]	22	[0.7737, 0.9460]
5	[0.1143, 0.3161]	14	[0.4371, 0.6770]	23	[0.8212, 0.9725]
6	[0.1468, 0.3593]	15	[0.4765, 0.7137]	24	[0.8713, 0.9935]
7	[0.1804, 0.4015]	16	[0.5164, 0.7498]	25	[0.9269, 1.0000]
8	[0.2149, 0.4429]	17	[0.5571, 0.7851]		

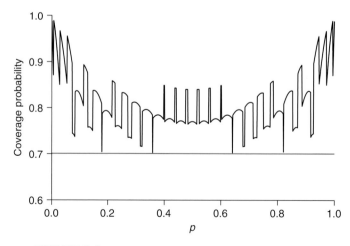

FIGURE 3.6   True coverage rates of exact 70% CI's, $n=25$.

here. It is an inevitable consequence, for discrete random variables, of the requirement that confidence intervals have *at least* $(1-\alpha)100\%$ coverage, for all values of the parameter.

3.2.3 Confidence Intervals – Summary

Like the scaled likelihood interval, CI's provide an expression of the uncertainty inherent in making statistical inference. They have somewhat greater appeal than the likelihood intervals, in that their length is related to probability rather than relative likelihood, which is a coin of unfamiliar currency.

CI's have several complications which are not well-known among practitioners. First, many familiar and simple forms such as $\hat{p} \pm z_{\alpha/2}\sqrt{\hat{p}(1-\hat{p})}$ are based on approximations of question-able value, even for fairly large sample sizes. Strictly speaking, these approximations *do not* yield CI's of specified confidence level, since it is required that the coverage probability of a CI be at least $(1-\alpha)$ for all values of the parameter. Second, when coverage probability of at least $(1-\alpha)$ is attained for all values of the parameter (as required by definition), the coverage probabilities may far exceed $(1-\alpha)$ for most values of the parameter. Our impression is that this phenomenon is not well-known among practitioners; while it is not an indictment of CI's per se, it is undoubtedly a factor mitigating the clarity of their interpretation.

Of the three intervals considered, only J_X is a legitimate 95% CI, for only it has coverage probability which is $\geq 95\%$ for every value of p. The guaranteed minimal coverage of 95% is obtained at a cost: greater coverage is attained by using a longer interval. We plot coverage probabilities for all three intervals in Fig. 3.7, noting that the inadequacies revealed occur despite reasonable sample size ($n=100$) and are more pronounced with smaller sample sizes.

Figure 3.7 was produced using a grid of 1000 values of p, evenly spaced over the interval from 0 to 1; the coverage probabilities on this grid are summarized in Table 3.3. We suspect that many practitioners would prefer the interval I_X^* as a representation of their intuitive concept of "90%

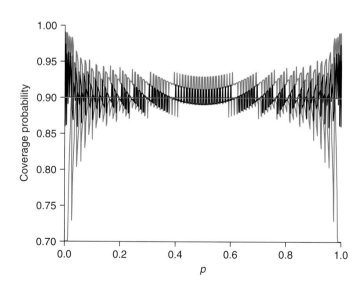

FIGURE 3.7 Coverage rates for exact 90% confidence interval J_X (green), and nominal 90% CI's, I_X^* (black) and I_X(blue), $n=100$.

TABLE 3.3   Summary of coverage properties (mean, standard deviation, percentiles, range) for two approximate 90% confidence intervals (I_X and I_X^*) and the exact confidence interval J_X, across a uniformly spaced grid of 1000 values on $[0, 1]$.

	Mean	sd	Min	2.5th	50th	97.5th	Max
I_X	0.881	0.038	0	0.707	0.891	0.912	0.914
I_X^*	0.902	0.015	0.859	0.875	0.900	0.941	0.974
J_X	0.926	0.016	0.902	0.906	0.923	0.976	0.990

confidence," being willing to sacrifice "minimal 90% coverage" for "typical 90% coverage." However, under the frequentist paradigm, there is no accounting for "typical" behavior across values of p but only a protection against worst-case scenarios. We turn our attention now to the construction of Bayesian interval estimates, which are designed and interpreted with a view to their typical behavior, and which we believe provide an intuitive and appealing portrayal of uncertainty about parameter values.

3.3 BAYESIAN INTERVAL ESTIMATION

The frequentist and Bayesian inferential systems are distinguished, primarily, by their treatment of unknown quantities. The distinction is often said to be that in frequentist thinking, parameters are "fixed but unknown constants," whereas in Bayesian thinking, parameters are random variables.

This epitome is slightly too brief. The fact is that both paradigms treat parameters as fixed and unknown. The frequentist view of parameters begins and ends thus. Nothing more is said; no statement is made about whether any particular values are more to be expected than others. No assessment of prior knowledge is made.

Under the Bayesian paradigm, parameters are fixed and unknown quantities, but are regarded as realizations of random variables, samples from probability distributions that are at least partially specified. These are called *prior distributions*, or simply *priors*. The prior distribution can be thought of as a summary of all that is known about the parameter of interest, without reference to the data. In the absence of prior knowledge about the parameter, or if we are interested in letting the data speak for themselves without the influence of our prior knowledge, a uniform distribution is often chosen.

3.3.1 Basics of Bayesian Inference

The prior distribution (for parameter θ) and the data distribution (for data X, given parameters θ) are all the ingredients needed for a Bayesian analysis. Using the bracket notation of Section 2.2.1, the data distribution is $[X|\theta]$, the prior distribution is $[\theta]$. Bayes' theorem combines these to produce the *posterior distribution* $[\theta|X]$:

$$[\theta|X] = \frac{[X|\theta][\theta]}{[X]}, \tag{3.6}$$

where

$$[X] = \int [X|\theta][\theta]\,d\theta. \tag{3.7}$$

The posterior distribution summarizes all that is known about the parameter, combining prior knowledge and information provided by the data. All Bayesian inference is based on the posterior distribution: if a point estimate is desired, one generally uses the posterior mean (or sometimes the mode, or the median). Interval estimates are constructed from percentiles of the posterior distribution, as described subsequently.

These three distributions, the data distribution, the prior, and the posterior, are the primary features of Bayesian analysis. The distribution $[X]$ is the *marginal distribution* of the data; as indicated by Eq. (3.7), it is the average value of the data distribution, averaged against the prior. This distribution is sometimes used for model checking but need not be computed in characterizing the posterior distribution of θ. That is, we may write Eq. (3.6) as

$$[\theta|X] \propto [X|\theta][\theta]; \tag{3.8}$$

knowing that $[\theta|X]$ is a distribution function, its integral with respect to θ must equal 1; it follows that the proportionality constant in Eq. (3.8) is $1/[X]$.

In defining the posterior distribution, the right-hand side of Eq. (3.8) treats $[X|\theta]$ as a function of θ, with X fixed, that is, as a version of the likelihood function, as discussed in Section 3.1. Equation (3.8) says that the posterior distribution is proportional to the product of the likelihood and the prior. Thus, the basis of inference is the product of information provided by the data and by the prior, summarized by the mnemonic

$$\text{Inferential Basis} = \text{Data} + \text{Prior Knowledge.} \tag{3.9}$$

3.3.2 Bayesian Inference for Binomial Success Rate

We return to our example, Shealer and Spendelow's (2002) observation of eight females among 10 kleptoparasitic roseate terns, to illustrate the basics of Bayesian inference. We have modeled the data as an observation X of a binomial random variable with index $n = 10$ and unknown success rate p; the data distribution is $[X|p] = \Pr(X = x|p) = B(x;n,p).$[8] Bayesian analysis of the data $X = 8$ requires nothing more than specification of a prior distribution $[p]$.

The choice of prior is, for some folks, the great Bayesian Bugaboo. The inferential basis being determined not only by the data but also by the choice of prior, some would caricature Bayesian inference as allowing a sort of intellectual anarchy: "choose your prior right, and you can get any answer you want."

---

8. Note that we no longer write $\Pr(X=x) = B(x;n,p)$, but, rather $\Pr(X=x|p) = B(x;n,p)$, in acknowledgment of the perspective that the data distribution $B(x;n,p)$ depends on a random quantity, p.

Indeed, there is great flexibility in the choice of prior: $[p]$ can be any distribution on $[0, 1]$. We see this as a strength, rather than a weakness of the Bayesian inferential system: Bayesian inference includes a formal mechanism for updating prior knowledge. This is a virtue, rather than a vice, when subject to the obvious proviso that one must report the prior as part of the analysis, and that one should assess the sensitivity of the conclusions drawn to the choice of prior. If one chooses a prior which expresses dead certainty about the value of the parameter, then the data will be ignored; however, if the prior expresses uncertainty about the parameter, then as sample size increases, the data will prevail in guiding inference.

For now, we will table the question of how one goes about choosing a prior distribution, and discussion of the implications of the choice, except to make note of two considerations: one practical and the other epistemological. The first of these is that certain families of prior distributions are very naturally chosen, on the grounds of convenience and expositional clarity. A good example is provided by the choice of a beta distribution

$$Be(p;a,b) = \frac{\Gamma(a+b)}{\Gamma(a)\Gamma(b)}p^{a-1}(1-p)^{b-1}, 0 < p < 1 \text{ and } a,b > 0, \tag{3.10}$$

as a prior for the binomial success parameter.[9] From Eq. (3.8), we find that the resulting posterior distribution is

$$[p\,|\,X] \propto B(x;n,p)\,Be(p\,;a,b) = \frac{\Gamma(a+b)}{\Gamma(a)\Gamma(b)}p^{a-1}(1-p)^{b-1}\binom{n}{x}p^{x}(1-p)^{n-x}$$

$$\propto p^{x+a-1}(1-p)^{n-x+b-1}. \tag{3.11}$$

The last line can be written as $p^{A-1}(1-p)^{B-1}$, with $A=a+x$ and $B=b+n-x$; we need only to add on the appropriate proportionality factor to put this in the form of the beta distribution as given in Eq. (3.10). The beta family of distributions is said to be *conjugate* for the binomial success parameter: the prior and posterior distribution are in the same family. The beauty of this is that the posterior from one study is ready made to serve as prior in the next. Also, the effect of data on parameter values provides a heuristic for understanding the quantification of uncertainty by the probability distributions: in the present example one may think of parameters a and b as running totals of previous numbers of successes and failures. Thus, a $Be(30,20)$ prior can be thought of as roughly equivalent to the knowledge acquired from previous experience of 30 successes and 20 failures in 50 previous trials. There is also a computational efficiency associated with using a conjugate prior: the posterior being of known form, there is no need to compute the integral in Eq. (3.7) which scales the posterior distribution defined by Eq. (3.6). We will speak more about conjugate priors in Chapters 4 and 6.

The second consideration is that analysts often wish to let the data speak for itself, as much as possible; they wish to substitute "0" for "*Prior Knowledge*" on the right-hand side of Eq. (3.9), to obtain an "Objective Bayesian analysis" in which the data alone provide the basis of inference. Technical difficulties in doing so are much trumpeted by non-Bayesian critics, and no single solution has been settled upon by Bayesian apologists. We believe that the issues are interesting but that their significance has been considerably overstated; we will return to discuss them in later chapters.

---

9. For details on the beta distribution, see Appendix B.10.

When an objective Bayesian analysis is desired for the binomial success rate p, a common choice of prior is the uniform distribution on [0, 1]. A uniform prior on p says that we have no more reason to believe it falls in any particular subinterval of [0, 1] than in any other of the same length. Note also that the $U(0,1)$ density is constant, hence the posterior distribution is simply a scaled version of the likelihood; thus, the desideratum of substituting "0" for "*Prior Knowledge*" on the right-hand side of Eq. (3.9) can be said to have been met.

The $U(0,1)$ distribution is a member of the beta family of distributions; it is $Be(a,b)$, with $a=b=1$. From the observations following Eq. (3.11), we conclude that the corresponding posterior distribution is $Be(X+1, N-X+1)$.

What of the kleptoparasitic terns? The data consist of eight successes in 10 independent Bernoulli trials. Thus given a uniform prior, the posterior distribution of p is $Be(9,3)$, plotted in Fig. 3.8. The mode of the distribution, from Appendix B.10, is 8/10, the same as the MLE as calculated in Section 3.1. This is no coincidence, but a consequence of the choice of the uniform prior on p, which leads to the posterior distribution being a scaled version of the likelihood. We will discuss similarities between "flat prior" Bayesian and frequentist analysis in subsequent chapters.

Of more immediate interest in the present context is the construction of Bayesian interval estimates. Since the 5th and 95th percentiles of the $Be(9,3)$ distribution are 0.530 and 0.921, respectively, we conclude that

$$\Pr\big(p \in (0.530, 0.921)|X=8\big) = 0.90.$$

Thus, under the Bayesian paradigm it is completely legitimate to say "there is a 90% chance that p is in the interval (0.530, 0.921)." This stands in marked contrast to the frequentist statement, based on the same data, that "the interval (0.493, 0.963) is an exact 90% confidence interval for p." The former has a simple and intuitive meaning, the latter is potentially misleading and commonly misunderstood.

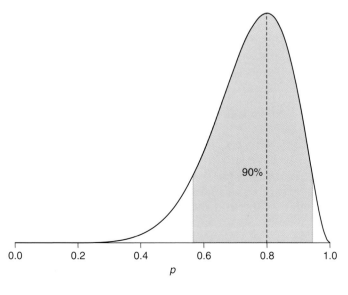

FIGURE 3.8 Posterior distribution for proportion females among kleptoparasites. Interval (0.566, 0.944) is the 90% HPDI, the shortest 90% credible interval.

Alternative interval estimates can be obtained from the same posterior distribution. Because the 10th and 90th percentiles of the $Be(9,3)$ distribution are 0.585 and 0.895, we may also conclude that

$$\Pr\big(p \in (0.585,1)|X=8\big) = 0.90$$

and

$$\Pr\big(p \in (0,0.895)|X=8\big) = 0.90.$$

We will refer to the intervals (0.530, 0.921), (0.585, 1), and (0, 0.895) as 90% *credible intervals* for p. We state the following definition:

Definition: Credible Interval for Binomial Success Rate

Suppose that $X|p \sim B(n,p)$. Then any interval B_X with the property that

$$\Pr(p \in B_X|X) = 1 - \alpha$$

is a $100(1-\alpha)\%$ credible interval for p.

We encourage the reader to compare the definitions of credible intervals and CI's. They are similar in that both involve probabilities for the event $p \in K_X$, where K_X is an interval calculated from the data X. The crucial difference is that the CI conditions on p, whereas the credible interval conditions on X. Bayesian and frequentist analyses are distinguished by whether one conditions on things unknown, or things known.

The three intervals presented may be distinguished as central, upper-tailed, and lower-tailed credible intervals. The lengths of these intervals are 0.391, 0.415, and 0.895. One might guess that the central credible interval would be the shortest possible, but in fact that distinction goes to the interval (0.566, 0.944), of length 0.378.[10] Some authors reserve the term "credible interval" for this particular interval, but since it is also distinguished as the *highest posterior density interval* (HPDI), we will apply the term "credible interval" to any interval containing $100(1-\alpha)\%$ of the posterior distribution's mass. It is often the case that the central credible interval is only marginally longer than the HPDI, and almost always the case that computation of the HPDI is substantially more difficult.

In the introduction to this text, we said that Bayesian inference is appealing because of its simplicity. The interpretation of interval estimates is a striking case in point. Consider once again the data on habitually kleptoparasitic roseate terns:

1. *Summary of Objective Bayesian analysis*: "In the absence of data, we know nothing about p, and suppose that it is equally likely to be anywhere in the interval [0, 1]. The data lead us to believe that there is a 90% chance that $0.566 \le p \le 0.944$."
2. *Summary of exact frequentist analysis*: "We have 90% confidence that p lies in the interval $J_X = (0.493, 0.963)$. Our confidence lies in the methods used to construct the interval: these methods succeed in producing an interval that includes p in at least (and possibly well over) 90% of their applications, regardless of the value of p."

---

10. See Fig. 3.8. This HPDI ranges from the 8th to 98th percentiles of the $Be(9,3)$ distribution. If the posterior distribution is symmetric and unimodal, the HPDI and the central credible interval coincide. The $Be(9,3)$ distribution is unimodal, but not symmetric; the distinction between central CI and HPDI is due to the skewness of the distribution.

In practice, exact analyses are seldom used, and a frank assessment of results would go something like this:

3. *Summary of approximate frequentist analysis*: "We have 90% confidence that p lies in the interval $I_X = (0.592, 1.008)$. Our confidence lies in the methods used to construct the interval: these methods attempt to produce an interval that includes p in at least 90% of their applications, regardless of the value of p. To be honest, we have some misgivings about the procedures, because we've used large sample approximations along the way in developing them, and it's not at all clear how well they work in this small sample setting."

3.3.3 Incorporating Prior Knowledge in Interval Estimate for Binomial Success Parameter

The $U(0, 1)$ prior for p we have used so far is chosen to represent prior ignorance about p, or to allow the data to speak without influence of prior knowledge. The Bayesian paradigm also provides a formal mechanism for updating prior knowledge. In Fig. 3.9, we illustrate the effect of prior on posterior distribution. The top panel displays a uniform prior distribution and four beta priors; the bottom panel displays the corresponding posterior distributions for the roseate tern kleptoparasite data, that is, an observation $X = 8$ from $X|p \sim B(10, p)$. For the purpose of illustration, we have chosen a prior mean of $\pi = a/(a+b) = 0.35$ for the beta distributions, a value in contradiction to the observed data. The concentration parameters for the beta distributions are $\theta = a + b = 2$ (red), 6 (blue), 24 (brown), and 120 (green). Clearly, the more specific the prior knowledge, the greater the influence of the prior on the posterior. The prior with $\theta = 120$ can be thought of as equivalent to previous experience of a data set consisting of 42 successes and 78 failures.[11] In light of such prior data, we would surely be inclined to discount the significance of eight successes in only 10 trials; the Bayesian calculus provides a formal mechanism for doing so.

A remarkable feature of this analysis is that the posterior mean is a weighted average of the prior mean and the maximum likelihood estimator. For $p \sim Be(a, b)$, $E(p) = a/(a+b)$. Thus, since the posterior distribution $p|X$ is $Be(a + X, b + n - X)$, the posterior mean is

$$E(p|X) = \frac{a+X}{a+b+n} = \frac{\theta}{\theta+n}\left(\frac{a}{a+b}\right) + \frac{n}{\theta+n}\left(\frac{X}{n}\right)$$

where $\theta = a + b$. Regarding θ as a prior sample size, we can see the weights on prior mean and data correspond to the relative sample sizes; if n is substantially larger than θ, the posterior mean will be essentially the same as the MLE, $\hat{p} = X/n$. If θ is much larger than n, the posterior mean will be closer to the prior mean than to the MLE. The Bayes estimate $\hat{\theta}^B = E(p|X)$ is often conceived as the result of moving a proportion $\theta/(\theta+n)$ of the way from the MLE along a line segment to the prior mean; this phenomenon of *shrinkage* is common in Bayesian analysis.

---

11. To be precise, we might say $Be(42, 78)$ is the posterior resulting from a uniform prior based on previous experience of 41 successes and 77 failures, analyzed using a uniform $Be(1, 1)$ prior.

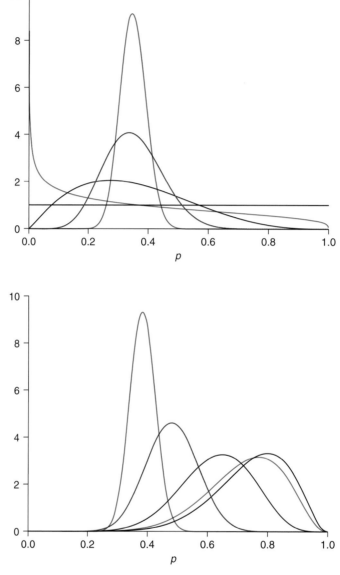

FIGURE 3.9   Top panel: uniform prior (black) and beta priors with mean 0.35 and $\theta = a + b = 2, 6, 24$, and 120 (red, blue, brown and green). Bottom panel: corresponding posteriors obtained from observation $X = 8$, for $X|p \sim B(10, p)$.

3.4 SUMMARY AND COMPARISON OF INFERENTIAL SYSTEMS

Frequentist methods generally, and CI's in particular, are defined with regard to hypothetical replicate data produced by a fixed set of parameters. No attempt is made to describe prior knowledge about the parameter. CI's are defined in terms of the worst-case scenario for coverage probability over all values of the parameter.

By contrast, Bayesian methods describe uncertainty about parameters using probability distributions. The prior distribution $[\theta]$ describes what is known or assumed about the parameter without regard to the data to be analyzed. The posterior distribution $[\theta|X]$ is obtained by combining the prior and the data distribution $[X|\theta]$ through Bayes' theorem, and is the basis of all Bayesian inference. The prior distribution can be chosen either as reflecting prior knowledge or with the goal of minimizing its influence on the analysis. Either way, all inference is based on the posterior distribution, conditioning on a fixed set of data rather than hypothetical replicate data sets.

We have focused on interval estimation, noting that Bayesian interval estimates are described using $\Pr(\theta \in C_X|X)$, whereas frequentist intervals are based on $\Pr(\theta \in C_X|\theta)$. More generally, we may describe frequentist methods as based on the data distribution $[X|\theta]$, and Bayesian methods as based on posterior distributions $[\theta|X]$.

Similar considerations govern the choice of point estimators. Frequentist estimation criterion rely on the the conditional distribution of hypothetical data, given a fixed set of parameters. Bayes estimates are based on the conditional distribution of parameters, given the observed data.

3.4.1 Point Estimation

Frequentist Estimation Criterion: Unbiasedness

A standard frequentist optimality criterion for estimators is unbiasedness. Estimator $\hat{\theta}(X)$ is unbiased for parameter θ if

$$E\left(\hat{\theta}(X)|\theta\right) = \theta,$$

the expectation being taken over hypothetical replicate data sets X. At first glance, the criterion appears reasonable: if over a lifetime of providing estimates, the statistician's values average out to the truth, everyone should be happy.

But a critic might say "I'm not interested in how you do over a lifetime of application, but in how you do *this time.*" The critic desires a different optimality criterion, not based on hypothetical replicates of the data set at hand, but only based on the present data set.

Such a criticism might not be all that unreasonable. For instance, suppose that X is a Poisson random variable with mean λ, and that an estimator of $\theta = \exp(-2\lambda)$ is desired. It can be shown that the unique unbiased estimator of θ is $(-1)^X$. The estimator equals ± 1, depending on whether X is even or odd. These values make no sense because $0 < \theta < 1$.

Many similar cases of "silly" unbiased estimators exist. Occasionally, an unbiased estimator of a variance component returns a negative estimate. Analysts often truncate or discard such estimates, not realizing that this practice effectively overthrows the basis of their selected optimality criterion, which depends on all outcomes, reasonable or otherwise, to evaluate typical performance.

Maximum Likelihood Oddity

Even the highly regarded criterion of maximum likelihood can fail, and spectacularly at that. Consider the following family of distributions:

$$f(x|\theta) = \frac{1/2}{\sqrt{\theta} + \sqrt{1-\theta}} \times \frac{1}{\sqrt{|x-\theta|}}, \tag{3.12}$$

for $0 < x < 1$, indexed by parameter $\theta \in (0,1)$. If you do the integration with respect to x, first over $x \in (0,\theta)$, then over $x \in (\theta,1)$, you will see that these distributions integrate to one, as they should.

However, the likelihood function is a mess, having an infinite spike at every observed value x_i. Figure 3.10 displays the likelihood function based on a sample $x_1 = 0.250, x_2 = 0.333, x_3 = 0.400$, and $x_4 = 0.500$.

So we would conclude that maximum likelihood is not a good criterion to use, in estimating θ. We would be forced to try something else. By contrast, Bayesian analysis always follows the same process: integrate the likelihood against a prior to obtain a posterior distribution, and base inference on the posterior distribution. Using a uniform prior on θ, the four observations lead to a posterior distribution with median 0.357; the interval $(0.182, 0.560)$ is a 90% credible interval.

Bayesian Estimation Criterion: Minimum MSE

Bayesians evaluate point estimators on the basis of their performance as measured by posterior distributions. For instance, the mean-squared error of $A = \hat{\theta}(X)$ (a function of data X) as an estimator of parameter θ, computed relative to the posterior distribution of θ, is

$$E\left((\theta - A)^2 | X\right).$$

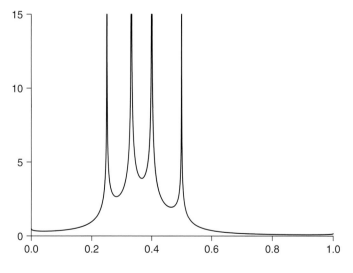

FIGURE 3.10 Likelihood function based on Eq. (3.12) for observations 0.250, 0.333, 0.400, and 0.500.

This MSE is minimized by setting $A = E(\theta|X)$, the posterior mean. If, instead, one wishes to minimize the mean absolute deviation,

$$E(|\theta - A||X),$$

the appropriate choice is the posterior median.[12] Either way, estimators are evaluated by their performance relative to the posterior distribution $[\theta|X]$.

3.4.2 Conclusion (peroration)

It has been widely observed that many classically trained statisticians use frequentist methods, but interpret them as though they were Bayesian. Statistics teachers take fiendish pleasure in catching their students out in thinking there's a 95% chance that the parameter is in their confidence interval, when the confidence is in the method, and is not a statement of probabilities. Similarly, we often hear (or read) analysts saying that they have an "unbiased estimate" of a quantity, when there is no such thing: unbiasedness is a property of estimators, of the machinery that produces estimates, rather than of the estimates themselves. The reality is that an unbiased estimator can produce really bad estimates.

The fact is, most folks are more naturally attracted to the Bayesian use of probability. We are confident that many of our readers will look at Table 3.3 and conclude that intervals I_X and I_X^* worked pretty well. And they do, from a flat prior Bayesian perspective. But they are dismal failures from a frequentist perspective. Looking again at Table 3.3, we imagine our readers will see the performance of interval J_X as rather unsatisfactory. And it is, from a Bayesian perspective. But of the three intervals, it is the only one satisfying the requirements for a frequentist CI.

The more one becomes aware of the complicated optimality criteria associated with frequentist thinking and the failures of such fundamental and apparently laudable criteria as unbiasedness, the more appealing the simplicity of the Bayesian way of thinking. The more one considers the approximate nature of many frequentist computations and the associated and unmeasured uncertainties, the more satisfactory the exact computation of posterior distributions becomes.

So then, our conclusion is this: if it is natural to look at things from a Bayesian perspective, why not go the whole way and embrace the Bayesian paradigm, with probability the tool and the basis for describing uncertainty? Subsequent chapters will highlight the simplicity of the inferential system, and the complexity of models that can be addressed with mathematical rigor, and without dubious distributional approximations.

---

12. See rules 16 and 17 in Appendix A.

Calculating Posterior Distributions

O U T L I N E

4.1	**Conjugacy**	**49**	**4.3**	**Markov Chain Monte Carlo**	**59**
	4.1.1 Hautapu Trout	49		4.3.1 Markov Chains and	
	4.1.2 Fisher's Tick Data	51		Stationary Distributions	59
	4.1.3 A Multivariate Posterior			4.3.2 Example: Standard Normal	
	Distribution	52		Stationary Distribution	61
				4.3.3 Metropolis–Hastings Algorithm	65
4.2	**Monte Carlo Methods**	**54**		4.3.4 Gibbs Sampling	67
	4.2.1 Cdf Inversion	55		4.3.5 MCMC Diagnostics	70
	4.2.2 Rejection Sampling	56			

As outlined in Chapter 3, Bayesian inference is based on straightforward use of Bayes' theorem. Inference about an unknown parameter θ is based on its posterior distribution,

$$[\theta|Y] = \frac{[Y|\theta][\theta]}{[Y]}. \tag{4.1}$$

In this setting, Eq. (4.1) is treated as having only one variable, θ. The data Y are fixed, and $[Y|\theta]$, rather than being the data distribution of Y for fixed θ, is the likelihood for θ given fixed Y.

Given the likelihood and the prior distribution $[\theta]$, the posterior distribution $[\theta|Y]$ is completely determined; it is the unique distribution function proportional to $[Y|\theta][\theta]$. The only role of the denominator $[Y]$ in defining the posterior is as a "normalizing constant," a scaling factor included so that the posterior distribution integrates to 1. Thus, Eq. (4.1) is often simply written as

$$[\theta|Y] \propto [Y|\theta][\theta]. \tag{4.2}$$

As far as defining the posterior distribution is concerned, the normalizing constant is irrelevant.

47

Of course, it is not enough to define the posterior distribution, to say it is the unique distribution function proportional to $[Y|\theta][\theta]$. We want more details. We want to be able to report its mean, its variance, its percentiles; we may want to know specific probabilities, such as $\Pr(\theta > 0|Y)$. While we do not need to know $[Y]$ to define $[\theta|Y]$, we do need to know it if we are to directly calculate these features. The normalizing constant is found by averaging the data distribution $[Y|\theta]$ against the prior $[\theta]$, as

$$[Y] = \int [Y|\theta][\theta]d\theta. \tag{4.3}$$

Unfortunately, evaluating this integral is almost always difficult and frequently impossible. It is not stretching the matter too far to say that the calculation of $[Y]$ is and has been the primary obstacle to implementation of Bayesian methods. In this chapter, we describe a variety of methods for getting around this difficulty, so that we may evaluate posterior distributions.

We begin in Section 4.1 with the simplest solution. In special cases, the prior combines with the likelihood to produce a posterior distribution of similar form to the prior: the prior is said to be *conjugate* to the likelihood. If one can choose a conjugate prior, the form of the posterior is known. Conjugacy completely solves the computational problem, but there are relatively few cases where it applies.

When simple solutions based on conjugacy are not available, most Bayesian applications examine posterior distributions by random sampling. If we can draw a sample of θ from $[\theta|Y]$, we can use features of the sample as estimates of corresponding features of the posterior distribution. The sample proportion of values $\theta > 0$ estimates $\Pr(\theta > 0|Y)$; the sample mean value of θ estimates the posterior mean value of θ. These estimates can be made as precise as we want by drawing large enough samples. Simulation methods for studying probability distributions are known generally as *Monte Carlo* methods; we consider these in Section 4.2.

Monte Carlo methods typically involve draws of independent samples from the distribution being studied. The most straightforward approach is through inversion of cumulative distribution functions (Section 4.2.1) but for posterior distributions this requires that we be able to compute the integral in Eq. (4.3). Another approach, which avoids this requirement, is *rejection sampling*, which we describe in Section 4.2.2.

It is usually not a straightforward matter to draw independent samples from posterior distributions, even while using rejection sampling and similar techniques. Bayesian statistics has been revolutionized by the development of techniques for drawing *dependent* samples from the posterior distribution. These samples can be used in a similar fashion to independent samples. A broad class of techniques, collectively referred to as *Markov chain Monte Carlo* (MCMC) has been developed since the mid-twentieth century and is introduced in Section 4.3. We describe the Metropolis–Hastings (MH) algorithm (Section 4.3.3) and Gibbs sampling (Section 4.3.4), two of the most important techniques used in MCMC. A thorough knowledge of the mathematics behind them is beyond the scope of this text, but also beyond the scope of what is needed by the practitioner. However, some skill is needed in using MCMC; a basic understanding of the process is crucial. There is a certain amount of art involved in the conduct and evaluation of simulations. We conclude with some guidelines for evaluating results in Section 4.3.5.

4.1 CONJUGACY

Sometimes, identifying posterior distributions is a piece of cake. The best case is when we can identify a family of distributions, which includes both the prior and posterior distributions. Distributions in this "conjugate family" are identified by a hyperparameter ψ. We can summarize what was known before data collection by a particular value ψ_0, and what is known afterwards, by a new value ψ_1. The process of updating knowledge by data boils down to updating ψ, and the transition from ψ_0 to ψ_1 involves simple summaries of the data.

The existence of a conjugate family depends on the form of the likelihood function. There are not many cases where conjugate families exist, but some of them correspond to common and useful models.[1] We present three examples, first involving the binomial likelihood (Section 4.1.1), second involving the Poisson likelihood (Section 4.1.2), and third involving the normal likelihood (Section 4.1.3). The first two of these are straightforward; the third illustrates the additional complexities of posterior analysis associated with multivariate parameters.[2]

4.1.1 Hautapu Trout

In 1988, the Wellington Fish and Game Council carried out a study to estimate the proportion of fish counted in drift dive counts of brown trout in the Hautapu River. Divers floated downstream in line-abreast, counting trout. In this study, 13 fish were tagged and released into a section of river that had been netted off to prevent fish from moving in or out. During three replicate dives 10, 7, and 8 of the 13 tagged fish were seen.

On each dive, each fish is either seen or not seen; we model these events as Bernoulli trials. We will suppose that the probabilities of seeing fish are constant, varying neither among dives nor among fish. If we assume further that the individual trials are independent, the data distribution for the total number of sightings Y is binomial, $Y \sim B(N, p)$, with $N = 3 \times 13 = 39$, and $Y = 10 + 7 + 8 = 25$, with p denoting the unknown sighting probability.

Suppose that our prior knowledge about p can be represented by a beta distribution with parameters a_0 and b_0. Then,

$$[p|Y] \propto [Y|p] \times [p]$$
$$\propto p^{25}(1-p)^{39-25} \times p^{a_0-1}(1-p)^{b_0-1}$$
$$= p^{a_1-1}(1-p)^{b_1-1},$$

where $a_1 = 25 + a_0$ and $b_1 = 39 - 25 + b_0$. Thus, the posterior distribution is proportional to a beta distribution, but with parameters a_0 and b_0 replaced by $a_1 = Y + a_0$ and $b_1 = N - Y + b_0$. Since the posterior distribution must integrate to 1, it must actually *be* a beta distribution; thus,

---

1. All members of the exponential family of distributions, which includes most of the commonly used distributions, have a conjugate prior distribution.
2. ... and might be found disturbing by folks who do not like lots of Greek letters. If that is you, do not fuss the details; it is the basic ideas that count.

we deduce that

$$[p|Y=25] = \frac{\Gamma(a_0+b_0+39)}{\Gamma(a_0+25)\Gamma(b_0+14)}p^{a_0+25-1}(1-p)^{b_0+14-1}.$$

Given the binomial likelihood, and a beta prior, we obtain a beta posterior: the beta family of priors is conjugate for the binomial likelihood. The hyperparameter $\psi = \{a,b\}$ is updated by adding the number of successes to a_0, and the number of failures to b_0.

How do we choose values a_0 and b_0? In the absence of specific knowledge, the choice $a_0 = 1$ and $b_0 = 1$ may be reasonable, since it describes a uniform prior on p. The probability that p lies within any subinterval of [0, 1] is the same for all equal-length subintervals; this choice is often made when one wishes to let the data speak for themselves, in an attempt to express complete lack of prior knowledge about p.

Given the uniform prior, we have $a_1 = 26$ and $b_1 = 15$. The posterior density, illustrated in Fig. 4.1, has its mode at

$$\frac{a_1-1}{a_1+b_1-2} = 25/39 = 0.641.$$

It is noteworthy that the posterior mode is the same as the ordinary maximum likelihood estimator, in consequence of the uniform prior on p.

The posterior distribution for p has 95% of its mass located between 0.488 and 0.777. Thus, from an initial position of being unwilling to assign any particular range of values for p as more likely than any other range of values we can now assert that there is a 95% probability that p lies in the range 0.488 to 0.777. From this posterior density, we can also express the belief that we are 96% certain that more than half the fish are seen on a dive.

The mean of the posterior density is $a_1/(a_1+b_1)=26/41=0.634$ and as discussed in Section 3.3.3, can be found as a weighted average of the prior mean (0.5) and the sample proportion of $25/39 = 0.641$, with a weight of $n/(a+b+n) = 39/41 = 0.95$ given to the sample proportion. With $n = 39$ and a $Be(1,1)$ prior, the posterior is dominated by information from

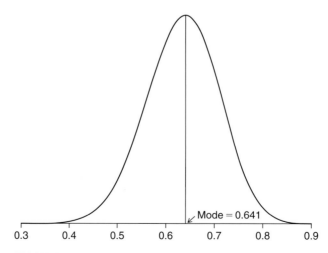

FIGURE 4.1   Posterior density of p for the Hautapu trout study.

the experiment – 95% weight is given to the sample proportion in the posterior mean, so there is relatively little shrinkage toward the prior mean.

4.1.2 Fisher's Tick Data

In a famous study analyzed by R. A. Fisher, data were collected on the numbers of ticks found on 60 sheep. As the data are counts, a useful starting model is the Poisson distribution.

Let $Y = \{y_1, \ldots, y_n\}$ denote the observed values in a random sample drawn from a Poisson distribution with parameter λ, and suppose that the prior for λ is a gamma distribution with parameters α_0 and β_0. We find that the posterior distribution $[\lambda|Y]$ is given by

$$[\lambda|Y] \propto \prod_{i=1}^{n} \frac{e^{-\lambda}\lambda^{y_i}}{y_i!} \times \frac{\beta_0^{\alpha_0}}{\Gamma(\alpha_0)} \lambda^{\alpha_0-1} e^{-\beta_0\lambda}$$

$$\propto \left(e^{-n\lambda}\lambda^{n\bar{y}}\right) \times \left(\lambda^{\alpha_0-1} e^{-\beta_0\lambda}\right)$$

$$= \lambda^{\alpha_1-1} e^{-\beta_1\lambda}, \tag{4.4}$$

where $\alpha_1 = n\bar{y} + \alpha_0$ and $\beta_1 = n + \beta_0$. Equation (4.4) is the *kernel* of a gamma distribution (i.e., all but the normalizing constant that makes the function integrate to 1). Thus, we can deduce that the posterior distribution is in the same family as the prior; both are gamma distributions. The gamma family of priors is conjugate for the Poisson likelihood. The hyperparameter $\psi = \{\alpha, \beta\}$ is updated by adding the total count to α_0, and the sample size to β_0.

For Fisher's tick data, the observed mean number of ticks per sheep was 3.25. A vague gamma prior, chosen to allow the data to speak for themselves and to specify little or no prior knowledge of λ, is one where α_0 and β_0 are both small. Setting $\alpha_0 = 0.0001$ and $\beta_0 = 0.0001$, a 95% credible interval from the posterior distribution (Fig. 4.2) indicates a value for λ somewhere in the range 2.81 to 3.72, with expected value 3.25.

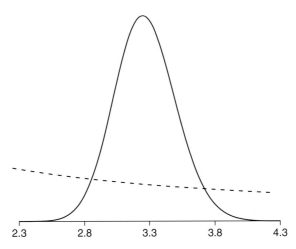

FIGURE 4.2   Posterior distribution for the Poisson rate parameter λ for Fisher's tick data. Also shown is a portion of the gamma(0.0001, 0.0001) prior, multiplied 10,000-fold to allow comparison.

I. PROBABILITY AND INFERENCE

With just one parameter, λ, the Poisson often provides a relatively poor fit to replicated count data; typically one finds evidence of excess variation. Recall that for independent Poisson random variables, the variance among replicates should equal the mean of the replicates. In this example, it seems likely that the Poisson rate parameter should vary from sheep to sheep. For example, larger sheep might be expected to carry more parasites than smaller sheep. Some animals because of their habits may have higher exposure to parasitism than others. One way to account for these missing covariates is to model the Poisson rate parameter as varying among sheep, with λ_i for sheep i being sampled from a prior distribution, itself determined by higher-level parameters.

Alternatively, we might specify λ_i as a parametric function of observed covariates (if these are available). Either way, the analysis proceeds hierarchically, with new parameters governing higher order relations. Model fitting becomes more complicated, but is easily handled under the Bayesian paradigm. We revisit these data later, illustrating how to go about investigating such relations.

4.1.3 A Multivariate Posterior Distribution

With more than one unknown, inference is based on a multivariate (joint) posterior distribution. This distribution is of the same form as Eq. (4.1), but now the parameter $\boldsymbol{\theta}$ is vector valued, and $[\boldsymbol{\theta}|Y]$ is a multivariate distribution.

Sometimes the multivariate characteristics of this joint distribution are of interest; one might take interest in the posterior correlations among parameters as a measure of association among estimates. More often, inference focuses on one parameter at a time. This is especially the case in models where there are many *nuisance parameters*, that is, parameters that are needed in order to correctly describe the sampling model but that are of little direct interest.

Models with large numbers of nuisance parameters are very common in ecological statistics. For example, capture probabilities p are needed in mark-recapture models, but inference usually focuses on demographic parameters such as survival probabilities or population size. The parameters p really are a nuisance: there may be nearly as many p's as parameters of interest, they may provide no insights into population features, and estimating them may introduce substantial cost, both in terms of design and analysis. But without including them, estimates of parameters of interest are likely to be biased.

For a vector-valued parameter $\boldsymbol{\theta} = (\theta_1, \theta_2, \ldots, \theta_k)'$, Eq. (4.2) becomes

$$[\boldsymbol{\theta}|Y] \propto [Y|\boldsymbol{\theta}][\boldsymbol{\theta}],$$

that is,

$$[\theta_1, \ldots, \theta_k|Y] \propto [Y|\theta_1, \ldots, \theta_k][\theta_1, \ldots, \theta_k].$$

Typically, but not always, the joint prior distribution is the product of independent priors for the individual θ_i.

To make inference about a single parameter, we examine the marginal posterior for the parameter of interest. This is found by integrating the remaining parameters from the joint posterior distribution. For example, the marginal posterior distribution for parameter θ_1 is found by,

$$[\theta_1|Y] = \int \cdots \int [\theta_1, \ldots, \theta_k|Y] d\theta_2 \ldots d\theta_k.$$

These integrals, like many encountered in Bayesian inference, are almost always analytically intractable. So inference for a single parameter in a multiparameter setting involves two potentially difficult calculations: first, of the joint posterior, second, of the marginal posterior from the joint. Fortunately, these two tasks are easily accomplished by the simulation techniques presented in this chapter.

There are cases where conjugate priors exist for multivariate parameters. This makes the first task, of identifying the joint posterior, feasible. An interesting example is when $Y = (Y_1, Y_2, \ldots, Y_n)'$ is a sample of a normal distribution, $Y_i \sim N(\mu, \sigma^2)$. Suppose that we wish to make inference about μ, and that the variance is an unknown. The parameter vector is $\theta = (\mu, \tau)'$, where $\tau = 1/\sigma^2$ (the precision) is a nuisance parameter. Inference for the mean is to be based on its marginal posterior distribution

$$[\mu | Y] = \int [\mu, \tau | Y] d\tau;$$

we must first calculate the joint posterior distribution $[\mu, \tau | Y]$, then integrate over τ. For this problem a conjugate family is defined in terms of a gamma distribution for τ and a normal distribution for μ, given τ. That is,

$$[\theta] = [\mu, \tau] = [\mu | \tau][\tau],$$

where $[\mu | \tau] = N(\eta, 1/(\kappa \tau))$ and $[\tau] = \text{Ga}(\alpha, \beta)$. As before, we have a specified set of hyperparameters $\psi = (\alpha, \beta, \eta, \kappa)'$ indexing the conjugate family of distributions.

Suppose that the prior distribution has parameter vector $\psi_0 = (\alpha_0, \beta_0, \eta_0, \kappa_0)'$. Then the joint posterior distribution is

$$[\mu, \tau | Y] \propto \tau^{n/2} \exp\left[-\frac{\tau}{2} \sum_{i=1}^{n} (Y_i - \mu)^2 \right]$$

$$\times (\kappa_0 \tau)^{1/2} \exp\left[-\frac{\kappa_0 \tau}{2} (\mu - \eta_0)^2 \right] \times \tau^{\alpha_0 - 1} \exp(-\beta_0 \tau). \tag{4.5}$$

After a bit of algebraic thrashing around, we can rewrite Eq. (4.5) as

$$[\mu, \tau | Y] \propto (\kappa_1 \tau)^{1/2} \exp\left[-\frac{\kappa_1 \tau}{2} (\mu - \eta_1)^2 \right] \times \tau^{\alpha_1 - 1} \exp(-\beta_1 \tau), \tag{4.6}$$

where

$$\alpha_1 = \alpha_0 + n/2,$$

$$\beta_1 = \beta_0 + \frac{(n-1)S^2}{2} + \frac{n\kappa_0 (\bar{y} - \eta_0)^2}{2(n + \kappa_0)},$$

$$\kappa_1 = \kappa_0 + n,$$

and

$$\eta_1 = \frac{n}{n + \kappa_0} \bar{y} + \frac{\kappa_0}{n + \kappa_0} \eta_0;$$

here $\bar{y}$ and S^2 are the sample mean and variance. We have established that the normal-gamma family of distributions, indexed by hyperparameter vector $\psi = (\alpha, \beta, \eta, \kappa)$ is conjugate for the parameter vector $\theta = (\mu, \tau)'$. Given a prior in the family, described by hyperparameter ψ_0, the posterior distribution is also in the family, and its hyperparameter ψ_1 is obtained by a simple formula involving ψ_0, the sample size n, and the sample mean and variance of the data.

It turns out that we can also integrate Eq. (4.6) as a function of τ to obtain the marginal posterior distribution $[\mu|Y]$.[3] Alternatively, we can use our knowledge of the conditional distributions $[\mu|\tau,Y]$ and $[\tau|Y]$ to sample pairs from the joint posterior distribution. First, we sample τ from a $Ga(\alpha_1,\beta_1)$ distribution. We then use the sampled τ in drawing μ from a $N(\eta_1,1/(\kappa_1\tau))$ distribution. Taken alone, the μ are a sample of the marginal posterior distribution.

4.2 MONTE CARLO METHODS

Markov chain Monte Carlo (MCMC) has revolutionized data analysis, breaking down the largest barrier to Bayesian analysis, that of computation. Describing all unknown quantities using probability distributions, always working in accord with the basic laws of probability, reasoning from prior to posterior distributions — all of these may be great ideas, but if the calculations are prohibitively difficult, their appeal is lost. MCMC makes the process possible.

Although the basic techniques of MCMC were first developed in the early 1950's, they were paid scant attention among statisticians until the late 1980's, and even less among wildlife and ecological data analysts prior to 2000. Since then, the use of MCMC has burgeoned among ecologists, largely due to the appeal of being able to fit complex hierarchical models.

Indeed, MCMC's ubiquity, the marvelous analyses produced using it, and the availability of free software and vast amounts of computational power are leading to a blackbox view manifest by a common sloppy misuse of the phrase "MCMC analysis," as in "I did an MCMC analysis of the life history data." This says nothing: MCMC is simply a tool for evaluating probability distributions, like Newton–Raphson is a tool for numerical optimization of functions, such as likelihoods. One would never say "I did a Newton–Raphson analysis of the life history data," at least, one would say "I used maximum likelihood …" although even this would sound strange, without reference to the particular model involved. Care is needed in using numerical optimization techniques, more perhaps than many users would realize, because of problems with local extrema. But even more care is required in using MCMC: one can easily be convinced that the procedure is working satisfactorily when it is not, or to believe that approximations are more precise than they really are. Using MCMC well requires some familiarity with how it works.

This chapter culminates with descriptions of various MCMC methods, and the caveats needed in using them. We begin by describing ordinary Monte Carlo (simulation) methods, useful in their own right, but also as stepping stones to understanding MCMC.

The basic idea of simulations is that we can study features of a probability distribution F by examining corresponding features of a sample $X_1, X_2, \ldots, X_n$ from F. Suppose we want to learn about the ratio of the largest lifespan to the average lifespan in samples of size 25,

---

3. The quantity $\sqrt{\kappa_1\alpha_1/\beta_1}(\mu - \eta_1)$ has a t-distribution with α_1 degrees of freedom. Taking limits as $\alpha_0, \beta_0, \kappa_0$, and η_0 approach zero, we obtain a $(1-\alpha)100\%$ credible interval

$$\bar{y} \pm \sqrt{\frac{n-1}{n}}\, t_{n,\alpha/2}\, \frac{s}{\sqrt{n}},$$

somewhat reminiscent of the frequentist confidence interval. The frequentist interval, based on $n-1$ degrees of freedom and without the multiplier $\sqrt{(n-1)/n}$, corresponds to a nonconjugate improper prior $[\mu,\tau] \propto 1/\tau$.

assuming exponential life lengths. We could attempt to do so analytically, based on the form of the exponential distribution, but the calculations are difficult. Alternatively, we can repeatedly generate samples of size 25 and calculate the ratio in each sample. Doing so 10 million times and summarizing the results takes less than 8 seconds on a 3.2 GHz laptop, and is sufficient to guarantee two decimal place accuracy in the mean (3.82) and the standard deviation (0.99).

Ordinary Monte Carlo methods are based on independent samples, making simple the evaluation of precision in their summaries. For instance, independence of the 10 million ratios leads to the conclusion that their mean has standard error $\sigma/\sqrt{n} \approx 0.99/\sqrt{10,000,000} = 0.001$. However, MCMC produces a *dependent* sequence of values. We discuss MCMC in detail, but first we describe two methods of ordinary Monte Carlo simulation.

Readers tempted to hurry on to materials on MCMC might be tempted to skip Sections 4.2.1 and 4.2.2. However, this material is useful in its own right and provides important background. Indeed, it would be worthwhile to try reproducing the examples we present. Generally, all that will be needed is software capable of producing samples of a uniform random number generator to produce $U \sim U(0,1)$.

4.2.1 *Cdf* Inversion

Given a uniform random number generator, many probability distributions are easily sampled.

For scalar (univariate) random variables, the easiest method is *cdf inversion*. Suppose that X is a continuous random variable with cumulative distribution function (*cdf*) $F(t) = \Pr(X \leq t)$. $F(t)$ is monotone increasing, hence invertible: there exists a function F^{-1} such that $F^{-1}(F(t)) = F(F^{-1}(t)) = t$.

Since $\Pr(X \leq t) = F(t)$,

$$\Pr(F(X) \leq t) = \Pr(X \leq F^{-1}(t)) = F(F^{-1}(t)) = t,$$

for all $t \in [0, 1]$. So $F(X)$ has the same distribution as a uniform random variable $U \sim U(0,1)$, from which it follows that $F^{-1}(U)$ has the same distribution as X.

Thus, if we want to obtain samples of a continuous random variable X, and we can calculate the inverse of X's *cdf*, $F^{-1}(t)$, all we need to do is calculate $X = F^{-1}(U)$, where U has a uniform distribution on $[0, 1]$.

For example, the exponential distribution with density $\lambda \exp(-\lambda x)$ has *cdf* $F(x) = 1 - \exp(-\lambda x)$. Solving $U = F(X)$, we obtain $X = -\log(1-U)/\lambda$. We can simplify matters a bit: if $U \sim U[0,1]$, then $1 - U$ also has a $U[0,1]$ distribution. Thus, $X = -\log(U)/\lambda$ has an exponential distribution. Recalling that the mean of an exponential distribution is $\mu = 1/\lambda$, we conclude that μ times the negative logarithm of $U \sim U[0,1]$ is an exponential random variable with mean μ.

In simulating the ratio of the maximum lifetime to the average lifetime, as described earlier, we note that the mean μ cancels out in the ratio. Thus it does not matter which value of μ we use, so we may as well use $\mu = 1$. We need only to compute $X_i = -\log(U_i)$ for $i = 1, 2, \ldots, 25$, then $M = \max(X_1, X_2, \ldots, X_{25})$, then $\bar{X}$, then $R = M/\bar{X}$, and repeat the process until we have a sample which is as large as we deem necessary to consider the distribution of R. Moderately large simulations, say 10,000 replications, can be carried out in a spreadsheet such as *Excel*. R code is given in Panel 4.1.

```
reps=100000;n=25
U=matrix(runif(reps*n),reps,n)
y=-log(U)

biggest=apply(y,1,max)
average=apply(y,1,mean)

stat=biggest/average
c(mean(stat),sd(stat))
```

4.2.2 Rejection Sampling

Rejection sampling is a simple technique for changing a sample from a distribution $c(x)$ into a (smaller) sample from another distribution $t(x)$. Imagine a histogram in marble of sampled values from distribution $c(x)$: rejection sampling chisels away at the marble, leaving behind a histogram of sampled values from distribution $t(x)$.

Rejection sampling has two uses. First, suppose that we have performed a simulation based on samples from distribution $c(x)$, and want to evaluate what simulation results would have been like had we instead sampled from $t(x)$. Rather than conducting an entirely new simulation based on $t(x)$, we might be able to use rejection sampling to obtain a subset of the results based on $c(x)$, a subset which can be treated as based on a sample from $t(x)$. For example, after sampling a collection of X's, we may decide that we are interested in the expected value of $X|X>0$. Rather than drawing a new sample from the conditional distribution of $X|X>0$, we simply discard the sampled X's ≤ 0 and compute the mean of the remaining values. In Section 5.5.2, we show how that, given a prior, data, and a sample of the posterior distribution, rejection sampling can be used to sample a posterior distribution resulting from the same data but a different prior.

A second use of rejection sampling is when it is easy to directly sample from $c(x)$ but difficult to directly sample $t(x)$. In this case, we use rejection sampling as an indirect method of drawing samples from $t(x)$. We refer to $c(x)$ as the *candidate* distribution and $t(x)$ as the *target* distribution. In this setting, the candidate distribution is of no interest *per se*, except as a means to the end of obtaining samples from the target distribution $t(x)$.

Here is how it is done: First, we choose an easily sampled candidate distribution $c(x)$, ensuring that there is a number $M<\infty$ such that over the entire range of $t(x)$, $t(x) \leq Mc(x)$. (We will explain why subsequently.)

Next, we draw a sample $X \sim c(x)$ and calculate $w(X)=t(X)/(Mc(X))$. Note that $0 \leq w(X) \leq 1$. We conduct a Bernoulli trial with success probability $w(X)$; if the trial is successful, the value X is accepted as an observation from $t(x)$, otherwise, the value is discarded. The collection of all X sampled from $c(x)$ is the original block of marble, the values accepted are the finished sculpture after chipping away the rejects. Simulation efficiency dictates that our choice of $c(x)$ should be guided by a desire to maintain high acceptance probabilities across the range of $t(x)$.

To illustrate, we will generate a sample of standard normal random variables from a sample of random variables having a triangular distribution. Triangular distributions are easily sampled: the sum of two independent, identically distributed uniform random variables has a

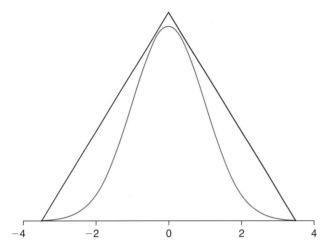

FIGURE 4.3 Scaled candidate distribution $Mc(x)$ (blue) and target normal distribution $t(x)$ (red).

triangular distribution. So, provided we have access to a uniform random number generator, we can easily sample from a triangular distribution and use these values to generate standard normal variates. To make things easy, we will cheat a little bit and restrict the range of the normal distribution to $(-3.49, 3.49)$.[4]

Now, let U_1 and U_2 be sampled from a $U(-7/4, 7/4)$ distribution; then $X = U_1 + U_2$ has a triangular distribution on $[-3.5, 3.5]$. The density function for X is

$$c(x) = (2/7) - 4|x|/49,$$

for $|x| \leq 3.5$. It can be verified that if $M = 15/(4\sqrt{2\pi})$, then

$$t(x) = \frac{1}{\sqrt{2\pi}} \exp(-x^2/2) \leq Mc(x)$$

for $|x| \leq 3.49$ (see Fig. 4.3). Thus, we sample X from the triangular distribution, rejecting immediately any of the few values $|x| > 3.49$ and accepting the remainder with probability $w(x) = t(x)/(Mc(x))$. The acceptance probability ranges from a low of about 6.5% at ± 3.2 to a high of about 97.6% at ± 0.3. In total, the acceptance rate is about 2/3, so that a sample of 9000 X's yields approximately 6000 standard normal observations.[5]

It is fairly easy to see why rejection sampling works. A sample from the triangular distribution can be portrayed as a histogram, with bars having heights approximating a triangle, as in Fig. 4.4. Think of each bar as a bin, into which we pour grains of sand, each grain representing

4. This bit of cheating is made necessary by the requirement that $Mc(x) \geq t(x)$ for all x, and by our desire to use a triangular distribution as a starting point. The normal density function never quite reaches zero, but any triangular density does, so that there inevitably will be x's with $c(x) = 0 < t(x)$. For such values, we simply can not cook up a big enough M. The Procrustean solution of truncating the normal distribution at ± 3.49 is not entirely satisfactory; however, less than 1 in 2000 values of a standard normal random variable fall outside of this range.

5. Of course, this is not the most efficient way of generating standard normal random variables; we include it as a simple illustration. A far better means of generating standard normal from uniform random variables is based on the Box–Muller transformation, given in Appendix B.3 (Eqs. B.3 and B.4).

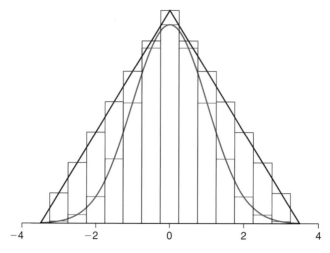

FIGURE 4.4 $Mc(x)$ and $t(x)$, with approximating histograms. The height ratio for histogram blocks centered at x is $w(x)$, the acceptance probability.

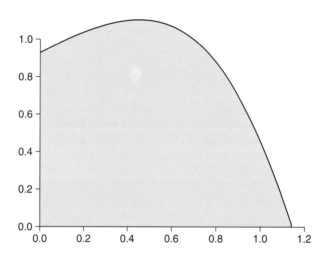

FIGURE 4.5   Density function $t(x) = C\sin(\exp(x))$, $0 < x < \log(\pi)$.

a sampled value. Random sampling from $c(x)$ ensures that the bins will fill up according to the shape of the candidate distribution. The lowest $100\,w(x)\%$ of the bar at each x represents accepted values. The proportion $w(x)$ is chosen to ensure that the histogram of accepted values conforms to the shape of the target distribution $t(x)$.

Here's one for the reader to try: let $t(x) = C\sin(\exp(x))$, $0 < x < \log(\pi)$, as plotted in Fig. 4.5. Try using rejection sampling to draw samples, to estimate the mean and the 90th percentile of the distribution. In working through the example, it will become apparent that one does not need to know the value of C (≈ 1.104) to implement rejection sampling. All that we need to know is that $t(x)$ is a legitimate density function, that is, that it is nonnegative and integrates to one. A solution is given in the footnote.[6]

4.3 MARKOV CHAIN MONTE CARLO

MCMC is a simulation technique for examining probability distributions. Devised in the early 1950's (Metropolis *et al.*, 1953), MCMC received little attention in statistical applications until the 1980's with the publication of important applications to image processing by Geman and Geman (1984) and Besag (1986). While it is tempting to speculate on the role of Besag's intriguingly titled "statistical analysis of dirty pictures" as contributing to the subsequent explosion of interest in the topic, the real explanation lies elsewhere: MCMC has taken the world by storm because of its usefulness and its relatively easy implementation.

4.3.1 Markov Chains and Stationary Distributions

A k-th order Markov chain is a sequence of random variables $X_1, X_2, X_3, \ldots$, with the property that, given all of the previous values, the probability distribution of the next value depends only on the last k values. That is,

$$[X_t | X_{t-1}, X_{t-2}, \ldots, X_1] = [X_t | X_{t-1}, X_{t-2}, \ldots, X_{t-k}].$$

Many Markov chains describe natural processes evolving through time, hence the index t is often referred to as "time," and the values of X_t are referred to as "states" (of nature).

Think of a gambler entering a casino with a stake of X_1 dollars and making a series of dollar bets at roulette. The roulette wheel has 38 equally likely outcomes, 18 of which win the gambler a dollar, 20 of which lose a dollar. The gambler's stake after t gambles is X_t. If the gambles are independent, then X_t, $t = 1, 2, \ldots$ is a first-order Markov chain. X_t is either $X_{t-1} + 1$ (with probability $p = 18/38$) or $X_{t-1} - 1$ (with probability $1 - p = 20/38$) regardless of the preceding history $(X_1, X_2, X_3, \ldots, X_{t-2})$.

Suppose that the gambler's original stake is $X_1 = 20$, and that she will gamble until she doubles her money or goes broke. Her stake X_t is always in the set $S = \{0, 1, 2, 3, \ldots, 40\}$, referred to as the *state space*. States 0 and 40 are called *absorbing* states: if $X_t = 0$, then $X_{t+k} = 0$ for $k = 1, 2, \ldots$; if $X_t = 40$, then $X_{t+k} = 40$ for $k = 1, 2, \ldots$.

If after $t - 1$ bets, the gambler's stake is $X_{t-1} = \$5$, then $X_t = \$4$ or $\$6$, with probabilities $20/38$ and $18/38$, regardless of the sequence of outcomes leading up to X_{t-1}. We might be intrigued by the knowledge that she has lost 20 bets in a row, and watched her stake dwindle from $25, or that she has narrowly avoided going broke and won four in a row, but the additional information provides no insights into the probability she wins her next bet. Similarly, if $X_{t-1} = 0$, then $X_t = 0$, or if $X_{t-1} = 40$, then $X_t = 40$; these are true regardless of the preceding

---

6. Let $k = \log(\pi)$ denote the upper endpoint of the range of $t(x)$. We will use a $U(0,k)$ density as the candidate distribution; this density is $c(x) \equiv 1/k$. The maximum value of $\sin(t(x))$ is 1, hence the maximum value of $t(x)$ is C. Thus, $Mc(x) \geq C$ ensures that $Mc(x) \geq t(x)$. $Mc(x) \geq C$ means $M(1/k) \geq C$; we would like the smallest M possible (why?), so set $M = kC$. Then, $w(x) = t(x)/(Mc(x)) = C\sin(\exp(x))/(kC(1/k)) = \sin(\exp(x))$. A quick simulation yielded 7.9 million accepted values out of 10 million; their 90th percentile was 0.888, and their mean was 0.489.

history. Her stake at time t depends on $(X_1, X_2, X_3, \ldots, X_{t-1})$ only through X_{t-1}. The sequence X_t is a first order Markov chain.

Some Markov chains have stationary distributions, probability distributions satisfying

$$\pi(A) = \Pr(X_t \in A),$$

for each subset A of the sample space. The important feature of this probability distribution is that it is "stationary" with respect to time; the probability that X_t is in a particular state or set of states does not depend on t.

Not all Markov chains have stationary distributions. Our gambler's chain, for example, does not. Consider $\Pr(X_t = 19)$. At time $t = 2$, the only possible states are $X_2 = X_1 \pm 1 = 19$ or 21, depending on whether she won her first gamble, thus $\Pr(X_2 = 19) = 20/38$. Eventually her chain will either reach 0 or 40, and remain there, so that $\Pr(X_t = 19)$ approaches zero as t gets large. The existence of a stationary distribution would require that $\Pr(X_t = 19)$ not change through time.[7]

Stationary distributions of Markov chains are the basis of MCMC. Suppose that we would like to sample distribution f, but that standard methods for producing independent samples, such as *cdf* inversion (Section 4.2.1) or rejection sampling (Section 4.2.2), are not feasible. In such situations, we might nonetheless be able to construct a Markov chain X_t, with stationary distribution f. The sampled values will not be independent, but they will still be a sample of the distribution we wish to investigate.

Users of MCMC should be aware of the ergodicity theorem, which says that a *positive recurrent* and *aperiodic* Markov chain has a stationary distribution $\pi(A)$ satisfying

$$\pi(A) = \lim_{n \to \infty} \Pr(X_n \in A | X_1),$$

for subsets A of the sample space. We define the terms positive recurrent and aperiodic subsequently, noting first the implications of the theorem. It not only guarantees the existence of a stationary distribution, but also states that the starting value X_1 does not affect the asymptotic behavior of the chain. Regardless of the starting value of the chain, it eventually settles into a pattern of visiting A with specified probability, $\pi(A)$. This is a useful observation for implementation of MCMC; in practice, we must specify starting values. The ergodicity theorem ensures that these do not matter too much. In practice, we typically discard some of the early observations from our chain, "burn-in values," which might not be representative of the stationary distribution.

We noted that our gambler's chain does not have a stationary distribution. In the first place, it is not aperiodic: it has a period of 2. The gambler can only return to a stake $0 < i < 40$ in an even number of gambles (why?). The period of a state is defined, somewhat obscurely, as the greatest common divisor of the possible return times. If the period of a state is 1, the state is said to be aperiodic. A Markov chain is aperiodic if all of its states are aperiodic.

Our gambler's Markov chain is also not recurrent. Suppose that $X_1 = i$, and let $T_i = \min\{n > 1 : X_n = i\}$; T_i is the time of first return to state i. State i is said to be recurrent if $\Pr(T_i < \infty) = 1$,

---

7. Despite the odds being close to even on each bet, the chance that the gambler will go broke before reaching \$40 is 89.2%. At the outset, her expected gain is thus $0.892 \times (-\$20) + 0.108 \times \$20 = -\$15.68$. If she had started with \$50, the chance of going broke before doubling her money would be 99.5%, and her expected gain $-\$49.49$. If you've got to gamble, make sure you own the casino.

and positive recurrent if $E(T_i) < \infty$. Thus, a state is recurrent if one can be certain that it will be revisited, and positive recurrent if it will not take too long. These descriptions are applied to the entire chain if applicable to every state. Loosely speaking, a positive recurrent chain moves efficiently through the state space, visiting all states with reasonable frequency. The gambler's chain is not recurrent because of the existence of absorbing states 0 and 40. From any state $0 < i < 40$, there is always a chance of reaching the absorbing state before returning to state i, so $\Pr(T_i < \infty) < 1$.

In applying MCMC, periodic behavior is rarely an issue but nonrecurrence can be. We turn to some examples.

4.3.2 Example: Standard Normal Stationary Distribution

Here is a simple example of a Markov chain that can be used with a uniform random number generator to produce samples of a standard normal distribution. The chain is defined with a tuning parameter A: any value $A > 0$ will work, but some will work better than others.

Algorithm 1

Let $X_0 = 0$. Then for $t = 1, 2, \ldots$, generate X_t according to the following rules:

Step 1: Generate two independent $U(0, 1)$ random variables, say u_1 and u_2.
Step 2: Calculate a candidate value, $X_{\text{cand}} = X_{t-1} + 2A(u_1 - 1/2)$.
Step 3: Calculate

$$r = \frac{\exp\left(-\frac{1}{2}X_{\text{cand}}^2\right)}{\exp\left(-\frac{1}{2}X_{t-1}^2\right)}$$

Step 4: If $u_2 < r$, set $X_t = X_{\text{cand}}$, otherwise set $X_t = X_{t-1}$.

At each time step, the chain either remains at its current value or moves incrementally to a randomly generated candidate value. The increment in Step 2 has a $U(-A, A)$ distribution, so the candidate value is sampled uniformly over an interval centered at the current value, that is, $X_{\text{cand}}|X_{t-1} \sim U(X_{t-1} - A, X_{t-1} + A)$. The values X_t are clearly a first-order Markov chain: the distribution of X_t given all of the previous values $X_{t-1}, X_{t-2}, \ldots, X_1$ depends only on the the most recent value X_{t-1}.

Note that Step 4 involves a Bernoulli trial, with success probability $= \min(r, 1)$, where r is the ratio of the (target) standard normal density evaluated at the candidate and current values of the chain. This success parameter is referred to as the acceptance or movement probability.

Algorithm 1 is an instance of the Metropolis–Hastings method, about which we will have more to say later. For now, we suggest it as a simple example to experiment with in order to gain insights about operational considerations for MCMC.

Effect of Tuning Parameter and Starting Value

The *history plot* for Markov chain X_t is obtained by plotting X_t against t. Figure 4.6 gives history plots for the first 1000 values of four Markov chains generated according to Algorithm 1,

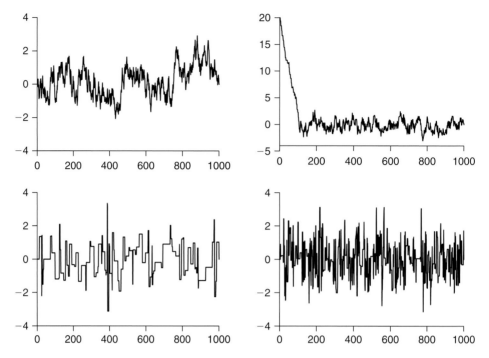

FIGURE 4.6   Plots of X_t against t, for Markov chains with standard normal stationary distribution. Reading clockwise from top left, these have tuning parameter $A = 0.5, 1.0, 3.7$, and 15.

three of which start with $X_0 = 0$ and one starting with $X_0 = 20$. Clockwise from top left, the chains have $A = 0.5, 1.0, 3.7$, and 15, respectively.

All four chains have a standard normal stationary distribution, but their history plots reveal obvious differences. The chains with $A = 0.5$ and 1.0 move slowly over the range of the standard normal distribution, taking many small steps. The chain with $A = 1.0$ and starting value $X_0 = 20$ moves slowly but deliberately down into the range of typical values of a standard normal random variable. Its first 100 values would typically be discarded as a "burn-in," values not representative of the stationary distribution. The chain with $A = 15$ takes occasional large steps but stalls frequently; for example, it stopped at a value of approximately 1 for t from 700 to 750.

These differences are reflected in the strength of association among values of X_t. For $h = 1, 2, \ldots$, the correlation $\rho(X_t, X_{t+h})$ between X_t and X_{t+h} is called the *autocorrelation at lag h*, and $R(h) = \rho(X_t, X_{t+h})$ is called the *autocorrelation function (ACF)*. Figure 4.7 gives the autocorrelation functions for three of the chains.

With $A = 3.7$, the autocorrelation drops off quickly, falling below 0.01 by lag 9; one could treat every ninth observation as nearly independent. The same is true for $A = 15$ at lag 35. The chain with $A = 0.5$ is highly autocorrelated, with the same value at lag 50 as the $A = 3.7$ chain at lag 3.

The consequence of high autocorrelation is diminished accuracy and precision in estimating features of the target distribution. The ergodicity theorem guarantees that sample features of

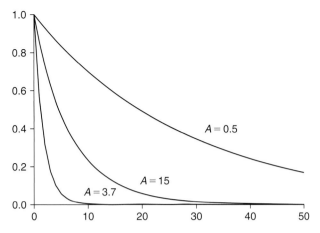

FIGURE 4.7 Autocorrelation functions for Markov chains with standard normal stationary distribution, and tuning parameter $A = 0.5, 3.7$, and 15.

the three Markov chains will approximate corresponding features of the normal distribution as chain length, n, gets larger. The sample mean will approach zero, the sample standard deviation will approach 1, and the sample 95th percentile will approach 1.960. For finite n, however, the sample features are only estimates of the features of the stationary distribution and may be biased or imprecise. The bias and precision of the various chains depends on the choice of A.

Not surprisingly (given Fig. 4.7), the chain with $A = 3.7$ is much more efficient in producing estimates of these quantities because its sampled values are less highly correlated than those of the other two chains. Table 4.1 reports an evaluation of Algorithm 1 using chains of length 1000, with $A = 0.5, 3.7$, and 15. For comparison, we include similar evaluations of independent samples of size 1000. So, for instance, the 97.5th percentile of the standard normal distribution is 1.960. Chains of length 1000 with $A = 3.7$ produced estimates that averaged 1.964, with a standard deviation of 0.158.

Chains with greater autocorrelation ($A = 0.5$ and 15) have substantially less precision in their estimates. Furthermore, there is a "small-sample" bias associated with the estimates of standard deviation and the 97.5th percentile: chain length of 1000 is insufficient to overcome this bias when $A = 0.5$ or 15. It is noteworthy that even with independent samples, $n = 1000$ is insufficient for high precision or accuracy. One would wish to use much larger samples.[8]

One would also wish to use methods that produce chains with as little autocorrelation as possible or to compensate for high autocorrelation by using longer chains. In Table 4.1, the standard deviations of the estimated parameters stand in ratios of approximately 6:2:4:1 for $A = 0.5, 3.7, 15$, and independent samples. Working on the principle that standard errors

---

8. We note a tendency in summaries of analyses using MCMC to report far more decimal places than justified. See Flegal *et al.* (2008) for comments and recommendations.

TABLE 4.1 Estimates of features of standard normal distribution, based on Markov chains of length 1000 according to algorithm 1 (for $A=0.5, 3.7$, and 15) and on independent samples of 1000 values. Values reported in the last two columns are summaries over 100,000 replicates.

	A	Mean	SD
Mean (=0)	0.5	0.000	0.234
	3.7	0.000	0.060
	15	0.000	0.118
	Independent samples	0.000	0.032
SD (=1)	0.5	0.966	0.118
	3.7	0.998	0.045
	15	0.990	0.093
	Independent samples	1.000	0.022
Median (=0)	0.5	0.000	0.239
	3.7	0.000	0.080
	15	0.000	0.170
	Independent samples	0.000	0.039
97.5th (=1.960)	0.5	1.892	0.389
	3.7	1.964	0.158
	15	1.937	0.292
	Independent samples	1.967	0.085

tend to be proportional to the square root of sample size, one might regard the chains with $A=0.5, 3.7$, and 15 as producing chains worth $1/36, 1/4$, and $1/16$ the value of independent samples.

We identified the value $A=3.7$ as reasonable by comparing its lag 1 autocorrelation with those of chains with other values of A. Figure 4.8 plots lag 1 autocorrelations and movement probabilities of chains produced with values of A ranging from 0 to 15. Small values of A result in small increments for the candidate values and high acceptance probabilities because nearby values have nearly identical probability; hence, r is close to 1. The chain moves slowly, and there is high autocorrelation. Large values of A lead to larger increments for the candidate values and lower acceptance probabilities. The lag 1 autocorrelation drops off as A increases, until a point where the acceptance rate is low enough that frequently $X_t = X_{t-1}$, once again resulting in high autocorrelation. The autocorrelation of the chains is minimized for A near 3.7, where the acceptance probability is approximately 0.42 and the lag 1 autocorrelation is about 0.56. Conventional wisdom is that acceptance rates in the range 30–50% are near optimal.

In working through this example, we have had a bit of a look under the hood. We imagine that many of our readers will just want to drive the car, and may find these details daunting. We wish to allay fears: readers should be aware of the availability of high quality software like

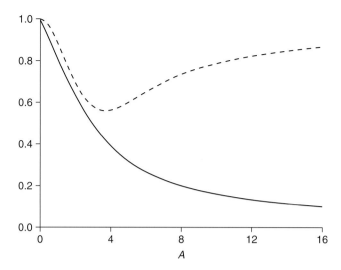

FIGURE 4.8 Lag 1 autocorrelation (dashed line) and movement probability (solid line) as function of tuning parameter A.

program WinBUGS, which can be used to implement MCMC for many models.[9] However, even using such software, some skill is required; familiarity with the autocorrelation function and other diagnostic tools (Section 4.3.5) is necessary for evaluating results.

We encourage readers to try writing their own MCMC code, beginning with simple models. Writing one's own code is a great way for developing intuition about MCMC performance and is occasionally a necessity, for analysis of complex models.

The primary tool for conducting MCMC is the Metropolis–Hastings (MH) algorithm, which we describe in Section 4.3.3. The MH algorithm is omnibus and easily implemented, but can require substantial tuning (tweaking, similar to the selection of A in the foregoing discussion) especially in application to multivariate distributions.

Gibbs sampling (Section 4.3.4) is a refinement of the MH algorithm for multivariate distributions. It can be remarkably efficient.

4.3.3 Metropolis–Hastings Algorithm

Algorithm 1, used to draw samples of the standard normal distribution, is a specific instance of the Metropolis–Hastings (MH) algorithm, which we now present.

Metropolis–Hastings Algorithm

Suppose that we wish to draw samples from target distribution $f(x)$. Let $j(x|y)$ be candidate generating distributions, describing probabilities for candidate values x, given current value y.

9. We will use BUGS as a generic term for either WinBUGS or the open-source version OpenBUGS. These programs are available for free download from http://www.mrc-bsu.cam.ac.uk/bugs/ and http://mathstat.helsinki.fi/openbugs/, respectively. The main practical difference between the two is that script files (for batch execution) are written differently. Almost all of the BUGS code presented in this text will run under OpenBUGS 3.0.3 or WinBUGS 1.4; exceptions are noted in the text.

Fix a value X_0. Then for $t = 1, 2, \ldots$, generate X_t according to the following rules:

Step 1: Generate a candidate value, X_{cand} by sampling from $j(x|X_{t-1})$.
Step 2: Calculate

$$r = \frac{f(X_{\text{cand}})j(X_{t-1}|X_{\text{cand}})}{f(X_{t-1})j(X_{\text{cand}}|X_{t-1})} \tag{4.7}$$

Step 3: Generate $U \sim U(0,1)$
Step 4: If $U < r$, set $X_t = X_{\text{cand}}$, otherwise set $X_t = X_{t-1}$.

The first thing to notice about the MH algorithm is that the target distribution is only involved in calculating r, and that it occurs in both the numerator and denominator. Think of what this means for Bayesians: if f is a posterior distribution (Eq. 4.1) that cursed [Y], the normalizing constant that caused so many headaches for earlier generations of Bayesians, cancels out!

Notice that the MH algorithm allows great latitude in selection of the candidate generating distribution. Limitations on $j(x|y)$ relate to the requirement that Markov chains be positive recurrent in order to have stationary distributions: every state must be reachable from every other state. Beyond this minimal requirement, there is the practical requirement that the chain move freely enough to have reasonably low autocorrelation.

We draw attention to two features of Algorithm 1 considered in Section 4.3.2. First, the candidate generating distribution is symmetric in its arguments. A uniform distribution centered at y and of length $2A$ has density function

$$j(x|y) = \frac{\mathbf{I}(|x - y| < A)}{2A},$$

where $\mathbf{I}(\cdot)$ is the indicator function. Clearly $j(x|y) = j(y|x)$, so that r in Step 2 simplifies to

$$r = \frac{f(X_{\text{cand}})}{f(X_{t-1})}.$$

The use of a symmetric candidate generator simplifies calculations and saves computation time.

We also draw attention to the tuning parameter A in Algorithm 1. Selection of an appropriate value can be done on the fly as the chain is generated. Starting with $A = 1$, we multiply A by (say) 1.01 each time a candidate value is accepted, and divide by 1.01 each time a candidate value is rejected. After (say) 5000 steps, we fix the value of A at the present value, discarding the previously sampled values. This process aims at an acceptance rate of about 50%; a lower acceptance rate is attained by multiplying by 1.01 following acceptance, and dividing by a smaller value (say, 1.007) following rejection. Following this approach, we settle on $A = 3.48$ and acceptance probability 0.44, close to the optimal value of 3.7.[10]

---

10. The value 1.007 was chosen with the goal of obtaining an acceptance rate near the optimal 42%, and is the solution of

$$p \times c + (1 - p)/d = 1,$$

with $p = 0.42$ and $c = 1.01$. The tuning process could be further refined by using a sequence of values c which gradually decrease to 1, and corresponding values $d = (1 - p)/(1 - pc)$.

4.3.4 Gibbs Sampling

Gibbs sampling is designed for multivariate posterior distributions. Let $\boldsymbol{\theta} = (\theta_1, \theta_2, \ldots, \theta_k)'$ represent unknown quantities, and X represent data. The goal is to draw a sample from $[\boldsymbol{\theta}|X]$.

Let $\boldsymbol{\theta}_{(-j)}$ denote the vector of length $k-1$ made up of all the elements of $\boldsymbol{\theta}$, but omitting θ_j. The *full conditional distribution* for θ_j is

$$[\theta_j|\boldsymbol{\theta}_{(-j)}, X];$$

this is the distribution of the jth component of $\boldsymbol{\theta}$, having fixed the values of all the other components, and having been informed by the data. Like the posterior distribution $[\boldsymbol{\theta}|X]$, it is proportional to $[X|\boldsymbol{\theta}][\boldsymbol{\theta}]$, the difference being that the normalizing constant is $[\boldsymbol{\theta}_{(-j)}, X]$ rather than $[X]$.

It turns out that full conditionals $[\theta_j|\boldsymbol{\theta}_{(-j)}, X]$ are often easily identified by inspection of $[X|\boldsymbol{\theta}][\boldsymbol{\theta}]$, when marginal posterior distributions $[\theta_j|X]$ or joint posterior distributions $[\boldsymbol{\theta}|X]$ are not. This convenience, and the ability to sample one component of the parameter vector at a time, make Gibbs sampling attractive.

Gibbs Sampling Algorithm

Suppose that we wish to draw samples from a joint posterior distribution $[\boldsymbol{\theta}|X]$.

Fix a value $\boldsymbol{\theta}^{(0)} = \left(\theta_1^{(0)}, \theta_2^{(0)}, \ldots, \theta_k^{(0)}\right)'$. Then for $t = 1, 2, \ldots$, generate $\boldsymbol{\theta}^{(t)}$ according to the following rules:

Step 1: Sample $\theta_1^{(t)}$ from the full conditional $\left[\theta_1 \big| \boldsymbol{\theta}_{(-1)}^{(t-1)}, X\right]$.

Step 2: Sample $\theta_2^{(t)}$ from the full conditional $\left[\theta_2 \big| \boldsymbol{\theta}_{(-2)}^{(t-1)}, X\right]$.

$\ldots$

Step k: Sample $\theta_k^{(t)}$ from the full conditional $\left[\theta_k \big| \boldsymbol{\theta}_{(-k)}^{(t-1)}, X\right]$.

Step $k+1$: Set $\boldsymbol{\theta}^{(t)} = \left(\theta_1^{(t)}, \theta_2^{(t)}, \ldots, \theta_k^{(t)}\right)'$.

Note: It is also acceptable to sequentially update $\boldsymbol{\theta}^{(t)}$ after each step in the preceding algorithm, and to use the partially updated $\boldsymbol{\theta}^{(t)}$ in sampling subsequent full conditionals. For instance, $\theta_3^{(t)}$ could be sampled from the full conditional distribution

$$\left[\theta_3 \big| \theta_1^{(t)}, \theta_2^{(t)}, \theta_4^{(t-1)}, \ldots, \theta_k^{(t-1)}, X\right],$$

rather than

$$\left[\theta_3 \big| \theta_1^{(t-1)}, \theta_2^{(t-1)}, \theta_4^{(t-1)}, \ldots, \theta_k^{(t-1)}, X\right].$$

Example

Suppose that we observe a sample $Y = (6, 7, 9, 9, 12, 13, 14, 15, 17, 18)'$ of $n = 10$ binomial random variables with index $N = 25$ and success rates sampled from a logit-normal distribution with mean μ and precision τ. We write $\phi_i = \text{logit}(p_i)$; inference will be based on a posterior distribution $[\boldsymbol{\theta}|Y]$, where $\boldsymbol{\theta} = (\mu, \tau, \phi_1, \phi_2, \ldots, \phi_{10})'$.

To complete the model specification, we require a prior on the hyperparameters μ and τ. For reasons that will become apparent subsequently, we choose the normal-gamma prior discussed in Section 4.1.3: that is, we specify priors $[\mu|\tau] = N(\eta_0, 1/(\kappa_0\tau))$ and $[\tau] = Ga(\alpha_0, \beta_0)$.

The posterior distribution $[\theta|Y] \propto [Y|\theta][\theta]$ is

$$\propto \prod_{i=1}^{n}\left\{\operatorname{expit}(\phi_i)^{Y_i}\left[1 - \operatorname{expit}(\phi_i)\right]^{25-Y_i} \times \tau^{1/2}\exp\left[\frac{-\tau}{2}(\phi_i - \mu)^2\right]\right\}$$

$$\times \tau^{1/2}\exp\left[\frac{-\kappa_0\tau}{2}(\mu - \eta_0)^2\right] \times \tau^{\alpha_0-1}\exp(-\beta_0\tau); \qquad (4.8)$$

here "expit" is the inverse of the logit transformation, so $\operatorname{expit}(\phi_i) = p_i$.

Calculation of this joint posterior distribution or of any of the marginal posterior distributions is out of the question; $[\theta|Y]$ is a mess. Gibbs sampling, however, is fairly straightforward. We need to identify full conditional distributional distributions for μ, τ, and the ϕ_i's. All of these are proportional to Eq. (4.8).

For example, the full conditional for τ will be proportional to all of the terms in Eq. (4.8) involving τ's. Using a handy notation for full conditionals, we have

$$[\tau|\cdot] \propto \tau^{n/2+1/2+\alpha_0-1}\exp\left\{\left[-\frac{1}{2}\sum_{i=1}^{n}(\phi_i - \mu)^2 - \frac{\kappa_0}{2}(\mu - \eta_0)^2 - \beta_0\right]\tau\right\};$$

that is, the full conditional distribution $[\tau|\cdot] = Ga(\alpha_1, \beta_1)$, with $\alpha_1 = (n+1)/2 + \alpha_0$, and

$$\beta_1 = \beta_0 + \frac{1}{2}\sum_{i=1}^{n}(\phi_i - \mu)^2 + \frac{\kappa_0}{2}(\mu - \eta_0)^2.$$

Similarly, the full conditional for μ will be proportional to all of the terms in Eq. (4.8) involving μ. We have

$$[\mu|\cdot] \propto \exp\left\{-\frac{\tau}{2}\left[\sum_{i=1}^{n}(\mu - \phi_i)^2 + \kappa_0(\mu - \eta_0)^2\right]\right\},$$

which can be wrestled into the form

$$[\mu|\cdot] \propto \exp\left\{-\frac{\kappa_1\tau}{2}(\mu - \eta_1)^2\right\},$$

with $\kappa_1 = \kappa_0 + n$ and

$$\eta_1 = \frac{n}{n+\kappa_0}\bar{\phi} + \frac{\kappa_0}{n+\kappa_0}\eta_0.$$

Thus, the full conditional for μ is $N(\eta_1, 1/(\kappa_1\tau))$.

Note that the prior and full conditional distributions for the pair $(\mu, \tau)'$ are both of the normal-gamma form, $[\mu|\tau] = N(\eta, 1/(\kappa\tau)), [\tau] = Ga(\alpha, \beta)$. The normal-gamma distribution is not conjugate: conjugacy requires the prior and posterior distributions to be of the same family. However, our knowledge of conjugate forms has led to a choice of prior for which the full conditional distributions are easily identified. As a general rule, Gibbs sampling can be expedited through the use of conjugate families.

All that remains is to identify full conditional distributions $[\phi_i| \cdot]$. Once again, we harvest the relevant terms from Eq. (4.8), obtaining

$$[\phi_i| \cdot] \propto \mathrm{expit}(\phi_i)^{Y_i}[1 - \mathrm{expit}(\phi_i)]^{25-Y_i} \exp\left[\frac{-\tau}{2}(\phi_i - \mu)^2\right]. \tag{4.9}$$

We ignore all of the other bits of Eq. (4.8), which do not involve ϕ_i; regarding Eq. (4.8) as a function of ϕ_i, they are merely constants, absorbed in the normalizing constant. Unfortunately, what's left as Eq. (4.9) does not correspond to any easily sampled, well-known distribution. The full conditionals for μ and τ are identified, and easily sampled, but $[\phi_i| \cdot]$ is not.

But we can still implement Gibbs sampling. We just need to figure out some way of sampling from the distribution shown in Eq. (4.9). One possibility is to use rejection sampling. Using the change of variables theorem (Section 2.2.4) we note that if $p \sim Be(Y, 25 - Y)$, then $\phi = \mathrm{logit}(p)$ has density

$$j(\phi) \propto \mathrm{expit}(\phi)^Y (1 - \mathrm{expit}(\phi))^{25-Y}. \tag{4.10}$$

The only difference between Eqs. (4.10) and (4.9) is the term $\exp[\frac{-\tau}{2}(\phi_i - \mu)^2]$, which is bounded above by 1. Hence, we can implement rejection sampling by drawing candidate values $p \sim Be(Y_i, 25 - Y_i)$, calculating $\phi = \mathrm{logit}(p)$, and accepting with probability $r = \exp[\frac{-\tau}{2}(\phi - \mu)^2]$.

The rejection sampling scheme for sampling $[\phi_i| \cdot]$ has one failing, illustrated in our analysis of the sample of size 10: occasionally, the values of μ and τ are such that the acceptance probability is very low. 34% of the time a single candidate was enough; 49% of the time the first or second candidate was accepted, but in about 1 in 300 cases over 1000 candidates were required. The result was a mean of 25.6 candidates, and long computation time.

A useful alternative is to use the M-H algorithm to sample $[\phi_i| \cdot]$, within the context of Gibbs sampling. Generating candidate values ϕ_{cand} using Eq. (4.10) and setting Eq. (4.9) as the target distribution, the MH acceptance probability (Eq. 4.7) is simply

$$r = \exp\left[\frac{-\tau}{2}\left[(\phi_{\mathrm{cand}} - \mu)^2 - (\phi_{\mathrm{curr}} - \mu)^2\right]\right].$$

One might anticipate that this *Metropolis within Gibbs* algorithm generates chains that are inferior to those based on rejection sampling within Gibbs sampling, just as MH samples are inferior to independent samples. This is confirmed in Fig. 4.9, where autocorrelations function for $\sigma = 1/\sqrt{\tau}$ are displayed for the sample data. The ACF using Metropolis within Gibbs (dashed line, top) tails off much more slowly than the ACF using rejection within Gibbs (solid line, middle). However, the computation time for rejection within Gibbs was about 16 times greater. Thus, a more relevant comparison is to produce a chain 16 times longer using Metropolis within Gibbs, and to "thin" the chain by a factor of 16, discarding all but every 16th observation. The autocorrelation for the thinned chain is substantially less than for the chain produced using rejection within Gibbs.[11]

---

11. There is a common misconception that MCMC output should be routinely thinned. Why would one throw out data, even if simulated? We recommend the practice only when it is necessitated by memory or storage limitations, or for comparison of methods, as in this section. One might also consider thinning chains to the point where values can be treated as nearly independent in order to obtain a rough and conservative estimate of the precision with which features of the target distribution are estimated.

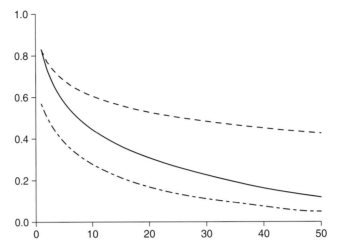

FIGURE 4.9   Autocorrelation functions for parameter $\sigma = 1/\sqrt{\tau}$ based on rejection sampling within Gibbs (solid line), Metropolis within Gibbs (dashed line), and Metropolis within Gibbs thinned by factor of 16 (dots and dashes).

PANEL 4.2   BUGS code for the logit normal model.

```
model{
 # Likelihood
     for (i in 1:10){
          x[i] ~ dbin(p[i],25)
          logit(p[i]) <- logitp[i]
          logitp[i] ~ dnorm(mu,tau)
     }
 # Priors
     mu ~ dnorm(0,tau.mu)
     tau.mu <- 0.001*tau
     tau ~ dgamma(0.001,0.001)
     sd <- 1/sqrt(tau)
 }
```

Having worked through the details of this section, we anticipate that many a reader will be pleased to note the simplicity of BUGS code required for analysis of the sample data (Panel 4.2). Compared with what is required to write one's own MCMC code, it seems almost presumptions to refer to this as "code"; all of the heavy lifting is done automatically. BUGS "code" is more like a whiteboard on which we sketch out notes describing a model, with no thoughts to specific MCMC methods.

4.3.5 MCMC Diagnostics

All numerical methods are subject to failure. Consider the Newton–Raphson algorithm for numerical maximization, which is often used for finding maximum likelihood estimators (MLE's). To maximize a function f of one variable, we start with a value x_1, then for $n = 1, 2, \ldots,$

calculate

$$x_{n+1} = x_n - \frac{f'(x_n)}{f''(x_n)};$$

here f' and f'' are the first and second derivatives of f. If all is well, the sequence $\{x_n\}$ converges to the maximizer of f. For instance, applying the method to the density function of $X \sim N(\mu, \sigma^2)$, $\{x_n\}$ should converge to the mode μ. And so it does, provided that we are fortunate enough to have chosen a starting value x_1 that is within $(\sqrt{2}/2)\sigma$ of μ. If $x_1 = \mu + (\sqrt{2}/2)\sigma$, the sequence bounces back and forth between $\mu \pm (\sqrt{2}/2)\sigma$ forever; if $x_1 > \mu + (\sqrt{2}/2)\sigma$, the sequence wanders off to $\pm\infty$.

Fortunately, such failings of the Newton–Raphson technique are usually obvious.[12] MCMC can fail too, and the dangerous part is that one might not notice it.

At the very least, one should have a look at history plots for the chains. If they look "grassy" like the first panel of Fig. 4.10, they are probably, but not certainly, working well. If they look like the second panel of Fig. 4.10, something is wrong. Generally, the first thing to check is whether the model has been correctly specified, looking to see whether there are redundant parameters. It may be that you have not properly tuned your Markov chain generator, or if you are using canned software, that the problem you are attempting is beyond its capability. It may be that you have chosen bad starting values for the chains; this is especially likely when using canned software that generates initial values automatically, and if you are using diffuse priors. Sometimes it pays to do a preliminary run using fairly informative priors, and to use a sampled value from the resultant chain as a reasonable starting value in the desired noninformative analysis.

Even grassy histories like the left-hand panel of Fig. 4.10 are not a guarantee that all is well. The authors once wrote a long and complicated piece of code that took hours to run and produced lovely grassy history plots. The only problem was that when we re-ran the code, the results were slightly different; the discrepancies were not huge, but were too large to be swept under the carpet. After several frustrating days of poring over our code, we found the problem: it turned out that a stray keystroke in our code had commented out the updating of a nuisance parameter, one for which we had assigned a random starting value. Thus, the nuisance parameter was fixed at a different value in each run; it being one of many nuisance parameters, we had not bothered to look at the samples from its posterior distribution.

We had, however, accidentally carried out good MCMC practice. It is a good idea to run multiple chains with distinct and diffuse starting values, and to compare the results. The Brooks–Gelman–Rubin diagnostic (Brooks and Gelman, 1998) implemented in BUGS provides one such comparison. Central $100(1 - \alpha)\%$ credible intervals are computed using data from individual chains and compared to intervals created using data pooled from multiple chains. The degree of agreement between the results is taken as a measure of convergence of the chains to their stationary distribution; sequential evaluation provides guidelines for choosing an adequate burn-in period.

It is always a good idea to produce long Markov chains, the longer the better. This is because features of the chains more closely approximate features of the posterior distribution

---

12. But not always: Newton–Raphson can get stuck at local maxima, when we're looking for global maxima. It is also perfectly satisfied to wind up at minima if such exist.

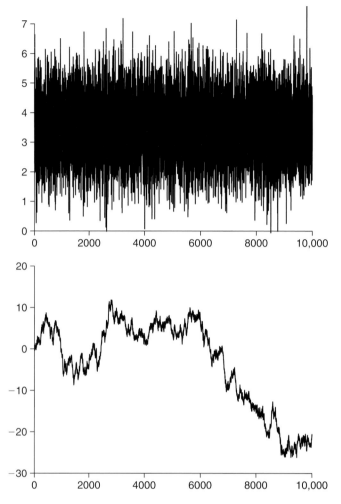

FIGURE 4.10   History plots, X_t against t, for two Markov chains.

as chain length increases.[13] While there are no simple rules for evaluating the precision of the approximations, a couple of simple options may provide insights. First, if the ACF tails off quickly enough that observations at lag k are essentially independent, a chain $\{X_i\}$ of length N can be thinned to a chain of length N/k for which the variance of the sample mean is approximated by $\sigma^2/(N/k) = k\sigma^2/N$, where $\sigma^2 = \mathrm{Var}(X_i)$, providing a conservative measure of precision for the mean of the entire chain.

13. While it is technically correct to describe MCMC as "estimating" features of posterior distributions, we prefer to use the term "approximating," because estimation implies finite resources, such as a given data set or sample size. Approximation, however, is only limited by our time and patience.

Another approach is to approximate the ACF by a geometric curve. If the autocorrelation at lag t is ρ^t, then the variance of the sample mean approximates

$$\left(\frac{1+\rho}{1-\rho}\right)\frac{\sigma^2}{N},\tag{4.11}$$

with σ^2 consistently estimated by the sample variance of the chain.

Yet another approach to estimating the precision with which a Markov chain's mean has been approximated is through the use of "batch means" (Roberts, 1996), implemented in BUGS). Suppose that $\{X_i\}$ is such that the average of n consecutive values has mean μ and variance v/n; suppose further that our chain of length N can be broken into K batches of length n, with nearly independent batch means $\hat{\mu}_k$. Then the sample standard deviation among the K values $\hat{\mu}_k$ is a consistent estimator of v/n as $K \to \infty$. Thus, we may estimate v by

$$\hat{v} = \frac{n}{k-1}\sum_{k=1}^{K}(\hat{\mu}_k - \bar{\mu})^2,$$

where $\bar{\mu}$ is the sample mean for the entire chain. We may then estimate $\mathrm{Var}(\bar{\mu})$ by $\hat{V}/N$. The square root of this quantity is reported as "MC Error" in BUGS.[14]

Illustration

Haramis *et al.* (2007), in a study of the dietary importance of horseshoe crab *Limulus polyphemus* eggs for migrating shorebirds, fit a nonlinear model relating stable isotope (SI) ratios $y = \delta^{15}N$ to body mass (x) of captured birds. Measured values were assumed to follow the relation

$$y_i = A\left[1 - b\exp(-c\,x_i)\right] + \epsilon_i,$$

where ϵ_i are independent mean zero normal random variables with variance $\sigma_i^2 = \alpha\exp(-\beta x_i)$. Assuming A, b, c and $\beta > 0$, the model posits SI ratios which increase toward an asymptote A as body mass increases, and that the variability among individuals decreases as the asymptote is reached.

We generated a Markov chain of length 1.025 million to sample the posterior distribution of the asymptote A, discarding the first 25,000 values as a burn-in; the process took slightly less than half an hour running BUGS on a 2.4 GHz processor. Owing to the complexity of the model, it is not surprising that the chain is highly autocorrelated. Its autocorrelation function to lag 250 is displayed in Fig. 4.11 (solid curve) along with an approximating geometric curve $f(t) = \rho^t$, with $\rho = 0.9856$, obtained by regressing the log of the ACF on the lag. Using Eq. (4.11), we have $\mathrm{Var}(\bar{\mu}) \approx 137.6\sigma^2/N$; substituting the sample variance 0.0431 of the one million values, we obtain $\mathrm{Var}(\bar{\mu}) \approx 0.0024^2$, and conclude that our estimate $\bar{\mu} = 14.3495$ is accurate to two decimal places. We report the posterior mean of A to be 14.35.

We note that BUGS reports "MC error" = 0.0029. The value based on batch means appears to be sensitive to the size of the batches. Using batches of size 100, 1000, and 5000, we obtain values

---

14. Note that we have used v rather than σ^2 in describing the batch means technique. We use σ^2 for the variance of individual X_i's in the chain; σ^2 is consistently estimated by the sample variance of the X_i's, regardless of the autocorrelation. The quantity v is different: in Eq. (4.11), $v = (1+\rho)/(1-\rho)\sigma^2$.

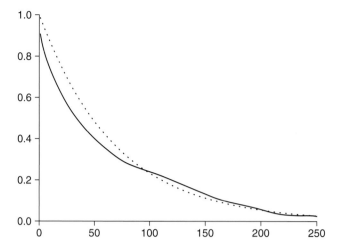

FIGURE 4.11 ACF (solid line) and approximating geometric curve (dashed).

0.0016, 0.0029, and 0.0034, respectively. It is important to recognize that these values are really only indices to precision. All indicate considerably less precision than we would incorrectly conclude by treating the sampled values as independent, and calculating $s/\sqrt{N} = 0.0002$. The value $(1+\rho)/(1-\rho) \approx 137.6$ allows a somewhat more honest appraisal of the value of our original chain of length one million: it allows comparable precision to an independent sample of size 1e6/137.6 $\approx$ 7300. Looked at this way, the chain length of 1e6 seems considerably less extravagant.

PART II

THE BAYESIAN MĀRAMATANGA

5

Bayesian Prediction

O U T L I N E

5.1	**The Bayesian Māramatanga**	**78**
	5.1.1 Derived Parameters	*78*
	5.1.2 The Posterior Predictive	
	Distribution	*79*
5.2	**Estimating Mean Residual Lifetime**	**81**
	5.2.1 Frequentist Analysis of	
	Derived Parameter ψ	*81*
	5.2.2 Posterior for ψ via Change of	
	Variables Theorem	*82*
	5.2.3 Posterior Features for ψ	
	based on Posterior of p	*82*
	5.2.4 Comparison of Methods	*84*
	5.2.5 Prior Sensitivity	*84*
5.3	**Derived Parameters in Summary**	
	Analyses: BBS Case Study	**85**
	5.3.1 Digression: When Are Count	
	Surveys of Value?	*86*
	5.3.2 Derived Parameters and	
	the BBS	*87*
	5.3.3 BBS Trend Summaries for	
	Grassland Birds	*89*

	5.3.4 Conclusion	*93*
5.4	**Derived Parameters and Out**	
	of Sample Inference in a Dose–	
	Response Study	**93**
	5.4.1 Background	*93*
	5.4.2 Laboratory Dose–Response	
	Study	*94*
	5.4.3 Derived Parameters	*95*
	5.4.4 Out of Sample Prediction	*95*
5.5	**Product Binomial Representation**	
	of the CJS Model	**98**
	5.5.1 CJS with Uniform Prior on λ	
	and τ	*100*
	5.5.2 CJS with Uniform Prior on p	
	and ϕ	*101*
	5.5.3 CJS Model: Summary	*104*
5.6	**Posterior Predictive Model**	
	Checking	**104**
	5.6.1 Bayesian and Frequentist	
	p-Values	*106*

5.1 THE BAYESIAN MĀRAMATANGA

Our approach in writing this book is to stress the usefulness of the Bayesian way of looking at things, but we cannot help but mention its beauty, as well.

"Way of looking at things" seems a fairly lame way of saying what we want to say. But our teeth are set on edge and our stomach churns when we hear the word "paradigm"; we have been negatively conditioned by the faddish and vacuous overuse of a perfectly serviceable word, meaning "a conceptual model underlying the theories and practice of a scientific subject" (OED).

We need a new word, and suggest the term "māramatanga," which is about as close as we could come in Maori, and sounds great. The Bayesian māramatanga is not only useful, but simple, exact, and coherent, as described in Chapter 1, and hence beautiful. In this chapter, we focus on one aspect of its simplicity, namely that *there are only two types of quantities in Bayesian inference, the **observed** and the **unobserved***. Observed quantities are data; these are the same whether one is frequentist or Bayesian. The difference comes when we turn our attention to unobserved quantities: these include parameters, missing data, future observations, even hypothetical outcomes which may have occurred in the past, or may yet occur in the future; all are treated equivalently, no distinction is made. Thus in the Bayesian setting there is no difference between estimating a parameter, and predicting a future observation. We do not need a whole raft of different techniques for predicting future observations, for dealing with missing data, for estimation; all are done in essentially the same way. The word *prediction* is used to describe all inference about unobserved quantities.

The examples presented in this chapter have to do with two types of unobserved quantities: functions of parameters (*derived parameters*) and hypothetical replicates.

5.1.1 Derived Parameters

As practicing statisticians consulting with quantitative biologists, we are frequently asked questions like the following: "How do I compute the variance of $-\log(\hat{\theta})$?" The better informed will sometimes indicate knowledge of a commonly used technique, and phrase the question as "What's the δ-method approximation to the variance of $-\log(\hat{\theta})$?"

The quick answer is that, under certain conditions, if $\hat{\theta}$ is unbiased for θ and has variance V, and if $\psi = g(\theta)$, then the variance of $\hat{\psi} = g(\hat{\theta})$ is approximately $V \times [g'(\theta)]^2$. This is the answer we give if we are in a hurry, or if we do not like the client.

A better response is to ask why the variance of $g(\hat{\theta})$ is desired, and perhaps to steer the client in another direction. Typically, the reason is that a confidence interval of the form

$$\hat{\psi} \pm z\sqrt{\frac{V(\hat{\psi})}{n}} \tag{5.1}$$

is to be calculated, using a standard normal multiplier z. In this case, it seems only fair to point out that the client is going to *estimate* an *approximate* variance, and use it in *approximate* interval estimate ... and to point out, by the way, that the δ-method might not work all that well.[1]

Biologists routinely need to estimate functions of parameters. To take a simple example, suppose that monthly survival is p, and can be modeled as independent of age and other effects. We may then estimate residual lifetime as a geometric random variable with parameter p, so that $\Pr(X = x|p) = (1 - p)p^x$, for $x = 0, 1, 2, \ldots$. Then the expected residual lifetime is $p/(1 - p)$; an estimator of p may need to be converted to an estimate of mean lifetime via this transformation. We work through this specific example in Section 5.2.

Parameters calculated as functions of other parameters are often referred to as *derived parameters*. Frequentist attempts at assessing the uncertainty associated with derived parameters are often ad hoc and approximate. The delta method can be especially problematic; generally, if $[\hat{\theta}_L, \hat{\theta}_U]$ is an interval estimate for θ, and $\psi = g(\theta)$ is a strictly increasing function of θ, the interval estimate

$$\left[g\left(\hat{\theta}_L\right), g\left(\hat{\theta}_U\right) \right] \tag{5.2}$$

is more reliable than applying δ-method approximations to (5.1).[2] But an altogether better approach, to our mind, is to adopt the Bayesian māramatanga.

Bayesian inference is ideally suited to dealing with derived parameters. Given the posterior distribution for θ, the posterior distribution for $g(\theta)$ is completely specified. If we are using a sampling-based assessment of the posterior distribution, such as MCMC, matters are made quite simple. Given a sample $\theta_1, \theta_2, \theta_3, \ldots, \theta_M$ from the posterior distribution $[\theta|X]$, and supposing that $\psi = g(\theta)$, we need only calculate $\psi_1 = g(\theta_1), \psi_2 = g(\theta_2), \psi_3 = g(\theta_3), \ldots, \psi_M = g(\theta_M)$ to obtain a sample from the posterior distribution $[\psi|X]$. There is no approximation involved, except such as results from the sampling-based assessment, and this sample size is not limited by the available data.

We illustrate the ease with which analysis of derived parameters can be conducted by means of examples in Sections 5.2–5.5. First, however, we complete our introduction to Bayesian prediction by describing the posterior predictive distribution, which is used for describing unobserved replicate data.

5.1.2 The Posterior Predictive Distribution

It is often useful to think about what new data would look like, if generated according to the same model as we are using to describe existing data. If we have a lot of faith in the model, the new data could be used to describe missing values in the historical record, or to anticipate future values. If we have doubts about the validity of the model, we could compare existing data with the new data to see whether there are inconsistencies between the model and existing data.

---

1. Suppose that $X \sim U(0, \theta)$. Then a natural estimator of θ is $\hat{\theta} = 2X$; the corresponding estimator of $-\log(\theta)$ is $-\log(2X)$. By direct calculation, it can be shown that the variance of $-\log(2X)$ is identically 1, independent of θ. As it turns out, the δ-method approximation is also independent of θ, but equals 1/3.
2. If $g(\theta)$ is strictly decreasing, the transformed interval estimate is $\left[g\left(\hat{\theta}_U\right), g\left(\hat{\theta}_L\right) \right]$.

All Bayesian inference is based on posterior distributions, including predictions of hypothetical data. It is useful to take a moment to consider how one would predict a new observation based on a model, and existing data.

Suppose that existing data X are modeled as having been sampled from a distribution $[X|\theta]$, governed by parameter θ, itself sampled from prior distribution $[\theta]$. To make predictions about a new observation X^{new}, we assume that it is a replicate of X, having the same distribution, and the same parameter θ. It will usually be appropriate to suppose that, except for having shared parameter θ, the new observation is independent of X.

We proceed as usual, using the posterior distribution of all unobserved quantities as the basis of inference. Bayes' theorem yields

$$[\text{Unobserved quantities}|\text{Observed quantities}] \propto$$
$$[\text{Observed quantities}|\text{Unobserved quantities}][\text{Unobserved quantities}],$$

i.e.,

$$[\theta, X^{new}|X] \propto [X|\theta, X^{new}] \times [\theta, X^{new}]. \tag{5.3}$$

Note that the first term on the right-hand side of (5.3) can be simplified, because X and X^{new} are conditionally independent, and that the second term can be factored, yielding

$$[\theta, X^{new}|X] \propto [X|\theta] \times [X^{new}|\theta] \times [\theta]. \tag{5.4}$$

Now the first and last terms on the right-hand side of (5.4), taken together, are proportional to the posterior distribution $[\theta|X]$. Hence, we may express (5.4) as

$$[\theta, X^{new}|X] \propto [X^{new}|\theta] \times [\theta|X];$$

this is the joint posterior distribution of θ and X^{new}. If we are only interested in inference about X^{new}, we may find its marginal posterior distribution by integrating out θ, obtaining

$$[X^{new}|X] = \int [X^{new}|\theta] \times [\theta|X] d\theta. \tag{5.5}$$

Equation (5.5) defines the *posterior predictive distribution*. There is really nothing out of the ordinary about the distribution defined by Eq. (5.5). It is the posterior distribution of a particular type of unobserved quantity, viz., a hypothetical replicate X^{new} generated under the same conditions as produced X. But the distinctive nature and usefulness of this particular posterior distribution are such as to justify its being distinguished with a special name, the posterior predictive distribution.

Examining Eq. (5.5), one notes that the posterior predictive distribution is obtained by integrating the data distribution for a new observation against the posterior distribution of the unknown parameter. In other words, the posterior predictive is the average value of the data distribution, averaged against draws θ sampled from the posterior distribution. If we are using a sampling-based assessment of the posterior distribution, such as MCMC, this means that we need merely to calculate the data distribution of X^{new} over values θ from the posterior distribution and to average these distributions. The approach is very similar to what is attempted using the parametric bootstrap in frequentist analysis, but without reliance on asymptotic approximations.

II. THE BAYESIAN MĀRAMATANGA

The posterior predictive distribution is a very handy tool, a further instance of the appeal of the Bayesian māramatanga. Draws from it represent our best attempt to generate data according to the model for X, properly accounting for the uncertainty due to imperfect knowledge of the parameter θ. We illustrate its use for "out of sample" inference in Section 5.4 and for model checking in Section 5.6.

5.2 ESTIMATING MEAN RESIDUAL LIFETIME

We begin with a very simple example of a derived parameter, the mean residual lifetime for a population with homogeneous survival rates. Suppose that we observe 20 individuals over a 12-month period, that 17 of the individuals are alive at the end of the study, and that the other 3 individuals survive for 7, 11, and 11 months, respectively.

A simple model is that monthly survival is p, varying neither among individuals nor with age. The data can then be summarized as 236 Bernoulli trials, with $17 \times 12 + 11 + 11 + 7 = 233$ successes (monthly survivals) and three failures, so that the likelihood function is $L(p) \propto p^{233}(1-p)^3$. The MLE of p is $\hat{p} = 233/236 = 0.9873$, with approximate standard error $\sqrt{\hat{p}(1-\hat{p})/236} = 0.0073$. The usual approximate 95% confidence interval is $0.9873 \pm 1.96(0.0073) = (0.9730, 1.0016)$.

We think of individual lifetimes X as geometric random variables with parameter p, so that $\Pr(X = x|p) = (1-p)p^x$, for $x = 0, 1, 2, \ldots$. Then the mean residual lifetime is $\mathrm{E}(X - t|X \geq t) = p/(1-p)$, independent of t. The geometric distribution is the discrete data analog of the exponential distribution, with its "lack of aging" condition; no matter how old you are, you still have the same expectation of future survival. This expectation is a function of the model parameter, a simple case of a derived parameter. We represent the mean residual lifetime by $\psi = g(p)$, where $g(p) = p/(1-p)$.

5.2.1 Frequentist Analysis of Derived Parameter ψ

The MLE of ψ is $\hat{\psi} = g(\hat{p}) = 77.7$. The δ-method estimate of the standard error is $f'(\hat{p}) \times 0.0073 = 45.1$, leading to a confidence interval using (5.1) of $77.7 \pm 1.96 \times 45.1 = (-10.8, 166.1)$. This interval does not inspire confidence.

For one thing, we know on a priori grounds that the mean lifetime is going to be larger than zero; indeed, the mean residual lifetime for the 20 individuals in the data set is at least $233/20 = 11.7$. For another thing, it seems clear that values of p quite close to 1 cannot be ruled out, and these suggest a very long mean residual lifetime, possibly much greater than 166. It is as though the δ-method confidence interval has got its mass in all the wrong places.

If one could trust the original interval estimate for p, truncated to $(0.9730, 1.0)$, one could trust an interval obtained by applying $g(p)$ to its endpoints, using Eq. (5.2). This results in an interval estimate for the mean residual lifetime of $(36.0, \infty)$. This is intuitively more satisfying than the δ-method interval, but our confidence in it rests upon our confidence in the original interval estimate; the observations in Chapter 3 suggest that this interval is not terribly reliable, and probably does not attain nominal coverage rates.

The exact 95% confidence interval for p (described in Section 3.2.2) is $(0.9633, 0.9974)$. This interval is conservative: it uses a technique that is guaranteed, regardless of the value of p, to have at least 95% probability of successfully including p; for many values of p, the interval is longer than necessary. Applying $g(p)$ to its endpoints, once again using Eq. (5.2), one obtains an interval estimate of (26.2, 383.6). This interval is a legitimate frequentist 95% confidence interval.

The approach used to create this interval has several drawbacks. First, it does not easily generalize: defining exact confidence regions is a challenging problem, with solutions having to be derived on a case by case basis. Second the computations involved in obtaining the exact confidence interval are not easily performed, even for the simple problem of a single binomial success parameter. Third, there is the matter of its interpretation, as with confidence intervals generally, as discussed in Chapter 3.

5.2.2 Posterior for ψ via Change of Variables Theorem

If we specify a Be(1,1) (uniform) prior for p, then by conjugacy, the posterior distribution of p is $[p|\text{Data}] = \text{Be}(1+233, 1+3)$. Because $\psi = g(p)$, the posterior distribution of p determines the posterior distribution of ψ. The only challenge to be faced is how to evaluate this posterior distribution, and for this there are several options.

First, we can directly calculate the posterior distribution of ψ by analytical means. The change of variables theorem (Section 2.2.4) says that if random variable ψ is defined as a function of random variable p by $\psi = g(p)$, then the distribution of ψ is

$$f_\psi(\psi) = f_p\left(g^{-1}(\psi)\right)\left|J_{g^{-1}}(\psi)\right| \tag{5.6}$$

Here, g^{-1} is the inverse of the transformation g and J denotes the Jacobian determinant of the transformation (the derivative, in the univariate case). For the problem at hand, $f_p(p)$ denotes the Be(234,4) density. Solving $\psi = g(p)$ for p, we have $p = g^{-1}(\psi) = \psi/(1+\psi)$. Consequently, the Jacobian of g^{-1} is $1/(1+\psi)^2$. Thus using (5.6) we find that the posterior distribution of ψ is

$$f_\psi(\psi) = \frac{\Gamma(238)}{\Gamma(234)\Gamma(4)}\left\{\frac{\psi}{1+\psi}\right\}^{234-1}\left\{\frac{1}{1+\psi}\right\}^{4-1}\left\{\frac{1}{1+\psi}\right\}^2$$

$$= 512616780\,\frac{\psi^{233}}{(1+\psi)^{238}}, \quad 0 \le \psi < \infty. \tag{5.7}$$

This density function is displayed in Fig. 5.1; all inference about ψ is based on this density. The implications of model, prior knowledge and data about ψ can all be examined by mathematical analysis of Eq. (5.7). For example, if we wished to summarize this distribution by a single point, we might choose the posterior mode. Setting the first derivative of $\log(f(\psi))$ equal to zero and solving for ψ, the mode is found to be at $\hat\psi = 233/(238-233) = 46.6$.[3]

5.2.3 Posterior Features for ψ based on Posterior of p

The calculations involved in computing the posterior distributions of derived parameters and in evaluating these posterior distributions can be quite difficult. But knowledge of the posterior distributions of the original parameters suffices; we need not actually compute $[\psi|\text{Data}]$ if we know $[p|\text{Data}]$.

---

3. The posterior mode for ψ is not the same as the MLE, because in assigning a uniform prior on p, a nonuniform prior is implicitly chosen for ψ. We return to this point in Section 5.2.5.

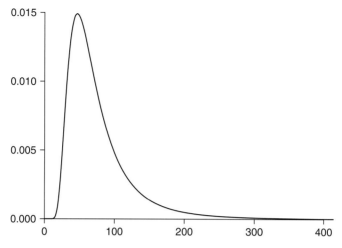

FIGURE 5.1   Posterior distribution of mean residual lifetime.

For example, the 95% highest posterior density interval (HPDI, i.e., shortest interval containing 95% mass) for p based on the Be(234, 4) distribution is (0.9668, 0.9969). This means that

$$Pr(0.9668 \leq p \leq 0.9969 \mid Data) = 0.95,$$

and because $\psi = g(p)$ is a monotone increasing function of p, that

$$Pr(g(0.9668) \leq g(p) \leq g(0.9969) \mid Data) = 0.95.$$

Thus, $\big(g(0.9668), g(0.9969)\big) = (29.1, 326.3)$ is a 95% credible interval for the mean residual lifetime, very much like the exact frequentist interval, though somewhat shorter.

Many analysts would not bother with exact calculations using the beta distributions, but instead use software like BUGS to produce a sample from the posterior for p, using the 2.5th to 97.5th percentiles as a 95% credible interval. The sample values would approximate the true 2.5th and 97.5th percentiles, yielding the interval (0.9635, 0.9954) for p, which on transformation as before yields a 95% credible interval of (26.4, 216.4) for ψ.

This latter interval is in fact shorter than the interval obtained by transforming the endpoints of the HPDI. Both are legitimate 95% credible intervals. Neither is the "right" answer, and neither the "wrong" answer; the posterior probability that ψ is in either interval is 95%. The second interval might be more appealing, however, because it is shorter. It is natural to ask whether it is possible to make an even shorter 95% credible interval.

Suppose that using BUGS or some other software we have a sample of a million values p^i from the Be(234, 4) distribution. Then $\psi^i = g(p^i), i = 1, 2, \ldots, 1,000,000$ is a sample from the posterior distribution of the mean residual lifetime. If we sort the million ψ^i, the interval from the ith to the $(950,000 + i)$th approximates a 95% credible interval. There are 50,000 such intervals, the shortest of which approximates the actual HPDI for ψ, namely (18.4, 173.8). This approach for finding the HPDI works provided that the posterior distribution is unimodal.

TABLE 5.1   Interval estimates of mean residual lifetime.

	Interval	
Frequentist	δ-method	$(-10.8, 166.1)$
	transformed endpoints of approx CI for p	$(36.0, \infty)$
	transformed endpoints of exact CI for p	$(26.2, 383.6)$
Bayesian	transformed endpoints of HPDI for p	$(29.1, 326.3)$
$[p] = U(0,1)$	$(2.5th, 97.5th)$ percentiles of ψ	$(26.4, 216.4)$
	HPDI for ψ	$(18.4, 173.8)$

5.2.4 Comparison of Methods

We have summarized the various interval estimates discussed in Table 5.1. The first two frequentist intervals are approximate, and self-evidently unsatisfactory. The third, based on the rarely used "exact" confidence procedure is conservative, and longer than the first of the three Bayesian credible intervals. All three of the Bayesian credible intervals have 95% posterior probability associated with them. The first was chosen for comparison to the exact frequentist interval, the second because it is the easiest to compute from simulation output, and the third because it is the shortest possible.

Of course, the best Bayesian summary of the information available about mean residual lifetime is the entire posterior distribution (Fig. 5.1), rather than one point estimate, or an individual interval estimate. Interval and point estimates are summaries of the density function (5.7), hence represent some loss of information.

The procedure for inference about a transformed parameter $\psi = g(p)$ is straightforward: knowledge about the posterior distribution of p translates directly into knowledge of the posterior distribution of ψ. This can be done directly, using the change of variables theorem (Section 2.2.4), or indirectly, by transforming sampled values of p, one by one, into sampled values of ψ. The latter approach is simple ... what's not to like?

5.2.5 Prior Sensitivity

What's not to like? Well ... there is always the matter of choosing priors. A uniform $U(0,1)$ prior might seem reasonable enough for studying parameter p, but consider the implications for studying parameter ψ. If p has a $U(0,1)$ distribution, then for $0 \leq t \leq 1$, $\Pr(p \leq t) = t$, hence

$$\Pr(\psi \leq x) = \Pr\left(\frac{p}{1-p} \leq x\right) = \Pr\left(p \leq \frac{x}{1+x}\right) = \frac{x}{1+x}.$$

Taking the derivative with respect to x yields the prior density for ψ, namely $1/(1+x)^2$ for $x > 0$. This distribution has infinite expectation, but still would seem to favor smaller values of ψ: its mode is at zero, and $x = 24, 49, 99$, and 999 are its 96th, 98th, 99th, and 99.9th percentiles. This choice of prior for ψ might seem unappealing.

Then, try another prior! Suppose that we wish to have a uniform prior on ψ, say $[\psi] = U(0,M)$ for a large value M. It can be shown that this uniform prior results from having a prior $[p] \propto 1/(1-p)^2$ on $(0, M/(M+1))$. We have two options for analysis.

First, we could use the likelihood $L(p) \propto p^{233}(1-p)^3$ in combination with the prior $[p] \propto 1/(1-p)^2$ on $(0, M/(M+1))$. At first glance, this prior might appear difficult to sample; it is not one that will appear in a listing of commonly used priors in software such as BUGS. The solution is to specify ψ as having the $U(0,M)$ distribution and then convert ψ's to p's by $p = \psi/(1+\psi)$. The resulting p have the desired prior distribution.

Alternatively, we may reparameterize the likelihood in terms of ψ, as

$$L(\psi) \propto \left(\frac{\psi}{1+\psi}\right)^{233} \left(\frac{1}{1+\psi}\right)^3,$$

again specifying ψ as having the $U(0,M)$ distribution. Done this way, we treat p as the derived parameter. It is not difficult to see that the two options are identical.

For $M = 10,000$, the resulting HPDI for ψ is $(22.1, 667.1)$. The HPDI is not changed by setting $M = 1,000,000$. This interval is similar to those previously obtained (Table 5.1), but the upper bound is higher; this observation is completely consistent with the foregoing comments on the prior for ψ induced by a uniform prior on p.

So then, which prior ought we to use? Should we place a uniform prior on p or on ψ? In either case, the induced prior on the other parameter is informative, and perhaps not to our liking. Although there is no absolute "right" answer, the Bayesian calculus is absolutely right, given the choice. The choice, if choice must be made, should be guided by prior knowledge. If, for instance, the $U(0,10,000)$ distribution were an appropriate expression of our uncertainty about the mean residual lifetime, then a 12-month study of monthly survival rates would probably not have been considered in the first place. The choice of a uniform distribution as an expression of prior uncertainty is generally most appropriate for the basic parameterization which guided model development and data collection protocols. Furthermore, we are not limited to these two choices; we are free to investigate the choice of various priors, and to see precisely what their effect is on inference. Despite differences in the inferred upper bounds for ψ, the inferred lower bounds are similar for the two priors considered here. We are free to evaluate as many priors as we might wish; honesty binds us to not choose a single "prior" on the basis of the resulting posterior, and to report the prior as part of the analysis.

In concluding this evaluation of the derived parameter ψ we reiterate our primary point: that derived parameters are naturally and consistently handled using posterior distributions.

5.3 DERIVED PARAMETERS IN SUMMARY ANALYSES: BBS CASE STUDY

Attend an ecological or wildlife meeting in North America, and you are likely to hear presentations drawing data from the North American Breeding Bird Survey (BBS); attend an ornithological meeting and the likelihood becomes near certainty.

The BBS has been conducted every summer since 1966, with the goal of monitoring bird population at large geographic scales. Data are collected for over 400 species of birds on more than 3500 roadside survey routes in the continental United States, southern Canada, and northern Mexico. A competent observer conducts 3-minute counts of birds seen or heard at each of 50 stops located at half-mile intervals along the route. The counts are carried out in accordance with protocols intended to reduce irrelevant sources of variation (Robbins *et al.*, 1986). Observers typically serve for 3 or 4 years, though some serve for far longer; on the other hand, roughly 25% only serve for a single year.

The BBS thus provides a vast collection of parameter estimates, a happy hunting ground in which to search for patterns and associations among parameters. Such investigations typically involve summaries of parameter values, summaries that are derived parameters. In this section, we focus on summaries of temporal trend parameters β_i, for species $i = 1, 2, ..., K$. By "trend" we mean the geometric mean rate of change over a specified time interval. We include this example as an appealing illustration of the Bayesian approach to derived parameters. Anticipating that some readers might have reservations about the quality of BBS data, we begin with a digression on the usefulness of count survey data.

5.3.1 Digression: When Are Count Surveys of Value?

The usefulness of BBS data for drawing inference is continually questioned, with legitimate concerns raised about how reliably variation in its counts can be associated with variation in population sizes. Can changes in bird populations sampled along roadsides be taken as representative of population changes in all of their habitats? After all, roadside habitats might be peripheral, low-quality habitats and hence indicate greater volatility in population sizes than exists in entire populations. Furthermore, features of roadside sites change through time as urbanization increases, and these changes might not be representative of the entire landscape.

BBS data also reflect variation among observers. There is strong evidence of trend in BBS counts associated entirely with changes in the pool of observers through time; new observers appear to be better than the observers they replace. Furthermore, there is strong evidence of temporal change in counts by individual observers, change that is unrelated to population change. It is thought that such change may be due to increasing familiarity with their routes and to changes in their ability to hear and to identify birds.

All of these questions about BBS data quality relate to the fact that count surveys only provide an index to population size: not all of the birds present along a survey route are counted. A useful though imperfect description of count data is that count C can be expressed as $C = Np$, where N is the size of the population studied and p is the proportion of individuals counted.[4] The problem with indices is that variation in C does not only reflect variation in N, but also variation in p. It can be shown that indices correlate well with population sizes if the coefficient of variation in detection rates is small, or if variation in detection rates can be explained through modeled effects of covariates.

The value of count survey data thus inevitably depends on our ability to determine and measure relevant explanatory variables, and on the validity of the models used to describe their effects. Many, many techniques have been developed for dealing with imperfect detection; for a review and critical commentary, see Johnson (2008). Despite its failings, the BBS and related surveys remain the only large scale source of data on population change for many species. It

---

4. The description is imperfect because it is unlikely that there ever is a closed population associated with a count.

is unreasonable to completely dismiss such data: better to make appropriately labeled weak inference than no inference at all.

5.3.2 Derived Parameters and the BBS

Suppose that we have a collection of trend estimates, $\hat{\beta}_i, i = 1, 2, \ldots, K$, such as presented in Table 5.2. What is the first thing we do when we look at the list? Most folks instinctively look

TABLE 5.2   BBS survey-wide estimates of 1966–99 trend for 28 grassland species.

	Species name	Trend	SE	N site
1	Upland Sandpiper	0.76	0.39	582
2	Long-billed Curlew	−0.77	1.01	222
3	Mountain Plover	−1.05	2.24	37
4	Greater Prairie-Chicken	−2.54	2.33	33
5	Sharp-tailed Grouse	−0.92	1.43	128
6	**Ring-necked Pheasant**	−1.06	0.32	1239
7	**Northern Harrier**	−0.80	0.40	935
8	**Ferruginous Hawk**	3.52	1.31	200
9	Common Barn Owl	−2.00	2.14	32
10	Short-eared Owl	−6.23	4.55	140
11	Burrowing Owl	1.00	2.74	278
12	**Horned Lark**	−1.89	0.22	1864
13	**Bobolink**	−1.25	0.31	1168
14	**Eastern Meadowlark**	−2.69	0.17	1984
15	**Western Meadowlark**	−0.75	0.17	1534
16	**Chestnut-col. Longspur**	−1.36	0.68	144
17	McCown's Longspur	−9.29	8.27	59
18	**Vesper Sparrow**	−0.61	0.24	1547
19	Savannah Sparrow	−0.34	0.29	1582
20	Baird's Sparrow	−2.04	1.48	120
21	**Grasshopper Sparrow**	−3.73	0.47	1443
22	Henslow's Sparrow	−4.82	2.50	149
23	LeConte's Sparrow	0.91	0.95	190
24	**Cassin's Sparrow**	−2.10	0.51	225
25	**Dickcissel**	−1.46	0.28	829
26	Lark Bunting	−3.74	2.30	339
27	**Sprague's Pipit**	−5.62	1.34	121
28	**Sedge Wren**	3.18	0.73	340

Species with trend estimate more than 1.96 standard errors away from zero in bold.

for the extreme values: "dang, wonder what is happening to those McCown's longspurs?" Next thing we do is to start making up stories to explain the extreme values. We get bit once or twice by the bug, but then get more cautious: "hang on, that trend for McCown's longspur is pretty badly estimated … it's not even significant."

Another stare at Table 5.2, and we get to noticing that there are a lot of negative trend estimates, 23 of the 28, in all. But then we reflect that only 12 of the 23 are significantly different from zero. Should we report that 12 of the species have significant declining trends? or perhaps, that of the 13 significant trends, 12 are negative? But then we get to thinking that reporting "significance" is more a statement about the quality of the data than it is about the parameters themselves, what is more, that our choice of $\alpha = 0.05$ was arbitrary, and we hesitate: perhaps we ought not to pass the parameter estimates through the filter of significance testing. Still, the nearly identical trend estimates for Western meadowlark (-0.75, SE $= 0.17$) and Long-billed curlew (-0.77, SE $= 1.01$) provide quite different support to the assertion that "grassland species are declining."

A related question occurs to us: "I wonder how many of these species have stable populations?" (Sauer and Link, 2002). To answer this, we must first define what we mean by a stable population; perhaps that the long-term trend is between -1% and $+1\%$. But again we find ourselves caught in the muddle of statistical significance. Sharp-tailed grouse have trend estimate of -0.92, but standard error of 1.43; the actual trend may well be less than -1; for that matter, it might be greater than $+1$. Are we to count them as having a stable population, or not?

Such easy questions that we would like to ask of the data, but classical training leaves us with no clear means of addressing them!

Each of the questions raised can be expressed in terms of simple functions of the parameters β_i.[5] Let $\mathbf{I}(statement)$ be the *indicator* function, equalling 1 if *statement* is true, equalling 0 if *statement* is false. Then $\mathbf{I}(\beta_i < 0)$ tells us whether species i is declining or not, and

$$D \equiv D(\beta_1, \beta_2, \ldots, \beta_K) = \sum_{i=1}^{K} \mathbf{I}(\beta_i < 0)$$

is the total number of declining species. Similarly, $\mathbf{I}(\beta_i \geq \beta_j)$ tell us whether the trend for species i is as large or larger than that for species j, and

$$R_i \equiv R_i(\beta_1, \beta_2, \ldots, \beta_K) = \sum_{j=1}^{K} \mathbf{I}(\beta_i \geq \beta_j)$$

gives the rank of species i in the group. The species with the most extreme decline is species m, defined as

$$m = \sum_{i=1}^{K} i \times \mathbf{I}(R_i = 1);$$

---

5. Functions of the *parameters*, not of the *estimates*.

the species with the largest trend is species M,

$$M = \sum_{i=1}^{K} i \times \mathbf{I}(R_i = K).$$

Finally, our criterion for population stability of species i is that $\mathbf{I}(|\beta_i| < 1) = 1$; the number of stable species among the K is simply

$$S \equiv S(\beta_1, \beta_2, \ldots, \beta_K) = \sum_{i=1}^{K} \mathbf{I}(|\beta_i| < 1).$$

Each of these quantities is a derived parameter, and a sample of its posterior distribution is easily obtained: letting θ_j denote the jth in a sample of the joint posterior distribution of all unknown quantities, we simply compute $D(\theta_j)$, $R_i(\theta_j)$, etc., to obtain samples from the posterior distributions of the derived parameters, then make inference using these posterior distributions.

5.3.3 BBS Trend Summaries for Grassland Birds

We treat the estimates $\hat{\beta}_i$ in Table 5.2 as conditionally independent unbiased estimates, normally distributed with precision $\tau(\hat{\beta}_i)$. We treat the estimated standard errors as mutually independent, and independent of the trend estimates, with distribution such that $SE_i^2 \sim$ $Ga(n/2, (n/2)\,\tau(\hat{\beta}_i))$, where n is the number of routes on which the species occurred; consulting Appendix B.9, it can be shown that this is equivalent to the familiar formulation

$$\frac{n\,SE^2}{\sigma_i^2} \sim \chi_n^2,$$

where $\sigma_i^2 = 1/\tau(\hat{\beta}_i)$.

We chose to treat the parameters β_i as a sample from a normal distribution with mean μ and precision $\tau(\beta)$. This imposition of group structure is important in ranking trends, because extreme trend estimates are likely artifacts of imprecise estimation.[6] We used a vague normal prior for μ, and vague gamma priors for all of the precision parameters. BUGS code is given in Panel 5.1. Subsequent summaries were based on Markov chains of length 10^6 after discarding a burn-in of length 10^5.

The posterior distribution of derived parameter D, the number of declining species, is given in Fig. 5.2. The posterior mode is at 23; the closest interval to a 95% HPDI is [20,25], with $\Pr(20 \leq D \leq 25\,|\text{Data}) = 0.968$. It happens that the number of negative trend *estimates* $\hat{\beta}_i$ is also 23, coinciding with the posterior mode for D. Although this is not entirely coincidental, it is not inevitable; what is more, a simple count of negative trend estimates provides no measure of the associated uncertainty.

Posterior modes, HPDI's and their coverage probabilities follow: For the number of stable species ($|\beta_i| < 1$) we predict $S = 10$, with $\Pr(7 \leq S \leq 14\,|\text{Data}) = 0.942$. We predict 15 unstable

---

6. The rank correlation between $|\hat{\beta}_i|$ and SE_i is 0.51.

PANEL 5.1   BUGS code for summary of BBS grassland bird trends.

```
list(delta=1,
betahat=c(0.76,-0.77,-1.05,-2.54,-0.92, -1.06,-0.80, 3.52,
 -2.00,-6.23,1.00,-1.89,-1.25,-2.69,-0.75,-1.36,-9.29,-0.61,
 -0.34,-2.04,-3.73,-4.82,0.91,-2.10,-1.46,-3.74,-5.62,3.18),
varhat=c(0.1521,1.0201,5.0176,5.4289,2.0449,0.1024,0.1600,
 1.7161,4.5796,20.7025,7.5076,0.0484,0.0961,0.0289,0.0289,
 0.4624,68.3929,0.0576,0.0841,2.1904,0.2209,6.2500,0.9025,
 0.2601,0.0784,5.2900,1.7956,0.5329),
n=c(582,222,37,33,128,1239,935,200,32,140,278,1864,1168,1984,
 1534,144,59,1547,1582,120,1443,149,190,225,829,339,121,340))

model{
        for(s in 1:28) {
                varhat[s] ~ dgamma(p[s],lam[s])
                p[s] <- n[s] / 2
                lam[s] <- p[s] * tau.betahat[s]
                tau.betahat[s] ~ dgamma(0.001,0.001)
                sd.betahat[s] <- 1/sqrt(tau.betahat[s])
                betahat[s] ~ dnorm(beta[s],tau.betahat[s])
                beta[s] ~ dnorm(mu,tau.beta)
                ranking[s] <- rank(beta[],s)
                rank.is.1[s]<- equals(ranking[s],1)
                rank.is.28[s]<- equals(ranking[s],28)
                stable[s] <- step(delta-abs(beta[s]))
                unstable[s] <- 1-stable[s]
                pos[s] <- step(beta[s])
                neg[s] <- 1-pos[s]
                speciesnum[s] <- s
        }
        numpos <- sum(pos[])
        numstable <- sum(stable[])
        unstabledown <- inprod(neg[],unstable[])
        unstableup <- inprod(pos[],unstable[])
        mu ~ dnorm( 0.0,1.0E-6)
        tau.beta ~ dgamma(0.001,0.001)
        sd.beta <- 1/sqrt(tau.beta)
        m <- inprod(rank.is.1[],speciesnum[])
        M <- inprod(rank.is.28[],speciesnum[])
        another.beta ~ dnorm(mu,tau.beta)
    }
```

and declining species (95.1% CI $= [12, 19]$) and we predict two unstable and increasing species (95.2% CI $= [1, 4]$).

Note that we have $10 + 15 + 2 = 27$ species in our summary, but there were 28 species in the study. Where has the extra species gone? Is this some sort of Bayesian deviltry, a tax collected by the authors of BUGS, or what? The "missing" species is an artifact of using posterior modes as point estimates. The posterior means were 10.21, 15.27, and 2.52, which add up to the desired 28, though we might scratch our heads and wonder what 0.21 species means; integer-valued estimates of integer-valued quantities are somewhat more appealing.

II. THE BAYESIAN MĀRAMATANGA

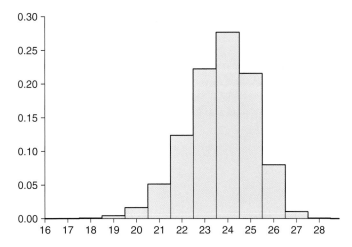

FIGURE 5.2 Posterior distribution of D, the number of grassland bird species with $\beta_i < 0$.

Our estimates of the number of stable species, etc., do not identify *which* species fall into the particular categories. Instead, our analysis provides posterior probabilities of stability for the individual species, namely $\Pr(|\beta_i| < 1|\text{Data})$. The distinction of being the most likely species to have a stable population goes to species 19, the savannah sparrow, for which we conclude $\Pr(|\beta_{19}| < 1|\text{Data}) = 0.987$. On the other hand, the data are almost perfectly ambiguous with regard to whether species 11, the burrowing owl, is stable or not: we have $\Pr(|\beta_{11}| < 1|\text{Data}) = 0.480$.

We return to our original observation of an apparently extreme trend for species 17, McCown's longspur. Our conclusions are based on two pieces of information about its trend: there is, on the one hand, the data ($\hat{\beta}_{17} = -9.29, \text{SE} = 8.27$), and on the other, the model, which says that β_{17} is sampled from the same distribution as the other β_i. Consider first the latter piece of information. In the BUGS code (Panel 5.1) we included a node `another.beta`, for a trend sampled from the same distribution, but not informed by data. This hypothetical value β tells us what the model says about trends; its posterior mean is -1.15, its standard deviation is 1.73. Thus, the model casts strong doubts on the point estimate $\hat{\beta}_{17} = -9.29$; trends sampled from the modeled distribution are extremely unlikely to be of such magnitude. For instance, the modeled probability of a $\beta < -6$ is less than $1/200$. Given that the variance associated with the data (SE^2) is nearly 33 times larger than the variance associated with the model, it is not surprising that the posterior mean for β_{17} is 96% of the way from $\hat{\beta}_{17}$ to the estimated group mean of -1.15, at -1.47, and that the posterior standard deviation for β_{17} (1.69) is nearly as large as the group standard deviation (1.73).

So where do we place McCown's longspur in a listing of Grassland species, sorted by trend? If we rank the species by the posterior mean values of β, McCown's longspur comes in 11th of 28. But to do so is to misleadingly portray the species as highly representative of the guild; its posterior mean is most highly adjusted toward the overall mean, simply because of the large SE associated with $\hat{\beta}_{17}$. The fact of the matter is that there really is no way of saying where this species ranks relative to the rest of the guild, as evidenced by the posterior distribution of its rank R_{17}, illustrated in Fig. 5.3.

Indeed, none of the species has clear title to the greatest decline in the group. The highest posterior probabilities are $\Pr(m=j|\text{Data}) = 0.413, 0.243, 0.084$, and 0.061 for $j = 27, 21, 22$, and 10. The posterior distribution for R_{27} (rank of trend for Sprague's pipit) is plotted in Fig. 5.4; we note in particular that $\Pr(R_{27} \leq 3) = 0.750$, so that we are 75% sure that the Sprague's pipit trend is one of the 3 worst. The largest trend most likely corresponds to the sedge wren ($\Pr(M=28|\text{Data}) = 0.691$) or the ferruginous hawk ($\Pr(M=8|\text{Data}) = 0.247$); for the remaining 26 species, $\Pr(M=j|\text{Data}) < 0.02$.

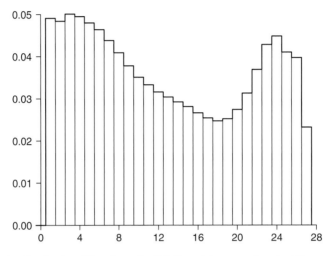

FIGURE 5.3   Posterior distribution of R_{17}, the rank of McCown's longspur (species 17) among the Grassland birds.

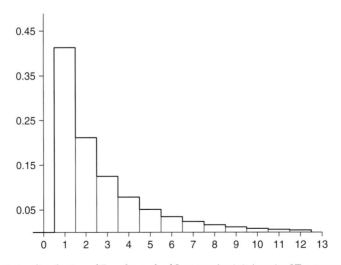

FIGURE 5.4   Posterior distribution of R_{27}, the rank of Sprague's pipit (species 27) among the Grassland birds.

5.3.4 Conclusion

Collections of parameter estimates such as Table 5.2 lend themselves to summary: we sort and rank the estimates, and make comparisons among them. Our intention, of course, is to sort, rank, and compare the latent parameter values themselves rather than the estimates. We have illustrated in this section that these comparisons are naturally expressed in terms of derived parameters, and easily handled in the Bayesian framework.

Analysts should give careful thought to the implications of the latent group structure imposed in the sort of analysis presented here. We noted that the posterior distribution for McCown's longspur trend, β_{17} is essentially uninformed by the data $\hat{\beta}_{17}$ and SE_{17}. Our conclusions about this species are, more than anything, a reflection of a model assumption, that McCown's longspur trend can be viewed as sampled from the same distribution as the other Grassland species. It should be noted that the summary analyses presented here do not require the assumption of the latent group structure. Instead of assuming that for all i, $\beta_i \sim N(\mu, \tau(\beta))$, we could assign independent vague priors to the β_i. The result is that the $\hat{\beta}_i$ become the posterior means (there is no shrinkage), the posterior median ranks more closely approximate the ranks of the raw estimates, and the credible intervals for individual rankings become slightly shorter. Either way, the approach outlined in this section provides legitimate summary analyses, properly accounting for all sources of variation in the data.

5.4 DERIVED PARAMETERS AND OUT OF SAMPLE INFERENCE IN A DOSE–RESPONSE STUDY

In this section, we describe a dose–response study investigating the effect on mallard drakes (*Anas platyrhynchos*) of exposures to sodium metavanadate ($NaVO_3$). The study was motivated by a die-off of Canada geese (*Branta canadensis*). A sample of the dead geese had unusually high levels of vanadium in liver and kidneys. Acknowledging the limitations of using mallards as surrogates for Canada geese, we might still reasonably inquire what sort of exposure levels, measured in terms of their lethality, would be associated with tissue concentrations observed in the geese. The analysis is of interest, first of all, in involving several derived parameters, and secondly, in using the posterior predictive distribution for out of sample inference.

5.4.1 Background

In January 2003, the Delaware Department of Natural Resources and Environmental Control (DDNREC) received a report of dead and dying Canada geese (*Branta canadensis*) at a petroleum refinery fly ash pond. The incident occurred during a period of extreme cold when most fresh water sources were frozen. The geese were apparently attracted to the fly ash pond because it was not completely frozen over.

The pond had been used for about 20 years as a basin for fly ash recovered from air scrubbers of the refinery's power plant and was found to have extremely high concentrations of vanadium. The DDNREC reported that composite samples of liver and kidney from two geese contained 57.3 and 226 parts per million (ppm) dry weight of vanadium, respectively, whereas values of other candidate toxicants (arsenic, cadmium, chromium, lead, mercury, nickel, selenium, and thallium) were low, and generally $< 0.5\,\mathrm{ppm}$ dry weight. Vanadium concentrations in liver and kidney rarely exceed $5\,\mathrm{ppm}$ dry weight in wild birds (Rattner *et al.*, 2006).

Some of the dead geese were examined by staff of the USGS National Wildlife Health Center, and found to be "in good flesh, with abundant body fat reserves [and without] lesions suggesting trauma"; healthy, except for congestion of lung, liver, and kidney tissues (Rattner *et al.*, 2006). The suspected cause of death was intestinal lesions and accompanying dehydration associated with vanadium toxicity.

5.4.2 Laboratory Dose–Response Study

Vanadium toxicity had been observed in human beings and studied in lab mammals, but little was known about toxic effects in birds and other wildlife. Thus, a dose–response study was conducted at the USGS Patuxent Wildlife Research Center, using adult mallard drakes (*Anas platyrhynchos*) as subjects. Subjects were administered sodium metavanadate ($NaVO_3$) in concentrations of 10, 18, 34, 62, 113, 208, 382, or 700 ppm body weight, and observed for 7 days. There were four subjects per dose level, and four controls (individuals sham-dosed with empty capsules). None of the control animals died; there were 0, 0, 0, 1, 4, 4, 3, and 4 deaths at the eight dose levels.

Vanadium concentrations were measured in the livers and kidneys of 15 of the 32 mallards in the dose–response study, to evaluate the relationship between exposure and tissue concentrations such as measured for the dead geese which motivated the study.

The data thus have two components: a binary response observed for all of the individuals, and a bivariate continuous response, observed for 15 individuals.

Such dose–response studies are a standard of toxicity studies. Mathematical models used to analyze dose–response data posit that animal i has a maximum tolerance level T_i, and that the animal exhibits a specific binary response (e.g. death) in response to exposure D if and only if $D > T_i$.[7] Toxicity is evaluated in terms of the population distribution of T. For example, the median tolerance is the dose D_{50} such that $Pr(T \leq D_{50}) = 0.50$. The dose level D_{50} is generally referred to as the EC50 (EC = "effective concentration"); when the response is death, it is referred to as the LD50 (LD = "lethal dose").

Let $d_i = \log(D_i)$, and let d^* denote the mean value of the d_i's. We modeled the death R_i of animal i, exposed to dose D_i, as a Bernoulli trial with parameter p_i, assuming that

$$\mathrm{logit}(p_i) = \alpha_R + \beta_R(d_i - d^*).$$

Liver and kidney concentrations for individual animals are likely associated, being dependent on shared covariates. The obvious shared covariate is dose level, but other factors such as the animal's condition and unique genetic features might be expected to play a role. Thus we model liver concentrations L_i and kidney concentrations K_i for individual i by a bivariate regression on dose level, with correlated residuals. Specifically, we suppose that

$$\log(L_i) = \alpha_L + \beta_L(d_i - d^*) + \epsilon_i$$

---

7. This statement is calm, nicely clinical and scientific; clearly not written by a duck.

and

$$\log(K_i) = \alpha_K + \beta_K(d_i - d^*) + v_i;$$

where (ϵ_i, v_i) has a bivariate normal distribution with zero mean vector and variance matrix Σ.

We assign vague normal priors to the regression coefficients, and a standard vague prior for Σ, namely the inverse-Wishart distribution with 2 degrees of freedom, and scale matrix equal to the identity. Data analysis is straightforward; BUGS data and code are in Panels 5.2 and 5.3.

5.4.3 Derived Parameters

Our analysis includes several derived parameters. There are σ_L, σ_K, and ρ, the residual standard deviations of liver and kidney measurements, and their correlation; these are all functions of the variance matrix Σ. Another derived parameter is the LD50, found by solving $p(D) = 50\%$, i.e.,

$$\text{logit}(p(D)) = \alpha_R + \beta_R(\log(D) - d^*) = \text{logit}(0.50) = 0$$

for D. The result is

$$D_{50} = \exp(d^* - \alpha_R/\beta_R).$$

In general, the xth percentile of the distribution of tolerances, the LDx, can be treated as a derived parameter, calculated as

$$D_x = \exp\left(d^* + (\text{logit}(x/100) - \alpha_R)/\beta_R\right).$$

5.4.4 Out of Sample Prediction

Recall that the liver and kidney measurements for geese recovered at the power plant were 57.3 and 226 µg/g. We might ask, "suppose that a duck in the lab study had liver and kidney levels of $L = 57.3$ and $K = 226$, and had died ($R = 1$), but that somehow we were unaware of its

PANEL 5.2   BUGS data for mallard $NaVO_3$ dose–response study.

```
dose = c(34,34,34,62,62,62,113,113,113,208,208,208,382,
     382,382,10,10,10,10,18,18,18,18,34,62,113,208,
     382,700,700,700,700),
resp = c(0,0,0,1,0,0,1,1,1,1,1,1,1,1,1,0,0,0,0,0,0,0,0,0,0,
     1,1,1,0,1,1,1),
log.liver = c(0.642,0.833,1.163,2.617,1.065,1.569,1.960,2.041,
     2.809,2.272,2.303,2.493,3.068,3.153,3.740,NA,NA,NA,NA,
     NA,NA,NA,NA,NA,NA,NA,NA,NA,NA,NA,NA,NA),
log.kidney = c(1.281,0.588,0.742,3.754,1.065,1.030,2.815,
     3.550,4.359,3.450,3.996,3.453,5.094,4.970,5.687,NA,NA,
     NA,NA,NA,NA,NA,NA,NA,NA,NA,NA,NA,NA,NA,NA,NA),
R=structure(.Data=c(1,0,0,1),.Dim=c(2,2))
```

PANEL 5.3   BUGS code for mallard NaVO₃ dose–response study.

```
model{
    ####### Data Transformations ######
    mean.log.dose <- mean(log.dose[1:32])
    for (i in 1:32){
        log.dose[i] <- log(dose[i])
        deve.dose[i] <- log.dose[i] - mean.log.dose }

    ####### Survival Model  ######
    for (i in 1:32){
        logitp[i] <- alpha.resp+beta.resp*deve.dose[i]
        logit(p[i]) <- logitp[i]
        resp[i] ~ dbern(p[i]) }

    ####### Tissue Model  ######
    for (i in 1:15){
        mu[i,1] <- alpha.liver+beta.liver*deve.dose[i]
        mu[i,2] <- alpha.kidney+beta.kidney*deve.dose[i]
        log.levels[i,1] <- log.liver[i]
        log.levels[i,2] <- log.kidney[i]
        log.levels[i,1:2] ~ dmnorm(mu[i,1:2],tau[1:2,1:2]) }

    ####### Priors ######
    alpha.resp ~ dnorm(0.0,1.0E-6)
    beta.resp ~ dnorm(0.0,1.0E-6)
    alpha.liver ~ dnorm(0.0,1.0E-6)
    beta.liver ~ dnorm(0.0,1.0E-6)
    alpha.kidney ~ dnorm(0.0,1.0E-6)
    beta.kidney ~ dnorm(0.0,1.0E-6)
    tau[1:2,1:2] ~ dwish(R[1:2,1:2],2)

    ####### Derived Parameters ######
    var[1:2,1:2] <-  inverse(tau[,])
    rho <- var[1,2]/sqrt(var[1,1]*var[2,2])
    sd.liver <- 1/sqrt(var[1,1])
    sd.kidney <- 1/sqrt(var[2,2])
    LD50 <- exp(mean.log.dose-alpha.resp/beta.resp)
}
```

exposure to NaVO₃. In terms of its lethality, i.e., the value LD_x, what level would we guess the duck had been exposed to?"

The solution is to assign a prior distribution for the hypothetical dose D^H, and to analyze hypothetical data $X^H = (L^H, K^H, R^H)'$ using the posterior predictive distribution. This, as it turns out, is much more easily *done* than *said*.

Here is the formal description of things: Let X denote the complete set of observed data, and θ represent the unknown parameters, $\theta = (\alpha_R, \beta_R, \alpha_L, \beta_L, \alpha_K, \beta_K, \Sigma)'$. Assuming that X^H is conditionally independent of X (i.e., given θ and D^H), and that the priors on θ and D^H are independent, it can be shown that

$$\left[D^H, \theta | X, X^H\right] \propto \left[X^H | D^H, \theta\right]\left[\theta | X\right]\left[D^H\right]. \tag{5.8}$$

Integrating both sides of Eq. (5.8) with respect to $\boldsymbol{\theta}$, we obtain the conditional distribution $[D^H|X, X^{New}]$. By inspection of the right-hand side of Eq. (5.8), this is seen to be proportional to the posterior predictive distribution $[X^H|D^H, X]$, weighted by the prior distribution $[D^H]$.

All of this is accomplished easily using BUGS, simply by appending the code in Panel 5.4 to the code in Panel 5.3. Nodes of the form $\theta.H$ defined using the "cut()" command are assigned the current value of θ, but are not allowed to influence the calculation of the posterior distribution of θ; thus nodes, whether stochastic (like `log.level.H`) or logical (like `P.H`) are drawn from the posterior predictive distribution.

The necessity of using the posterior predictive distribution arises, in this case, because the observations $X^H = (L^H, K^H, R^H)'$ are hypothetical; there was no duck in the lab study associated with these values. To treat them as observations would improperly influence the posterior distribution $[X^H|D^H, \boldsymbol{\theta}]$.

A hypothetical observation with $L^H = 57.3$, $K^H = 226$, and $R^H = 1$ leads to a posterior predictive distribution for `P.H` with mean of 0.976; its 5th percentile was 0.889. Thus, we can conclude that the Canada geese measurements were consistent with an exposure to the LD97.6 in the lab study; furthermore, that had we observed such measurements in a mallard drake in the lab study, we would have 90% confidence that the duck's exposure was at least as toxic as the LD88.9.

We find this a very satisfying analysis, for the usual reasons: there are no data limited approximations involved in the calculations, there is no resting on dubious asymptotics.

PANEL 5.4   BUGS posterior prediction code for NaVO3 study.

```
####### Posterior Predictive Elements ######
alpha.resp.H <- cut(alpha.resp)
beta.resp.H <- cut(beta.resp)
alpha.liver.H <- cut(alpha.liver)
beta.liver.H <- cut(beta.liver)
alpha.kidney.H <- cut(alpha.kidney)
beta.kidney.H <- cut(beta.kidney)
tau.H[1,1] <- cut(tau[1,1])
tau.H[1,2] <- cut(tau[1,2])
tau.H[2,1] <- cut(tau[2,1])
tau.H[2,2] <- cut(tau[2,2])

Dose.H ~ dunif(0,10000)
log.Dose.H <- log(Dose.H)
deve.Dose.H <- log.Dose.H-mean.log.dose
logitP.H <- alpha.resp.H+beta.resp.H*deve.Dose.H
logit(P.H) <- logitP.H
Resp.H <- 1
Resp.H ~ dbern(P.H)

mu.H[1] <- alpha.liver.H+beta.liver.H*deve.Dose.H
mu.H[2] <- alpha.kidney.H+beta.kidney.H*deve.Dose.H
log.Levels.H[1] <- log(57.3)
log.Levels.H[2] <- log(226)
log.Levels.H[1:2] ~ dmnorm(mu.H[1:2],tau.H[1:2,1:2])
```

Inference is based on clearly articulated premises and follows the overarching principle that all conclusions should be based on summaries of posterior distributions. Subject to the obvious limitations of comparing ducks and geese, the answers provided are as good as they could be.

5.5 PRODUCT BINOMIAL REPRESENTATION OF THE CJS MODEL

In this section, we demonstrate the use of derived parameters as a means of simplifying otherwise complex calculations. Our interest is not so much in the transformed parameters per se, but in their use in simplifying the description of a model.

The Cormack–Jolly–Seber (CJS) model is of enormous importance in wildlife studies; its development by Cormack (1964) and later extensions by Jolly (1965) and Seber (1965) are important milestones in the advancement of statistical methodology for estimating demographic parameters. The CJS model is covered in greater detail in Chapter 11; here, our interest is in computational efficiencies for Bayesian analysis based on the use of derived parameters.

The model applies when attempts are made on t sampling occasions to observe individual animals. The ability to identify individual animals is usually ensured by capturing them, placing permanent and uniquely distinctive marks on them, and releasing them to the general population.

The CJS model describes an open population: animals can become associated with the study area at any time during the course of the study (whether by birth or by immigration) and can leave the study area (by death or by emigration). In the simplest cases it is assumed that there is no temporary emigration.

The model describes events associated with animals as Bernoulli trials, which are referred to as survival, capture, and release events. Between sampling occasions an animal may die or permanently emigrate or survive and remain in the study area. For simplicity, the event that the animal survives and remains in the study area is referred to as *survival*. The probability that an animal alive at sampling occasion i survives until sampling occasion $i+1$ is designated ϕ_i. It is assumed that this probability is constant among marked animals, and for inference, that survival of marked and unmarked animals are the same. Marked animals available for capture at time i are recaptured with probability p_i.

The CJS model conditions on the numbers R_i, the total number of animals (previously marked or unmarked) that are released at occasion i. Identifiable parameters are summarized as

$$\theta = \{p_2, p_3, \ldots, p_{t-1}; \phi_1, \phi_2, \ldots, \phi_{t-2}, \lambda_{t-1}\},$$

where $\lambda_{t-1} = \phi_{t-1} p_t$. Cormack (1964) expressed the likelihood as

$$\prod_{i=1}^{t-1} \phi_i^{v_i} p_{i+1}^{a_{i+1}} \left(1 - p_{i+1}\right)^{v_i - a_{i+1}} \chi_i^{c_i};$$

here, v_i, a_i, and c_i are statistics, χ_i is a function of survival and capture parameters. Specifically, v_i is the number of animals first released on a sampling occasion prior to i and recaptured at a sampling occasion j, $i \leq j \leq t$; a_i is the number of animals first released on a sampling occasion prior to i and recaptured at sampling occasion i; and c_i is the number of animals released on sampling occasion i and never again recaptured at a sampling occasion $\leq t$. Parameters χ_i are defined recursively, with $\chi_t = 1$ and

$$\chi_i = (1 - \phi_i) + \phi_i(1 - p_{i+1})\chi_{i+1},$$

for $i = t-1, t-2, \ldots, 2, 1$.

The complicated form of the likelihood prevents analytic calculation of posterior distributions for ϕ_i and p_i; Bayesian analysis must proceed using MCMC.[8] Gibbs sampling is not an option, because full conditional distributions are also not available. Thus, posterior distributions must be examined using the Metropolis–Hastings algorithm. For example, one might place independent $U(0,1)$ priors on the components of θ and generate candidate values by adding $N(0, \sigma^2)$ to the logit of current parameter values. Some tuning of the algorithm is required (choosing appropriate values of σ^2), but the results will be satisfactory.

Here, we show how computational efficiencies may be gained through a reparameterization of the likelihood and the use of derived parameters.

Sufficient statistics for θ are summarized as

$$S = \{r_1, r_2, \ldots, r_{t-1}; m_2, m_3, \ldots, m_t\};$$

where r_i is the number of animals marked and released at i and subsequently recaptured on a sampling occasion $\leq t$, and m_i is the number of marked animals captured on sampling occasion i. It is convenient to define $T_2 = r_1$, and to recursively calculate

$$T_{i+1} = T_i - m_i + r_i,$$

for $i = 2, 3, \ldots, t-1$. T_i is the number of animals marked and released prior to i that are subsequently recaptured at some sampling occasion j, $i \leq j \leq t$.

Burnham (1991) showed that the rather complicated CJS likelihood can be rewritten in terms of sufficient statistics and derived parameters, as

$$\prod_{i=1}^{t-1} B(r_i; R_i, \lambda_i) \prod_{i=2}^{t-1} B(T_i; m_i, \tau_i); \tag{5.9}$$

here, the λ_i and τ_i are functions of the parameters ϕ_i and p_i. The λ_i and τ_i are calculated recursively from the ϕ_i and p_i, starting with $\lambda_t = 0$, then calculating

$$\lambda_i = \phi_i p_{i+1} + \phi_i(1 - p_{i+1})\lambda_{i+1}, \tag{5.10}$$

for $i = t-1, t-2, \ldots, 1$. Note that $\lambda_i = 1 - \chi_i$. Simultaneously, we calculate

$$\tau_i = \phi_{i-1} p_i / \lambda_{i-1}, \tag{5.11}$$

for $i = t-1, t-2, \ldots, 2$. We will write

$$\psi = g(\theta) = \{\lambda_1, \lambda_2, \ldots, \lambda_{t-1}; \tau_2, \tau_3, \ldots, \tau_{t-1}\}$$

for the alternative parameterization.

---

8. In Section 11.3, we adopt a different form of the likelihood which leads to a Gibbs sampler based on simple full conditional distributions.

The point we emphasize for now is the simplicity of (5.9); if we were interested in the parameters λ_i and τ_i themselves, and could place independent beta priors on them, the posterior distributions would be obtained by inspection, using conjugacy. If, for instance, $\lambda_i \sim Be(a,b)$ then the posterior distribution would be $[\lambda_i | Data] \sim Be(a+r_i, b+R_i-r_i)$. This observation and the ease with which derived parameters are handled under the Bayesian approach form the basis of two alternative means for analysis of the CJS model.

5.5.1 CJS with Uniform Prior on λ and τ

Typically, we would choose to place priors on the parameters ϕ_i and p_i; these are, after all, the basic and most relevant quantities in the model, and provide the most natural descriptions of the phenomena being studied. The derived parameter λ_i is seen from (5.9) to be the probability that an animal released at occasion i is recaptured after occasion i; parameter τ_i is the probability that an animal released prior to occasion i is recaptured on or after occasion i; both depend in rather complicated ways on capture and survival probabilities. We would probably have little basis for placing priors on λ_i and τ_i; further, it is not at all clear what the choice means in terms of priors on ϕ_i and p_i.[9]

However, it is often the case that "the data overwhelm the prior": given sufficient data, the choice of prior does not wind up having that much of an effect on the posterior. So then, if we choose to conduct our analysis using independent uniform priors on λ_i and τ_i, it follows immediately from (5.9) that the posterior distributions are independent, with

$$[\lambda_i | Data] = Be(r_i+1, R_i - r_i +1) \tag{5.12}$$

and

$$[\tau_i | Data] = Be(T_i+1, m_i - T_i +1). \tag{5.13}$$

Given readily available software, we can generate samples from these distributions. Then we can invert the calculations (5.10) and (5.11) to convert the sampled λ_i and τ_i to ϕ_i and p_i.

It can be shown that (5.10) and (5.11) imply

$$\phi_i = \left(\lambda_i/\lambda_{i+1}\right)\left(1 - \tau_{i+1}(1-\lambda_{i+1})\right), \tag{5.14}$$

for $i=1,2,\ldots,t-2$, and that

$$p_i = \left(\tau_i \lambda_i\right)/\left(1-\tau_i(1-\lambda_i)\right), \tag{5.15}$$

for $i=2,3,\ldots,t-1$. Applying (5.14) and (5.15) to λ_i and τ_i sampled according to (5.12) and (5.13), we obtain samples from posterior distributions of ϕ_i and p_i, given uniform priors on ψ.

Assigning uniform priors to the transformed parameters ψ is not the same as assigning uniform priors to θ. Because each set of parameters can be calculated from the other, we may think of uniform priors for ψ as inducing (nonuniform) priors on θ. We can then use rejection sampling to convert the sample of ϕ_i and p_i obtained according to this induced prior on θ to a sample obtained using uniform priors on θ, as we now describe.

---

9. It should be noted that $0 \leq \tau_i \leq 1$ and $0 \leq \lambda_i \leq 1$, for all i, is not sufficient to guarantee that $0 \leq \phi_i \leq 1$ and $0 \leq p_i \leq 1$ for all i. Thus, the uniform prior on τ and ϕ does not rule out parameter values that are inadmissible according to the biological model. Such inadmissible values can simply be rejected; we discuss this matter further in Section 5.5.2.

5.5.2 CJS with Uniform Prior on p and ϕ

Recall that rejection sampling (Section 4.2.2) is used to obtain samples from distribution $t(x)$ by first sampling from distribution $c(x)$ (think t for *target*, and c for *candidate*). Typically, $c(x)$ is easily sampled, but $t(x)$ is not. The requisite condition for rejection sampling is that there is a number M such that $t(x) \leq Mc(x)$ over the support of $t(x)$; thus $w(x) = t(x)/Mc(x) \leq 1$. Rejection sampling works by making a draw $X = x$ from $c(x)$, then accepting or rejecting the observation as representative of $t(x)$ on the basis of a Bernoulli trial with success probability $w(x)$.

Now suppose that $f_t(\beta|X)$ and $f_c(\beta|X)$ are posterior distributions for a parameter β, based on two distinct priors, $[\beta]_t$ and $[\beta]_c$ and data X. Then

$$\frac{f_t(\beta|X)}{f_c(\beta|X)} \propto \frac{[x|\beta][\beta]_t}{[x|\beta][\beta]_c} = \frac{[\beta]_t}{[\beta]_c}. \tag{5.16}$$

From Eq. (5.16), we see that if prior $[\beta]_t$ can be sampled by rejection sampling of prior $[\beta]_c$ using

$$w(\beta) = \frac{[\beta]_t}{M[\beta]_c}, \tag{5.17}$$

then $f_t(\beta|X)$ can be sampled by rejection sampling of $f_c(\beta|X)$, using the same $w(x)$. Note that M has been chosen so that $w(\beta) \leq 1$ over the support of $[\beta]_t$. It may be possible to improve on the efficiency of the rejection sampler for $f_t(\beta|X)$ by only requiring M to be large enough to guarantee $w(\beta) \leq 1$ over (most of) the support of $f_t(\beta|X)$; this results in a smaller value of M and higher acceptance probabilities.

For the CJS model with uniform priors on θ (i.e., the survival probabilities ϕ and detection rates p), the posterior distribution desired is $f_t(\theta|X)$ with $[\theta]_t \equiv 1$; the candidate generating distribution is the easily sampled $f_c(\theta|X)$, corresponding to the prior $[\theta]_c$ induced by placing uniform priors on ψ, as described in Section 5.5.1. It can be shown (Link and Barker, 2008) that

$$[\theta]_c \propto \frac{\lambda_{t-1}}{\lambda_1} \prod_{i=1}^{t-2} \phi_i, \tag{5.18}$$

which is easily calculated.

We apply these techniques to a famous set of data, the moth (*Gonodontis bidentata*) data reported on by Bishop *et al.* (1978), and subsequently analyzed by various authors (Crosbie, 1979; Crosbie and Manly, 1985; Link and Barker, 2005). The data consist of records for 689 male *Gonodontis bidentata* that were captured, marked, and released daily over 17 days at Cressington Park in northwest England. These moths were nonmelanic; demographic parameters were estimated as part of a larger study looking at comparative fitness of distinct color morphs. Sufficient statistics for the CJS model are given in Table 5.3.

An interesting feature of the Burnham parameterization (5.9) is that not all combinations of τ_i and λ_i correspond to admissible values of ϕ_i and p_i. That is, when we translate the parameters τ and λ into ϕ_i and p_i, we may wind up with values $\phi_i > 1$. A related problem in frequentist analysis occurs when a maximum likelihood estimator is inadmissible. In such cases, it helps to realize that the likelihood "knows" nothing about the biology of the problem, and is simply treating the parameters as symbols, measuring fit without regard to meaning;

TABLE 5.3   Sufficient statistics for CJS model of *Gonodontis* data.

i	1	2	3	4	5	6	7	8	9	10	11	12	13	14	15	16	17
u_i	15	44	37	45	14	71	45	55	32	71	12	69	69	42	40	28	0
r_i	10	22	14	15	13	12	16	17	12	12	3	25	17	14	7	4	–
R_i	15	52	54	62	29	84	51	74	43	85	15	81	93	59	60	37	8
m_i	–	8	17	17	15	13	6	19	11	14	3	12	24	17	20	9	8

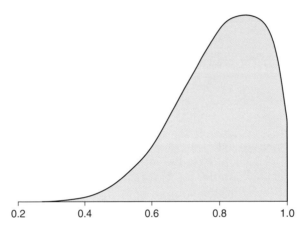

FIGURE 5.5   Posterior distribution of ϕ_1 for *Gonodontis*.

it may happen that the best fit occurs for parameter values that are biologically inadmissible though mathematically feasible.

In this case, it really makes sense to have informative priors, ones which rule out inadmissible ϕ or p (and the corresponding τ and λ). If we are to use the approach described in Section 5.5.1, of placing uniform priors on τ and λ, then converting to ϕ or p, we may simply reject any sampled sets with inadmissible ϕ or p. In effect, we have replaced the uniform priors over the unit intervals (for τ and λ) with uniform priors over an admissible range.

As it turns out, the posterior distributions for survival rates ϕ_i, under uniform priors, are fairly diffuse. An example is given in Fig. 5.5, for the first daily survival rate, ϕ_1. The posterior mode is at about 0.83, the 2.5th and 97.5th percentiles are 0.502 and 0.988; there is considerable mass near 1.0. It is not surprising that many sampled sets of τ and λ lead to inadmissible ϕ; indeed, we reject 94.5% of the sampled sets. So great is the gain in efficiency from direct sampling of the posterior distribution, that this "waste" still leads to a much more efficient sampling scheme than random walk Metropolis–Hastings.

For the *Gonodontis* data of Table 5.3, we generated τ and λ according to the β distributions (5.12) and (5.13) until we had 2,001,676 sets of admissible ϕ and p. These samples allow us to consider the "ψ-uniform" posterior distribution. Rejection sampling, described subsequently, reduced these to a sample of 302,424 from the "θ-uniform" posterior distribution. The computation time for the whole process was 1222 seconds on a 3.2 Ghz Pentium 4 system.

For time comparisons, we also generated a sample of size 302,424 from the θ-uniform posterior using program BUGS. The time for this sample, including a burn-in of length 10,000, was 2944 seconds, more than 2.4 times as long. What's more, the values generated by BUGS are autocorrelated, whereas the samples generated by rejection sampling were independent, meaning that the effective sample size from BUGS is smaller.

Posterior means and standard deviations of the survival estimates are given in Table 5.4. There is near perfect agreement between the values for the "θ-uniform" posteriors, whether obtained by rejection sampling or using BUGS (as one would expect). Furthermore, the results for the θ-uniform posterior are nearly identical (ϕ tend to be slightly larger, p slightly smaller). Use of this approximation allows a further 6-fold reduction in computation time $(2,001,676/302,424 = 6.6)$.

We conclude this example by describing the rejection sampling scheme used to convert the sample of the ψ-uniform posterior into a sample from the θ-uniform posterior. Substituting (5.18) in (5.17), we have

$$w(\theta) \propto \frac{\lambda_1}{\lambda_{t-1} \prod_{i=1}^{t-2} \phi_i}. \tag{5.19}$$

Note that we are considering $w(\theta)$ as a function of $\theta = \{\phi, p\}$, and that Eq. (5.19) is expressed in terms of λ simply for notational simplicity. We calculated the right-hand side of Eq. (5.19) for the sample of 2,001,676 values from the ψ-uniform posterior. The distribution of these is very highly skewed, with a few values much larger than the others. The largest value was

TABLE 5.4   Posterior mean and standard deviations for "ψ-uniform" posterior (Columns 1 and 2), "θ-uniform" posterior by rejection sampling (Columns 3 and 4), and "θ-uniform" posterior as implemented using BUGS (Columns 5 and 6).

i	1	2	3	4	5	6
ϕ_1	0.791	0.131	0.788	0.131	0.789	0.131
ϕ_2	0.752	0.132	0.741	0.135	0.741	0.135
ϕ_3	0.472	0.131	0.455	0.129	0.455	0.129
ϕ_4	0.345	0.090	0.335	0.088	0.335	0.088
ϕ_5	0.796	0.139	0.781	0.144	0.781	0.144
ϕ_6	0.404	0.123	0.386	0.120	0.386	0.119
ϕ_7	0.617	0.153	0.608	0.152	0.608	0.152
ϕ_8	0.621	0.156	0.607	0.156	0.606	0.156
ϕ_9	0.786	0.143	0.778	0.146	0.777	0.146
ϕ_{10}	0.602	0.185	0.595	0.186	0.596	0.186
ϕ_{11}	0.543	0.177	0.534	0.177	0.533	0.177
ϕ_{12}	0.731	0.141	0.724	0.142	0.723	0.142
ϕ_{13}	0.476	0.134	0.465	0.133	0.465	0.132
ϕ_{14}	0.627	0.177	0.614	0.178	0.613	0.178
ϕ_{15}	0.456	0.197	0.430	0.194	0.431	0.194

nearly four times as large as the 99.9th percentile. Scaling the right-hand side of (5.19) would be safest, but would result in an acceptance rate one fourth as large as scaling by the 99.9th percentile. Scaling by the 99.9th percentile has only the slightest effect on our evaluation of the posterior distribution.

5.5.3 CJS Model: Summary

The CJS model and its many extensions play an important role in the study of marked animal populations. Here, we have used alternative parameterizations of the CJS model to illustrate an appealing feature of Bayesian inference, especially when implemented using sampling-based investigations of posterior distributions (like MCMC): this is the ease with which one may make inference about functions of parameters. Given the posterior distribution of parameter α, the posterior distribution of parameter $\beta = g(\alpha)$ is easily obtained. If the posterior distribution of α is of known mathematical form, as $f_\alpha(\alpha)$, the posterior distribution of β is obtained analytically through the change of variables theorem (as in Section 5.2.2). More commonly, we may find ourselves examining the posterior distribution of α through simulation, and in this setting, if $\alpha_1, \alpha_2, \ldots, \alpha_B$ is a sample of $f_\alpha(\alpha)$, then $\beta_1 = g(\alpha_1), \beta_2 = g(\alpha_2), \ldots, \beta_B = g(\alpha_B)$ is a sample of $f_\beta(\beta)$. It's as simple as that.

Our example shows that straightforward use of derived parameters can lead to substantial computation efficiencies. Coupled with rejection sampling, so as to exactly reproduce the desired posterior distribution, the example documents a 59% reduction in computation time; using the posterior distribution based on a slightly different prior, the example documents a 94% reduction in computation time.

5.6 POSTERIOR PREDICTIVE MODEL CHECKING

G.E.P. Box famously said "all models are wrong, but some are useful." This fine bit of rhetoric reminds us that some models might not be useful; that failures of model assumptions might lead to poor inference. It is often said, and truly, that "statisticians, like artists, ought not to fall in love with their models." We need to evaluate whether our data are consistent with our models, and if not, whether the inconsistency is liable to prejudice our inference.[10]

The posterior predictive distribution is a handy tool for examining whether our data are consistent with our model. Suppose that we have data X, and a model M under which the data distribution of X is specified, except for an unknown parameter θ. It might be that we can describe some feature that is characteristic of M, a feature that is reflected in a statistic $f(X)$. We can calculate the statistic for our data, draw hypothetical replicate data sets X^{New} from the posterior predictive distribution $[X^{\text{New}}|X]$, and see how our value $f(X)$ matches up with hypothetical replicate values $f(X^{\text{New}})$.

---

10. On the other hand, Box's dictum that "all models are wrong", while catchy, is either tautologous or overstated. If one defines a model as an inexact approximation to reality, then there can be no quarrel with the statement, by definition. However, if one defines a model as a mathematical description of phenomena, then the possibility of the description being exact cannot be dismissed. We have heard Box's statement misused to trivially dismiss model-based inference; sometimes, to favor nonparametric analyses on the mistaken grounds that they "require no assumptions"; other times, to argue against the value of parsimony.

PANEL 5.5 BUGS code for posterior predictive analysis of Fisher's tick data.

```
list(x=c(0,0,0,0,0,0,0,1,1,1,1,1,1,1,1,1,2,2,2,2,
    2,2,2,2,3,3,3,3,3,3,3,3,3,3,3,3,3,4,4,4,4,4,4,4,4,5,
    5,5,5,5,6,6,6,6,7,7,7,9,10,10))

model{
    for(i in 1:60){
        x[i]  ~ dpois(lambda)
        x.new[i]  ~ dpois(lambda)
    }
    lambda  ~ dgamma(0.001,0.001)
    stat.x <- pow(sd(x[ ]),2)/mean(x[ ])
    stat.x.new <- pow(sd(x.new[ ]),2)/mean(x.new[ ])
    pvalue <- step(stat.x.new-stat.x)
}
```

A concrete example helps. In Section 4.1.2, we discussed Fisher's tick data, consisting of the number of parasites on each of 60 sheep. The data being counts, it is natural to consider the Poisson distribution as a starting point for analysis. The data and BUGS code are given in Panel 5.5. The Ga(0.001,0.001) prior for the rate parameter λ approximates the Jeffreys prior (Section 6.2.3).

We have calculated the ratio of variance to mean in the data as stat.x. Poisson random variables have variance and mean equal. Fisher's tick data have sample variance that is 1.89 times larger than the sample mean, suggesting a failure of the Poisson model. On the other hand, we recognize that the sample mean and variance are random variables, and thus would not have expected that stat.x should be exactly equal to 1. We need to determine whether the discrepancy might be within the bounds of natural variation.

Making this assessment is challenging. The data distribution of $f(X)$ depends on the unknown parameter λ, and is analytically intractable. So, we might fix a value of λ (using, say, the MLE), and approximate the data distribution through simulation. This "parametric bootstrap" seems reasonable enough, but does not account for our uncertainty about λ. Laird and Louis (1987) proposed bootstrap solutions to this inadequacy of the parametric bootstrap.

A Bayesian approach is to allow λ to vary in accordance with credible values; that is, by sampling values λ from the posterior distribution $[\lambda|X]$. Thus our draws X^{New} are not from a distribution $[X^{\text{New}}|\lambda]$ for fixed λ, but from the average value of this distribution against the posterior distribution. This is

$$\left[X^{\text{New}}|X\right] = \int \left[X^{\text{New}}|\lambda\right]\left[\lambda|X\right]d\lambda,$$

the posterior predictive distribution.

Once again, the sampling is easier done than said. Consider the code of Panel 5.5. For each value λ in the Markov chain sample of $[\lambda|X]$, the code samples a new data set X^{New}. The collection of all such data sets is thus a sample from the posterior predictive $\left[X^{\text{New}}|X\right]$. For each sampled value X^{New}, we compute the sample variance/mean ratio, and an indicator (the node `pvalue`) for whether the sampled data have larger variance/mean ratio than the original data.[11]

The node `pvalue` is so named because of a similarity between it and the frequentist p-value. Its mean value is $\Pr\{f(X^{\mathrm{New}}) \geq f(X) \mid X\}$, the probability of observing a variance/mean ratio as large or larger than the one in our data set. For Fisher's tick data, this *Bayesian p-value* (see Section 5.6.1 below) has a value less than 1 in 4000: all but 215 of 1 million sampled variance/mean ratios were smaller than the observed value of 1.82.[12] There are strong indications that the Poisson model is not justified.

Given that the variance/mean ratio is too large, the Poisson model will predict too many large or small values. Keeping track of the million hypothetical values `x.new[1]`, we find that 4.0% were zeros, and only 1.9% were bigger than 7. These probabilities are the ones we would use to make predictions about tick burdens of sheep not yet observed. However, these values are inconsistent with the rates of 11.7% and 5.0% observed in our data set. The Poisson model contradicts the plain testimony of the data.

5.6.1 Bayesian and Frequentist p-Values

Bayesian p-values are used with the same purpose as frequentist p-values, to assess the adequacy of a model M for describing data X. The idea, in both cases, is to compare the value of a relevant statistic $f(X)$ with the value of hypothetical replicate values generated under M. The statistic $f(X)$ is chosen as reflecting a specific characteristic feature of M, one which is not shared by alternative models.

In the case of the Poisson distribution, we considered the variance/mean ratio as an appropriate statistic, because the Poisson mean and variance are the same. However, Poisson distributions are not the only ones with this feature. The discrete uniform distribution on $\{0, 1, 2, \ldots, T\}$ has mean of $T/2$, and variance $T(T+2)/12$. These are equal when $T=4$. The point is that we must exercise some care in our choice of statistic, and recognize that not finding evidence against a model is not the same as establishing the adequacy of the model. Model building and assessment cannot be reduced to an automatic and unthinking process.

Another difficulty to be faced is how we are to evaluate hypothetical replicates under the model, especially if (as is often the case) their probability distributions depend on unknown parameters of the model. The frequentist solution is to require that probability assessments for hypothetical replicates be true for all values of the unknown parameter.[13] For instance, if we can say that

$$\Pr\left(f(X) > f\left(X^{\mathrm{New}}\right) \mid \boldsymbol{\theta}\right) \leq 0.023,$$

for all values $\boldsymbol{\theta}$, then 0.023 is a legitimate frequentist p-value.

---

11. Readers might wonder why it is not necessary to use the `cut()` command here, as in the example of Panel 5.4. There would be no harm in so doing, writing `x.new[i] ~ dpois(lambda.new)`, with `lambda.new <- cut(lambda)`. But this is not necessary. The node `x.new[i]` is not an observed value and is independent of the data which define the likelihood, thus sampled values `x.new[i]` have no effect on the posterior distribution $[\lambda \mid X]$.

12. Note also that the gamma prior on λ is conjugate for the Poisson distribution, with the result that BUGS can draw independent samples of the posterior distribution. Consequently the 1 million sampled values are independent, and we can use asymptotic theory to justify our statement that the p-value is less than 1 in 4000.

13. This requirement is the guiding principle. In practice, as with confidence intervals (Section 3.2), p-values are often based on approximations.

Bayesians, on the other hand, prefer to average across the uncertainty associated with the unknown parameter. Ideally, this means using the marginal distribution

$$\left[X^{\mathrm{New}}\right] = \int \left[X^{\mathrm{New}}|\boldsymbol{\theta}\right]\left[\boldsymbol{\theta}\right] d\boldsymbol{\theta} \qquad (5.20)$$

for model criticism and evaluation. This works well with informative priors, but objective analyses based on vague priors might lead to excessively diffuse marginal distributions of data, with little discriminatory power.

Use of the posterior predictive distribution $\left[X^{\mathrm{New}}\,|\,X\right]$ amounts to using $[\boldsymbol{\theta}\,|X]$ in place of $[\boldsymbol{\theta}]$ in Eq. (5.20). Instead of weighting by our uncertainty about $\boldsymbol{\theta}$ prior to data collection, we are weighting by the uncertainty remaining after observing X, and assuming M is true. There is something somewhat incestuous in this approach, perhaps prejudicing our inference in favor of M, and leading to excessively favorable evaluations of M. Recognizing this possibility, the posterior predictive remains a useful tool for screening unacceptable models: if a Bayesian p-value suggests model inadequacy, its testimony is to be trusted. Examples that illustrate use of the Bayesian p-value are in Sections 8.5.3 and 9.3.2.

CHAPTER

6

Priors

O U T L I N E

6.1	An Example where Priors Matter	110
	6.1.1 Hypergeometric Distribution	111
	6.1.2 Dusky Seaside Sparrow	113
6.2	Objective Bayesian Inference	115
	6.2.1 Uniform Priors and Transformation Invariance	116

	6.2.2 Improper Priors	117
	6.2.3 Jeffreys Priors	120
	6.2.4 Summary	122
6.3	Afterword	123

The philosophical appeal of Bayesian inference – its coherent use of probability to quantify all uncertainty, its simplicity, and exactness – all of this is set at nought for some by the necessity of specifying priors for unknown parameters. The aroma of subjectivity associated with choosing priors is the biggest barrier to the widespread use of Bayesian inference methods by scientists today.

To complicate matters, many Bayesian theorists insist that there is no such thing as "objective" probability, no Kilogram Standard, as it were, to serve as the basis of definition. All that matters is coherency: you have your probabilities, I have mine, and what matters is that we agree on the rules for manipulating them. Such theorists embrace the term "subjective probability," questioning whether there is or needs to be a "physical" or "material" reality associated with probability.[1] Thus, the word "subjective" has come to be linked with Bayesian analysis, though in a technical sense, and without the antiscientific shadings of dishonesty.

---

1. For readers inclined toward such philosophical musings, we recommend I. I. Good's entertaining collection of essays (Good, 1983).

Given that we have to choose prior distributions, how should we proceed to avoid the taint of subjectivity, in its nontechnical sense? We begin our discussion with the following observations:

Observation 1) The mnemonic

$$\text{Inferential Basis} = \text{Data} + \text{Prior Knowledge} \tag{6.1}$$

describes all inference, Bayesian or otherwise. Prior knowledge is inevitably used, if only to provide a context in which "Data" is more than a list of numbers or symbols. Legitimate inference always acknowledges and ponders its assumptions, thus reducing the chance of self-deceit as well as the appearance of biased advocacy. Under the Bayesian paradigm, prior distributions for parameters are assumptions; analysts can avoid even the semblance of dishonesty by reporting prior choice.

Observation 2) Thus, it should be recognized and acknowledged, without embarrassment, that Bayesian inference is not merely data analysis, but analysis of data and priors. The effects of priors on inference can be evaluated easily enough, simply by trying several of them.

Observation 3) Provided that prior distributions do not categorically rule out particular parameter values, data tend to "overwhelm" priors as sample size increases. If you have decent data, the choice of prior should not matter all that much. Having said that, we acknowledge that scientists are often interested in insights at the limits of what their data can support. If they have a lot of data, they will tend to fit complicated models and in such cases priors may still have some influence on inference. On the other hand, there is no such thing as a free lunch: all methods of statistical inference have shortcomings when we have few data.

Sometimes, Bayesian analysis is desirable precisely because the prior distribution may have an effect on inference. Prior knowledge of a parameter, whether from previous studies or informed common sense, might be quantified in terms of a probability distribution. This distribution is described as an "informative" prior. Posterior inference is the formal mechanism for incorporating prior knowledge with the information provided by data. We present an example in Section 6.1.

More often than not, however, analysts wish to let the data speak for themselves and to choose "noninformative" priors. Their desire is to minimize, as much as possible, the contribution of *Prior Knowledge* on the right-hand side of (6.1) and to produce an "objective Bayesian analysis."

Perhaps surprisingly, there is no automatic, simple, universally agreed-upon method of doing so: defining a prior distribution describing ignorance is more challenging than one might think. We explain why and describe general principles for objective Bayesian analysis in Section 6.2.

6.1 AN EXAMPLE WHERE PRIORS MATTER

The Dusky seaside sparrow (*Ammodramus maritimus nigrescens*) (Fig. 6.1) lived in salt marshes along the south central coast of Florida until it was driven to extinction by marsh

FIGURE 6.1   Dusky seaside sparrow. Photo credit P.W. Sykes, USFWS.

management practices conducted to control mosquito populations at the Kennedy Space Center. The pathetic story of events leading up to its extinction in 1987 is told by Walters (1992).

In 1979, attempts were made to capture the six remaining individuals, with the goal of captive propagation. Only five were captured; all were male. What, we might ask, is the probability that the one bird that escaped was female?

Given the small sample size involved, one would anticipate considerable sensitivity to the choice of priors on model parameters. Conducting posterior inference forces a careful statement of prior assumptions and their logical consequences. We begin by describing Bayesian inference for the hypergeometric distribution, which we will use as a model for the data at hand.

6.1.1 Hypergeometric Distribution

Consider n draws, without replacement, from a box containing a known number (N) of balls. There are m red balls in our sample of n; we wish to make inference about M, the total number of red balls in the box. The data distribution of m is hypergeometric, with *pdf*

$$f(m|M,N,n) = \frac{\binom{M}{m}\binom{N-M}{n-m}}{\binom{N}{n}}. \tag{6.2}$$

Inference will be based on a posterior distribution

$$\pi(M|m,N,n) \propto f(m|M,N,n)g(M|N,n). \tag{6.3}$$

Note that we require a prior distribution $g(M|N,n)$; this will probably not depend on n, but will most likely depend on N, so we will designate the prior as $g(M|N)$.

Given no prior information about M, we would likely choose a discrete uniform prior $g(M|N) = 1/(N+1)$ for $M = 0,1,2,\ldots,N$. Then, the posterior distribution required is

$$\pi(M|m,N,n) = \frac{\dfrac{\binom{M}{m}\binom{N-M}{n-m}}{\binom{N}{n}}\dfrac{1}{N+1}}{\sum_{H=0}^{N}\dfrac{\binom{H}{m}\binom{N-H}{n-m}}{\binom{N}{n}}\dfrac{1}{N+1}}$$

$$= \frac{\binom{M}{m}\binom{N-M}{n-m}}{\sum_{H=m}^{N-(n-m)}\binom{H}{m}\binom{N-H}{n-m}}. \tag{6.4}$$

Note that the terms $\binom{N}{n}$ and $\frac{1}{N+1}$ cancel; these are constants with regard to the distribution of M. Also, note that once simplified the summation begins at $H = m$, because $\binom{H}{m} = 0$ for $H < m$. This reflects the fact that $f(m|M,N,n) = 0$ for $m > M$: we couldn't have drawn out more red balls than there were in the first place. By the same token, the summation ends at $N - (n - m)$: we know there are at least $n - m$ nonred balls.

A sample calculation is given in Table 6.1. Suppose the box has $N = 30$ balls in it, we draw $n = 20$ and count $m = 18$ red ones. Before drawing from the box, we had no knowledge of

TABLE 6.1 Calculation of posterior for M based on $m = 18$, $n = 20$, and $N = 30$, using Eq. (6.4).

M	$\binom{M}{m}$	$\binom{N-M}{n-m}$	$\binom{M}{m}\binom{N-M}{n-m}$	$\pi(M|m,N,n)$
18	1	66	66	0.0000
19	19	55	1,045	0.0000
20	190	45	8,550	0.0002
21	1,330	36	47,880	0.0011
22	7,315	28	204,820	0.0046
23	33,649	21	706,629	0.0159
24	134,596	15	2,018,940	0.0455
25	480,700	10	4,807,000	0.1084
26	1,562,275	6	9,373,650	0.2113
27	4,686,825	3	14,060,475	0.3170
28	13,123,110	1	13,123,110	0.2959
		Total	44,352,165	1.0000

the number of red balls, and assigned equal probabilities to $N=0,1,2,\ldots,30$. Removing 20 of the balls and noting 18 red ones, we are certain that there are no fewer than 18 red, nor more than 28. Using the posterior distribution, we conclude with 99.4% certainty that there are at least 5 red balls among the 10 remaining in the box, since $\Pr(\#\text{Red remaining} \geq 5) = \Pr(M-m \geq 5) = \Pr(M \geq 23) = 1 - 0.6\%$.

6.1.2 Dusky Seaside Sparrow

Assuming equal capture probabilities for the final 6 Dusky seaside sparrows, we can model the number of male birds captured as a hypergeometric random variable m with *pdf* given by Eq. (6.2), with $n=5$ and $N=6$. Using the discrete uniform prior on the number of males M, Eq. (6.4) yields $\pi(M|m=5,N=6,n=5)=1/7$ for $M=5$, and $6/7$, for $M=6$. Given our choice of prior, we conclude that the odds are 6:1 in favor of the remaining bird being male.

In this case, an alternative prior on M might be reasonable: as Fisher (1935) said in describing nonparametric models, "an erroneous assumption of ignorance is not innocuous; it often leads to manifest absurdities." For instance, a binomial prior $g(M|N) = \binom{N}{M}(1/2)^M(1-1/2)^{N-M} = \binom{N}{M}(1/2)^N$ might more realistically portray expectations of a balanced sex ratio. Using this prior in Eq. (6.3) the posterior distribution is[2]

$$\pi(M|m,N,n) \propto \binom{M}{m}\binom{N-M}{n-m}\binom{N}{M}. \tag{6.5}$$

Substituting $m=n=5$ and $N=6$, the right-hand side of Eq. (6.5) equals 6 for both $M=5$ and $M=6$. Since these are the only possible values, the posterior probabilities for $M=5$ and $M=6$ are the same, i.e., 50%. The chance that the escaped bird was female increases from $1/7$ in the original analysis to $1/2$, by the choice of this alternative prior.

The difference between the two posterior probabilities makes perfect sense, on some thought. Each prior on M can be thought of as the consequence of treating gender as an independent Bernoulli trial with common success parameter $p = \Pr(\text{Male})$ for each of the N birds. If we were to treat p as unknown, with our absence of knowledge modeled by supposing p is sampled from a uniform distribution on $[0,1]$, then M would have distribution

$$g(M|N) = \int_0^1 \binom{N}{M} p^M(1-p)^{N-M} dp$$

$$= \frac{\binom{N}{M} \Gamma(M+1)\Gamma(N-M+1)}{\Gamma(N+2)} = \frac{1}{N+1},$$

for $M=0,1,2,\ldots,N$. This is the prior used in the first analysis.[3] In the second analysis, $p=0.50$ for all birds; each has a 50% chance of being male.

---

2. The terms $\binom{N}{n}$ and $(1/2)^N$ are constants with regard to the variable M, hence absorbed into the proportionality constant.

3. The integral is easily calculated by "integrating like a statistician": we recognize the integrand as nearly identical in form to a known density function, in this case the beta density, and multiply in the appropriate constant so as to make the integral evaluate to one.

This implicit difference in priors on p explains the difference in posterior probabilities that the missing sparrow was female. Let Y be the indicator variable for the event that the escaped bird was female. Then,

$$\Pr(Y=1)=\mathrm{E}\big(\Pr(Y=1\mid p)\big)=\mathrm{E}(1-p)=1-\mathrm{E}(p).$$

The genders of the 5 captured sparrows update a $Be(1,1)$ prior to a $Be(1+5,1+0)$ posterior for p. The mean of the $Be(6,1)$ distribution is $\mathrm{E}(p)=6/7$, hence the posterior probability that $Y=1$ is $1/7$, as under the first analysis. In the second analysis, the value p is the same for all birds, captured or not; the genders of the 5 captured sparrows provide no information about the one that was not caught. Hence, the posterior probability that $Y=1$ under the second analysis was $1/2$.

But which answer is correct, $\Pr(\text{Female})=1/2$ or $1/7$? The answer is "neither" and "both." *Neither*, in the sense that additional knowledge would lead to a different assessment of the prior probabilities for the missing bird and different conclusions, and *Both*, in the sense that each is obtained using a mathematically sound and transparent calculation. One might quibble with the assumptions made in obtaining either value, but there can be no objection to the results as mathematical evaluations of clearly articulated premises. Our view in the present case is that the value $1/7$ seems more reasonable, as there is no reason to assume that a population under severe stress should have a stable population sex ratio.

We can examine the effect of the assumption of equal catchability, as follows: Specify $[M\mid N,\phi]=B(N,\phi)$ (with known $\phi=1/2$, or with $\phi\sim U(0,1)$ as before), and suppose $[m\mid M]=B(M,\pi_M)$ and $[f\mid F]=B(F,\pi_F)$ (where $F=N-M$). The data are $m=5, f=0$. The previous analysis assumed that $\pi_M=\pi_F$ and made use of the fact that $[m\mid n=m+f,M]$ is a hypergeometric distribution. Instead, we might assign independent uniform priors to π_F and π_M. Results of the two previous and two new analyses are summarized in Table 6.2.

How are we to understand the differences among the inferences? Note first that the data suggest $\phi=\Pr(\text{Male})\geq 5/6$, arguing against the missing bird being female. Hence, setting $\phi=0.50$ increases the posterior probability that the missing bird is female. The data also suggest that $\pi_M=\Pr(\text{Captured}\mid\text{Male})\geq 5/6$. Supposing $\pi_M=\pi_F$, the indication is of a fairly high probability of capture for females. Thus, relaxing the constraint $\pi_M=\pi_F$ allows for lower capture probability of females and hence increases the posterior probability that the missing bird is female.

Four such disparate answers as given in Table 6.2 could be distressing, especially when it is realized that the data provide scant basis for modeling decisions.[4] On the other hand,

TABLE 6.2   Posterior probability for $F=1$ under four different models.

	$\pi_M=\pi_F$	$\pi_M,\pi_F\sim U(0,1)$
$\phi\sim U(0,1)$	$1/7=0.14$	$7/19=0.37$
$\phi=\frac{1}{2}$	$1/2=0.50$	$7/9=0.78$

4. We might consider Table 6.2 as summarizing results for four "models" but can also distinguish them as four prior specifications. Models with $\phi=1/2$ have a degenerate prior (a point mass distribution for ϕ), as do models with $\pi_M=\pi_F$ (a joint distribution on the unit square, concentrated on the line $y=x$).

each of the calculations is an exact summary of the data and assumptions; no ambiguity has been built in through the use of dubious analytical approximations. The differences among the results highlight the limitations of the data and force us to think about the validity of the various model assumptions. Some will throw up their hands and say that no conclusions can be drawn; others will be content with multiple answers. Others might informally combine the various results, assigning weights to the various answers: "I'll assign 70% weight to 1/7, 15% to 7/19, 10% to 1/2, and 5% to 7/9, and come up with a composite value of roughly 1/4." More formal approaches to weighted inference are discussed in Chapter 7.

One way or another, the analysis makes clear that the conclusions we draw about the escaped sparrow are largely a reflection of prior convictions. We now turn our attention to the question of objective Bayesian inference, asking how we might minimize the effects of prior knowledge on the conclusions we draw.

6.2   OBJECTIVE BAYESIAN INFERENCE

The posterior distribution of parameter θ, given data X, is

$$[\theta|X] \propto [X|\theta][\theta]; \qquad (6.6)$$

it is proportional to the product of likelihood and prior. It is easy to see how we might *maximize* the effect of our prior convictions: we choose a likelihood, and then choose a prior for θ with a very limited range. Suppose we choose a prior distribution for θ that is uniform on an interval $[\theta_0 - \epsilon, \theta_0 + \epsilon]$, for some value $\epsilon > 0$. Since the RHS of (6.6) is zero outside of the interval, the posterior distribution $[\theta|X]$ is restricted to the same interval. By choosing a very small value for ϵ, we can restrict the posterior distribution to values very close to θ_0, regardless of the data X. If our prior knowledge amounts to certainty, the data wind up being ignored.

It is conceivable that prior knowledge of *some* quantities could be so unassailable to suggest that we ignore the data's indications, but usually we want to allow the data to inform us. Indeed, if we were to move toward an imbalance in the effects of prior knowledge and data, the more desirable extreme would typically be to *minimize* the effect of prior knowledge. How do we do it? How do we conduct an "objective" Bayesian analysis, where prior knowledge contributes very little to the final inference? How do we define a noninformative prior?

A first step is to agree that "prior knowledge" should be limited, as much as possible, to the form of the likelihood $[X|\theta]$. Next, suppose that instead of a very restrictive uniform prior on θ, we choose a uniform prior over a large range. If our model's parameter space has finite range (e.g., $0 < p < 1$, for the success parameter p of Bernoulli trials, or $0 < \psi < 2\pi$ for an angular measurement ψ), a uniform prior over the entire range can be specified. This seems a reasonable specification of ignorance about θ: we are saying that no single value seems any more likely than any other. Thomas Bayes (1763) used a uniform prior distribution as an expression of prior ignorance in the original treatise on inverse probability, and hence the appropriateness of the choice is known as "Bayes postulate." Note that since the uniform prior has distribution $[\theta] = c$, for some constant c, it follows from Eq. (6.6) that the posterior distribution will be proportional to the likelihood. In particular, the posterior mode will be the maximum likelihood estimator.

Uniform priors thus seem an appealing basis for objective Bayesian analysis, but a few small defects need to be noted. First, a uniform prior for θ almost always implies a nonuniform prior for a transformed parameter $\psi = g(\theta)$. Thus if you and I analyze the same data under different parameterizations of the same model, we may get different answers. My expression of ignorance about θ amounts to an informative prior about your parameter ψ. So uniform priors lack "transformation invariance." The problem is usually not as bad as it sounds; we illustrate this in Section 6.2.1.

Another problem for using uniform priors as noninformative is that the range of θ may be infinite (e.g., $-\infty < \mu < \infty$, for the mean μ of a normal distribution). A uniform distribution over an infinite range is not possible, because the integral of a nonzero constant function over an infinite range is ∞, not 1; the "prior" $[\theta] \propto c$ is said to be "improper." Use of this improper prior amounts to treating

$$f_X(\theta) = \frac{[X|\theta]}{\int [X|\theta]d\theta}$$

as a posterior distribution, an approach that, though not strictly justifiable, nevertheless often yields sensible results.[5] Results based on $f_X(\theta)$ usually coincide with the results obtained using proper priors with very large but finite ranges. We discuss improper priors in Section 6.2.2.

Much effort has been expended in attempting to prescribe methods for constructing non-informative priors (Kass and Wasserman, 1996). Bernardo (1979) defined *reference priors* as those that maximize the expected Kullback–Lieblier divergence of the posterior from the prior; other authors (e.g. Kass and Wasserman, 1996) use the term in a nonspecific sense. Bernardo's reference priors can be mathematically complex; however, under conditions that guarantee asymptotic normality of the posterior distribution, they coincide with the Jeffreys prior, which we describe in Section 6.2.3. Jeffreys priors, unlike uniform priors, have the property of transformation invariance.

6.2.1 Uniform Priors and Transformation Invariance

Suppose that our data model is binomial, $X|p \sim B(N,p)$, for some $0 < p < 1$. A uniform prior $[p] = U(0,1) = Be(1,1)$ seems a reasonable candidate for a noninformative prior on p: we assert that all possible values of p are equally likely. Fine. But suppose we are asked about parameter $\psi = p^2$. Parameter ψ also takes values between 0 and 1, but if $[p] = U(0,1)$, then

$$\Pr(\psi \le t) = \Pr(p^2 \le t) = \Pr(p \le \sqrt{t}) = \sqrt{t};$$

thus, the density function of ψ is

$$f(t) = \frac{d\sqrt{t}}{dt} = \frac{1}{2\sqrt{t}}, \quad 0 < t < 1,$$

displayed in Fig. 6.2. The distribution of ψ is not uniform; it is $Be(1/2,1)$. There is a 50% chance that $\psi \le 0.25$, and the expected value of ψ is $1/3$. Thus if "noninformative" is defined

---

5. We have used notation $f_X(\theta)$ rather than $f(\theta|X)$ to emphasize that this function, though defined in terms of the data X, and though possibly a distribution function, is not a posterior distribution, there being no proper prior associated with it.

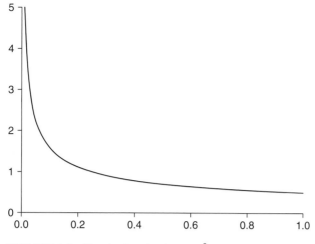

FIGURE 6.2   Density function for $\psi = p^2$, assuming $p \sim U(0,1)$.

as "uniform," a noninformative prior for p results in an informative prior for p^2. There's something unsettling about saying "I know nothing about p, but I know something about p^2."

The foregoing illustrates that uniformity of priors is not transformation invariant.[6] We do not regard this as a terribly serious problem. Suppose that we were to choose a $U(0,1)$ prior for $\psi = p^2$. Reasoning as before, this choice can be demonstrated to induce a $Be(2,1)$ prior on p. The resulting posterior distribution is $[p|X] = Be(X+2, N-X+1)$ rather than $Be(X+1, N-X+1)$. The two posterior distributions are plotted for $X = 15, N = 30$ in Fig. 6.3; 95% HPDI's are (0.33, 0.67) from the uniform prior and (0.35, 0.68) from the $Be(2,1)$ prior. Thus, even with a fairly small data set, inference is reasonably insensitive to the choice of prior.

6.2.2 Improper Priors

Uniform priors seem reasonable when the parameter has a finite range because no value is favored over any other. It would be nice to extend this idea to parameters with infinite range, but a uniform prior with infinite prior is impossible. We need to introduce the idea of an *improper prior*.

Given a data distribution $p(X|\theta)$ and a prior distribution $\pi(\theta)$, the posterior distribution is

$$f(\theta|X) = \frac{p(X|\theta)\pi(\theta)}{\int p(X|\theta)\pi(\theta)\,d\theta}. \tag{6.7}$$

Occasionally, it makes sense to substitute a nonnegative function $g(\theta)$ for $\pi(\theta)$ in Eq. (6.7), despite the fact that $g(\theta)$ has an infinite integral and cannot be scaled to be a probability distribution. The function $g(\theta)$ is referred to as an improper prior. It can happen that even

---

6. The frequentist criterion of unbiasedness also lacks transformation invariance. For instance, the unbiasedness of s^2 as an estimator of σ^2 guarantees that s is biased as an estimator of σ. The proof is straightforward: $0 < \mathrm{Var}(s) = E(s^2) - E(s)^2 = \sigma^2 - E(s)^2$, hence $E(s)^2 < \sigma^2$, so $E(s) < \sigma$.

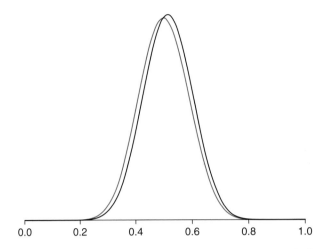

FIGURE 6.3   Distributions $Be(17,16)$ (black) and $Be(16,16)$ (red).

though $\int g(\theta)d\theta = \infty$, the integral $\int p(X|\theta)g(\theta)d\theta$ is finite, with the consequence that

$$f_X(\theta) = \frac{p(X|\theta)g(\theta)}{\int p(X|\theta)g(\theta)\,d\theta} \tag{6.8}$$

defines a perfectly legitimate probability distribution, which we use as though it were a posterior distribution. We have already mentioned one example, the uniform prior over an infinite range. Other examples arise in the context of conjugate priors.

Improper Priors in Conjugate Families: Binomial Success Rate

A family of distributions $[\theta] = g(\theta|\psi)$ is said to be conjugate for a likelihood $[X|\theta]$ if priors in the family combine with the likelihood to produce posteriors in the same family (Chapter 4). The prior has hyperparameter ψ_0, and the posterior has hyperparameter ψ_1. Typically, ψ_1 is easily calculated from ψ_0 and the data X. For example, given a binomial likelihood proportional to $p^X(1-p)^{N-X}$, the beta family of priors is conjugate, with hyperparameter $\psi = (\alpha, \beta)$. The updating formula is

$$\alpha_1 = \alpha_0 + X, \quad \beta_1 = \beta_0 + (N - X). \tag{6.9}$$

We discussed conjugacy in Chapter 4 as a computational efficiency. Conjugacy is also useful in that the updating formula may suggest a form for a noninformative prior. For example, the α and β parameters of the beta distribution are updated as running totals of successes and failures [Eq. (6.9)]. The uniform prior for p is obtained by setting $\alpha_0 = \beta_0 = 1$; we might interpret this as prior knowledge equivalent to one success in two trials. Why not enforce an even greater standard of ignorance and set $\alpha_0 = \beta_0 = 0$ in the prior; zero successes in zero trials. This $Be(0,0)$ prior was suggested by Haldane (1931).

The only problem is that there's no such thing as a $Be(0,0)$ distribution. The hyperparameters α and β must both be strictly positive. If there *were* such a thing as a $Be(0,0)$ density, it would be proportional to $g(p) = p^{-1}(1-p)^{-1}$ and would integrate to 1 over the interval $[0,1]$. However, the function $g(p)$ becomes infinite as $p \to 1$ or $p \to 0$ and does so rapidly, with the consequence that

$$\int_0^1 g(p)dp = \infty.$$

Consequently, there is no constant c, by which we might scale $g(p)$ to make it integrate to 1; $g(p)$, like a uniform distribution over an infinite range, is an improper prior.

Nonetheless, we may treat $g(p)$ as though it were a proper prior, as in Eq. (6.8). Provided that $X \neq 0$ or N, $f_X(\theta)$ is a legitimate distribution function; an improper prior leads to what may be treated as a proper posterior distribution. This "posterior" is also in the beta family, with updating of hyperparameters following the rule given by Eq. (6.9): $\alpha_1 = 0 + X = X$ and $\beta_1 = 0 + N - X = N - X$.

It is best to think of the improper prior as a limit of proper priors, $Be(0,0) = \lim_{\epsilon \to 0} Be(\epsilon, \epsilon)$, and the resulting "posterior" as the limit of the corresponding sequence of posteriors, $Be(X, N - X) = \lim_{\epsilon \to 0} Be(X + \epsilon, N - X + \epsilon)$.[7] Lest this seem like an excessive attention to detail, consider what happens when $X = 0$. The posterior probability of $p > 0$ approaches zero as ϵ gets smaller. If we use the improper prior and observe $X = 0$, our posterior indicates absolute certainty that $p = 0$, even with $N = 1$.

The foregoing demonstrates that improper priors might lead to improper posteriors. This problem is not inevitable, however, and improper *priors* can be quite useful. An example is the problem of estimating a normal mean based on a random sample.

Improper Priors in Conjugate Families: Normal Mean

Given $[\bar{X}|\mu] = N(\mu, \sigma^2/n)$, with σ known, the family of normal distributions is conjugate for μ. Thus, if $[\mu] = N(\mu_0, v_0)$, the posterior distribution $[\mu|\bar{X}] = N(\mu_1, v_1)$, with

$$\mu_1 = \left(\frac{\sigma^2/n}{v_0 + \sigma^2/n}\right)\mu_0 + \left(\frac{v_0}{v_0 + \sigma^2/n}\right)\bar{X} \qquad (6.10)$$

and

$$v_1 = \frac{v_0 \sigma^2/n}{v_0 + \sigma^2/n}. \qquad (6.11)$$

The improper prior with infinite variance seems a very reasonable description of prior ignorance. This prior is the limit of proper priors, with $v_0 \to \infty$. Inspection of (6.10) and (6.11) reveals that the improper prior $N(\mu_0, \infty)$ leads to a proper posterior $N(\bar{X}, \sigma^2/n)$, from which we calculate the $(1-\alpha)100\%$ credible interval

$$\bar{X} \pm z_{\alpha/2}\frac{\sigma}{\sqrt{n}}.$$

---

7. This view of improper priors is enforced in software BUGS by the requirement that hyperparameters α and β must both be strictly positive. One may choose $\alpha = \beta = 0.001$, but not $\alpha = \beta = 0$. However, if an improper prior leads to an improper posterior, a nearly improper prior does not let you off the hook – the resulting posterior will be nearly improper and result in poor inference.

This interval is the numerical equivalent of the standard frequentist confidence interval. The numerical coincidence of the two intervals should not obscure the difference in paradigms. The Bayesian interpretation is expressed in terms of probability for the parameter; the frequentist in terms of probability for the method.

6.2.3 Jeffreys Priors

We have considered uniform priors as candidates for the title of being noninformative. Given a uniform prior, the posterior distribution is simply a scaled version of the likelihood; no prior knowledge is brought to bear on inference other than the form of the likelihood. Uniform priors make perfect sense when the range of the parameter is finite, and the idea extends naturally enough when the range of the parameter is infinite, through the use of improper priors.

But what of the problem of transformation invariance? The material of Section 6.2.1 shows that if we equate "noninformativeness" with uniform distributions, then a noninformative prior for parameter θ implies an informative prior for a transformed parameter $g(\theta)$. Thus, reparameterizations of the same model could lead to distinct inferences about the same parameter. The differences might be slight, given decent data, but the problem is there nonetheless; it would be nice if we could define the quality of "noninformativeness" so as to avoid such unpleasantness.

Jeffreys (1946) suggested a clever solution, leading to a prior specification which bears his name. To understand his solution, we need to recall the change of variables theorem (Section 2.2.4), and the definition and properties of the Fisher information.

Fisher Information

Let $L(\theta|X) \propto [X|\theta]$ be the likelihood function for θ based on an observation X and define the *score function* to be the derivative of $\log L(\theta|X)$ with respect to θ, i.e.,

$$S(\theta|X) = \frac{d \log L(\theta|X)}{d\theta} = \frac{dL(\theta|X)/d\theta}{L(\theta|X)}. \tag{6.12}$$

The Fisher information, $I(\theta)$, can be defined as the expected value of the squared score function, taken over values of X with θ fixed.[8] That is,

$$I(\theta) = E_X\left(S(\theta|X)^2\right). \tag{6.13}$$

Suppose that we were to consider a reparameterization of the model in terms of ψ. We would need to be able to calculate ψ's from θ's, and vice versa, so we can write $\psi = \psi(\theta)$ and $\theta = \theta(\psi)$. To calculate the loglikelihood function for ψ, we need only substitute $\theta = \theta(\psi)$ in the loglikelihood for θ. Thus, the score function for ψ is

$$S(\psi|X) = \frac{d \log L(\theta(\psi)|X)}{d\psi} = \frac{d \log L(\theta(\psi)|X)}{d\theta}\frac{d\theta}{d\psi} = S(\theta|X)\frac{d\theta}{d\psi},$$

---

8. Fisher (1925b) defined the information as minus the expected second derivative of the data distribution with respect to the parameter, that is, $-E(d^2 f(X|\theta)/d\theta^2)$. The two definitions coincide under mild regularity conditions.

in consequence of the chain rule. Using Eq. (6.13) and recognizing that $d\theta/d\psi$ is a constant with respect to the data X, it follows that

$$I(\psi) = I(\theta) \left(\frac{d\theta}{d\psi} \right)^2. \tag{6.14}$$

So, why is $I(\theta)$ referred to as "information?" Note from Eq. (6.12) that for fixed X, the score function is a proportional rate of change in the likelihood as θ varies. Under mild mathematical conditions, the expected value of this rate of change (over data sets X) is zero, so the Fisher information is the variance of the score function. As such, it measures the potential of observations to effect change in the likelihood.

Jeffreys Prior

Jeffreys (1946) suggested that a prior distribution for θ be defined as $[\theta] \propto \sqrt{I(\theta)}$. It follows as an immediate consequence of Eq. (6.14) and the change of variables theorem that a reparameterization $\psi = \psi(\theta)$ will have prior distribution $[\psi] \propto \sqrt{I(\psi)}$. Jeffreys prior is therefore transformation invariant.

A likelihood function $\mathcal{L}(\theta|X)$ is said to be data-translated if it can be expressed in the form $\mathcal{L}(\theta|X) \propto g(\psi(\theta) - s(X))$.[9] For a data-translated likelihood, the data influence the position but not the shape of the likelihood. Box and Tiao (1973) argue that it is therefore appropriate to impose a uniform prior on the parameter $\psi(\theta)$. Using Jeffreys prior is equivalent to choosing a parameterization under which it is appropriate to assign a uniform prior; if the likelihood is data-translated, the approaches coincide.

Jeffreys Prior for Binomial Success Rate

We calculate Jeffreys prior for the binomial success rate as follows. First, the likelihood is $L(p|X) \propto p^X(1-p)^{n-X}$; hence, the loglikelihood is a constant plus $X \log(p) + (n-X) \log(1-p)$, and the score function is

$$S(p|X) = \frac{X}{p} - \frac{n-X}{1-p} = \frac{X-np}{p(1-p)}.$$

Given that $E(X) = np$, it is clear that $I(p) = \text{Var}(X)/(p(1-p))^2 = n/(p(1-p))$; hence Jeffreys prior is $[p] \propto 1/\sqrt{p(1-p)} = p^{-1/2}(1-p)^{-1/2}$. This prior is readily recognized as the $Be(1/2, 1/2)$ distribution.[10]

It is worth considering, for a moment, exactly what we mean by "transformation invariance." Suppose that instead of parameterizing our binomial distribution in terms of the success rate p, we decide that we would rather make inference in terms of the natural logarithm of the odds of a success, $\eta = \text{logit}(p)$. We could effect inference by taking a random sample from the posterior distribution $[p|X]$, then converting each sampled p to an η. The resulting posterior for η corresponds to a particular prior induced by the prior for p. Depending on our definition of "noninformativeness," a noninformative prior for p might not induce a noninformative prior

9. To illustrate, consider a sample of n normal random variables with mean zero, and unknown variance σ^2. The likelihood for σ^2 is data-translated with $g(t) = \exp(-nx - e^{-x})$, $\psi(\sigma^2) = -\log(\sigma)$, and $s(X) = \sqrt{\Sigma X^2}$.
10. Also known as the arcsine distribution, since its cumulative distribution is $1/2 + (1/\pi)\sin^{-1}(2t-1)$ for $0 \le t \le 1$.

for η. If we take Jeffreys prior as our definition of noninformativeness, no such problem arises. If we have used Jeffreys prior for p, drawing a sample of p's and transforming them to η's yields a posterior sample for η based on a Jeffreys prior for η.

Jeffreys Prior for Multivariate Parameters

Most of the time, our models involve multiple parameters: we have a vector-valued parameter $\boldsymbol{\theta}=(\theta_1,\theta_2,\ldots,\theta_k)'$. Jeffreys suggested two possibilities. First, we might obtain priors for each component θ_i according to his recommended approach in the univariate case, treating $\theta_j, j\neq i$ as known. We obtain a joint prior on $\boldsymbol{\theta}$ as the product of (independent) priors. This approach might be sensible in the case of normal data with unknown mean and variance but is not in the case of multinomial data with unknown cell probabilities, since these must sum to one (and therefore cannot be independent).

If we wish to retain transformation invariance, Jeffreys prior for $\boldsymbol{\theta}$ is proportional to the square root of the determinant of the Fisher information matrix, this being the matrix with $\{i,j\}$ element

$$-\mathrm{E}\left(\frac{\partial^2 L(\boldsymbol{\theta}|X)}{\partial\theta_i\partial\theta_j}\right).$$

Although the use of Jeffreys prior for single parameter problems is widely accepted as appropriate, some controversy remains about the multivariate Jeffreys prior (Kass and Wasserman, 1996).

6.2.4 Summary

We have mentioned uniform priors, priors based on conjugacy, and Jeffreys priors as possible noninformative priors. These and many other solutions have been offered to the question of how one is to produce an objective Bayesian analysis, one which removes the taint of subjectivity or arbitrariness from the process of inference. In the case of the binomial success parameter, the three approaches considered result in $Be(1,1)$, $Be(0,0)$ and $Be(1/2,1/2)$ priors. With all but small sizes, the results are practically indistinguishable. For example, in Chapter 3, we discussed the observation of Shealer *et al.* that 8 of 10 kleptoparasitic roseate terns were female. Posterior distributions based on the three choices of prior are given in Fig. 6.4. Even with the very small sample size, the results are quite similar. The 95% HPDI's for the success rate are (0.52, 0.96), (0.54, 0.98), and (0.57, 0.99); posterior probabilities for $p\leq 1/2$ are 0.033, 0.026, and 0.020 (results for uniform, Jeffreys and Haldane priors, respectively).

Jeffreys prior is sometimes improper but its performance is described as "somewhat magical" in that it "almost always yields a proper posterior distribution" (Yang and Berger, 1996). Its form is generally quite simple. For example, a location-scale distribution is one with the property that the density function satisfies

$$f(t|\mu,\sigma)=\frac{1}{\sigma}g\left(\frac{t-\mu}{\sigma}\right)$$

where $g(\cdot)$ is a baseline *pdf*; $-\infty\leq\mu\leq\infty$ and $\sigma>0$ are known as the location and scale parameter, respectively. The normal distribution provides a canonical example, with the mean in the role of location parameter, and the standard deviation as the scale. Provided that the range

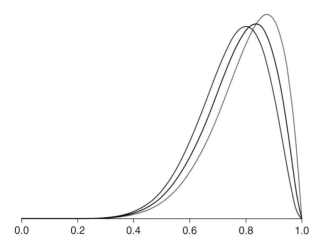

FIGURE 6.4 $X = 8$ females among 10 klep-toparasitic roseate terns, modeled as $X \sim B(10, p)$. Posterior distributions for p based on uniform $Be(1,1)$ prior (blue), Jeffreys $Be(1/2, 1/2)$ prior (black), and Haldane $Be(0,0)$ prior (red).

of $f(t|\mu, \sigma)$ does not depend on unknown parameters and that the Fisher information exists, Jeffreys prior for the location parameter is uniform, and Jeffreys prior for the scale parameter is proportional to σ^{-1}.

An interesting feature of Jeffreys priors is that they are also, at least approximately, "probability matching" priors. A probability matching prior is one that produces a credible interval that is identical to a corresponding frequentist confidence interval (Kass and Wasserman, 1996). A simple illustration is Bayesian inference for μ using a random sample $\boldsymbol{y} = (y_1, \ldots, y_n)'$ from a $N(\mu, \sigma^2)$ distribution when both parameters are unknown. If we adopt as our "prior" the improper $[\mu, \sigma] \propto 1/\sigma$, the resulting posterior distribution for μ has the property that $\sqrt{n}(\mu - \bar{y})/s$ has a t-distribution with $n - 1$ degrees of freedom. Thus, the HPDI for posterior inference about μ exactly matches the usual frequentist confidence interval for this problem.[11] In this particular example, the prior $[\mu, \sigma] \propto 1/\sigma$ corresponds to the product of Jeffreys prior for μ when σ is known and Jeffreys prior for σ computed when μ is known, and is the prior Jeffreys recommended himself (Kass and Wasserman, 1996).

6.3 AFTERWORD

Little (2006), while describing the strengths and weaknesses of Bayesian inference, complained that "the frequentist paradigm does not provide enough exact answers; [but that] with Bayes, there is an embarrassment of riches, because once the likelihood is nailed down, every prior distribution leads to a different answer!" We might add, "all of them correct."

---

11. There is much appeal in the idea that a Bayesian credible interval is well-calibrated against long-run frequencies in the sense that 95%, say, of our 95% credible intervals will contain the true value of the parameter in the long-run, no matter what value the parameter takes in that application. Unfortunately, the subset of cases where a Bayesian credible interval can be shown to be well-calibrated is small and frequency calibration remains a limited criterion for use in the search for noninformative priors.

Generally, two solutions exist to the problem of choosing priors. The first, embraced by subjective Bayesians, is to choose a prior by elicitation of informed opinion and to rejoice in the elegance of the process of updating this prior to a corresponding posterior. The alternative, favored by those who would wish to remove subjective judgments from the process, is to choose reasonable default priors expressing limited information in an effort to allow data to speak for themselves in context of a specified likelihood. Uniform priors for parameters with finite ranges, improper uniform priors for location parameters, and priors $[\theta] \propto 1/\theta$ for scale parameters, all merit consideration when we desire an objective prior specification. Parameters with range $\theta > 0$ might be assigned "vague" uniform priors, i.e., $U(0,L)$, where L is "large"; alternatively, we might consider a $Ga(\epsilon,\epsilon)$ prior, for a small value of ϵ as an approximation to the improper prior with $[\theta] \propto 1/\theta$. Regression coefficients and other parameters with range extending over the entire real line might be assigned mean zero normal priors, with large variance.

We caution, however, against unthinking reliance on default choices for priors. A proper prior always leads to a proper posterior; frequently an improper prior also leads to a proper posterior. However, an improper prior might lead to an improper posterior, which is a serious problem. In such a case, a proper approximation to the improper prior is no solution, because the resulting proper posterior may also approximate the improper posterior, leading to unstable inference. This instability may not be obvious in posterior inference based on simulations.

Gelman (2006) discusses the problem of inference about variance parameters in hierarchical models, a simple example being

$$y_{ij} \sim N(\mu + \alpha_j, \sigma_y^2) \ i=1,\ldots,n_j; \ j=1,\ldots J$$
$$\alpha_j \sim N(0,\sigma_\alpha^2), \sim j=1,\ldots,J, \tag{6.15}$$

where we easily have enough data for inference about μ and σ_y^2. If we could have observed $\alpha_1,\ldots,\alpha_J$, then we could use the conjugate inverse-gamma $IG(\alpha,\beta)$ prior for σ_α^2. In this case the noninformative $IG(0,0)$ "prior" would lead to a proper posterior. It is thus tempting to use the $IG(\epsilon,\epsilon)$ prior for σ_α^2 in the hierarchical model (6.15), with ϵ set to a small value to represent vague prior knowledge. However, Gelman (2006) shows that the $IG(\epsilon,\epsilon)$ prior does not lead to any proper limiting posterior distribution for σ_α^2 in the limit $\epsilon \to \infty$ for this hierarchical model. That is, the improper $IG(0,0)$ prior will lead to improper proper posterior which an $IG(\epsilon,\epsilon)$ prior will approximate, leading to unstable inference. For such problems, Gelman (2006) recommends a uniform $U(0,A)$ prior for σ and some large A. This prior yields a limiting proper posterior density as $A \to \infty$ as long as $J \geq 3$.

In the context of this hierarchical model problem, Gelman (2006) discusses the concept of "weakly informative priors": priors that are chosen so that the information they contain deliberately understates the prior information that is actually available. Thus, Gelman (2006) offers an alternative to the search for uninformative priors; instead, he advocates use of families of subjective but weakly informative priors.

There is much literature on the selection of priors with a particularly useful review given by Kass and Wasserman (1996). We agree with their conclusion that, "...the problems raised by the research on priors chosen by formal rules are serious and may not be dismissed lightly: When sample sizes are small (relative to the number of parameters being estimated), it is dangerous to put faith in any 'default' solution; but when asymptotics take over Jeffreys's rules and their variants remain reasonable choices."

Whether the goal is to conduct a subjective or objective Bayesian analysis, the solution to the "problem" of choosing priors is to be found in the observations with which we began this chapter. One ought always to report the prior as part of the analysis; one ought always to conduct sensitivity analyses to examine the influence of prior choice. Having done these things, further worrying about priors may simply be a straining at gnats.

7

Multimodel Inference

O U T L I N E

7.1	**The BMI Model**	**129**
	7.1.1 Example: BMI for Two Fully Specified Models	*131*
	7.1.2 Example: BMI with Unknown Parameters	*132*
7.2	**Bayes Factors**	**134**
	7.2.1 Bayes Factors and Likelihood Ratio Statistics	*134*
	7.2.2 Bayes Factors are Multipliers of Odds	*135*
	7.2.3 Updating Bayes Factors	*135*
	7.2.4 Bayes Factors as Measures of Relative Support	*136*
	7.2.5 Problems with Vague Priors on Parameters	*137*

7.3	**Multimodel Computation**	**139**
	7.3.1 Multimodel Inference in BUGS	*139*
	7.3.2 Reversible Jump Markov Chain Monte Carlo	*144*
	7.3.3 Bayesian Information Criterion	*148*
7.4	**Indices of Model Acceptability: AIC and DIC**	**149**
	7.4.1 Akaike's Information Criterion	*150*
	7.4.2 Deviance Information Criteria	*153*
	7.4.3 Example: Trout Return Rates	*153*
7.5	**Afterword**	**157**

Inference about ecological processes is almost inevitably model based. No matter how much planning has gone into our investigation prior to collecting data, no matter how carefully we have designed our study, no matter how familiar we are with the system we are studying, there comes a point where we must describe our observations using a mathematical model. The model will have components related to the processes we are studying, and components related to the acquisition of data. Some of these components will be structural, amenable to computation given knowledge of covariates and parameters, other components will be stochastic, describable only as random noise. The model is fully specified, except for unknown parameters estimable from our data.

The process of inference often goes no further than estimating unknown parameters. There is no acknowledgment of model uncertainty, save perhaps a "goodness of fit" test to see whether our observations are consistent with the model.[1] Although we might not always be careful to mention it, we know that our inference is conditional on the model chosen, and might have been different had we posited a different model.

In practice there are usually several, even many plausible models that we could consider for our data. Each has distinct structural features and stochastic components. We are uncertain about which components are necessary. Inference based on the selection of a single model may sweep this uncertainty under the carpet.

The choice of a model is particularly important when candidate models involve complicated structure, and many parameters. We distinguish "parameters of interest" and "nuisance parameters," the former being the objects of our inquiry, the latter being required only so as to avoid a distorted view of the parameters of interest. We do not wish to waste inferential resources by including unnecessary nuisance parameters, nor do we wish to risk being misled by neglecting necessary ones. We also wish to choose the "right" set of parameters of interest, and not some that merely happen to be correlated with the right set.

Instead of conditioning our inference on the choice of a single model, we might wish to include model uncertainty as part of the inferential process. The goal is to produce a composite inference reflecting the uncertainties within and between models. To give a concrete example, suppose that we have analyzed a single data set under K distinct models, obtaining estimated survival rates $\hat{\phi}_k$ and associated standard errors $s(\hat{\phi}_k)$. Suppose further that we can quantify our relative confidence in the models, so as to produce model weights $w_k \geq 0$ satisfying $\sum w_k = 1$. Buckland *et al.* (1997) suggested combining model specific estimates using

$$\tilde{\phi} = \sum_{k=1}^{K} w_k \hat{\phi}_k \qquad (7.1)$$

and a composite measure of uncertainty

$$s(\tilde{\phi}) = \sum_{k=1}^{K} w_k \sqrt{s(\hat{\phi}_k)^2 + (\hat{\phi}_k - \tilde{\phi})^2}.$$

Estimates (and standard errors) 0.74(0.13), 0.70(0.12), 0.76(0.10), and 0.65(0.08) with weights 0.4, 0.3, 0.2, and 0.1 are thus combined to $\tilde{\phi} = 0.72$ and $s(\tilde{\phi}) = 0.12$. The idea is that $\tilde{\phi}$ is an estimate based on all of the models, and that $s(\tilde{\phi})$ incorporates uncertainties within and between models.

The two problems of multimodel inference are thus model selection, and model weighting. Model selection is an attempt to choose a best model from a set of candidates; inference is then conditioned on that selection. Model weighting attempts to combine model specific inferences in a way which acknowledges the relative degree of trust we place in models, as well as the uncertainties associated with model specific inferences.

---

1. Goodness of fit tests are usually conducted with fingers crossed and in fervent hope that the *p*-value will come out larger than 0.05. In which case, the null hypothesis of model adequacy is treated as having been established – in flat contradiction to the philosophy of hypothesis testing.

II. THE BAYESIAN MĀRAMATANGA

Unfortunately, there is currently no consensus on how one ought to acknowledge and account for such uncertainty … and there probably never will be.[2] There are many competing ideas and methods in the literature, enough to leave the practitioner boggled, astounded, and disheartened. The Bayesian approach to multimodel inference does not, alas, finally settle all of the complicated issues surrounding the topic. In some measure, multimodel inference must remain at the intersection of art and science.

However, Bayesian multimodel inference (BMI) has a strong philosophical appeal; like Bayesian inference generally, it retains the features of simplicity, exactness, and coherency described in Chapter 1. Indeed, BMI is a very natural extension of the basic Bayesian technique: one makes inference about unknown quantities (in this case, *models*) based on their posterior distributions, given data. Posterior model probabilities are used for combining model-specific estimates in the spirit of Eq. (7.1), and to combine model-specific inferences. And as for model selection, if a single model is desired, posterior model probabilities provide an objective basis for choice.

In this chapter we provide an overview of BMI, with comments on model weights, Bayes factors, the Bayesian information criterion (BIC), and the deviance information criterion (DIC). We also discuss computational issues, describing reversible jump Markov chain Monte Carlo (RJMCMC) and simple implementations of BMI in program BUGS.

One of the challenges for BMI, perhaps the most serious, is the selection of vague priors for parameters. When dealing with a single model, given adequate data the choice of vague prior has little influence on inference. Unfortunately this is not the case in the multimodel setting. We describe the problems and some possible solutions.

In recent years, Akaike's information criterion (AIC) has been heavily promoted among wildlife statisticians (Burnham and Anderson, 2002) as a basis for model selection and model weighting. AIC has taken the field by storm, and has been uncritically accepted by many practitioners. Its performance is reasonably evaluated in a Bayesian context; we do so in Sections 7.4.1 and 7.4.3.

7.1 THE BMI MODEL

BMI is really no different than any other sort of Bayesian inference: all quantities are treated as random, the only distinction being between quantities that are known or unknown. Inference is made using posterior distributions of unknown quantities given known quantities.

Thus BMI begins by considering a random variable *Model*, drawn from some collection called the *model set*. We might conceive of Nature blindfolded, making a single draw from a bucket containing K models (Fig. 7.1). If the selected model is fully specified (i.e., there are no unknown parameters in the model specification), Nature generates *Data* according to *Model*; if there are unknown parameters, Nature draws parameters for the particular model from a prior distribution specific to *Model*, and then generates *Data*.

Much of statistical inference is done pretending that we know with certainty which model Nature has drawn. Multimodel inference acknowledges model uncertainty, supposing that *Model* could have been any one in the model set.

---

2. Nor perhaps *should* there be, for reasons to be explained at the end of this chapter.

FIGURE 7.1   A bucket of models.

More formally, we conceive of *Model* as a multinomial random variable with index 1 and cell probabilities $\pi_1, \pi_2, \ldots, \pi_K$. If we have no a priori reason for favoring one model over another, we might set $\pi_1 = \pi_2 = \cdots = \pi_K = 1/K$; otherwise, we may choose prior model probabilities reflecting prior beliefs. These beliefs may reflect specific knowledge about the system studied and might also reflect convictions about desirable model features, such as parsimony.

Objections!

We anticipate two objections. First "that's not how things work! *Nature* drawing from a bucket?" But the multinomial draw is merely a model of our uncertainty; a mathematical convenience rather than an exact depiction of reality. The BMI framework is a meta-model, a model about models. Thus like all models, it need not be an exact depiction of reality, merely one that is useful. Subsequently, we describe the process of updating prior model probabilities to posterior model probabilities, based on information provided by data. The usefulness of this meta-model lies in its providing a mathematically formal and coherent system of assessing the support provided by data to various competing models.

Another objection: "But Truth isn't in your bucket!" Much unnecessary ink has been spilled on this topic, with declamations about Science and Truth and Knowledge that can leave our head spinning: is there such a thing as Truth? is it even possible for a Model to be Truth? But entertaining as such philosophical ramblings might be, they have no bearing on the issue at hand: it does not matter whether Truth is in the model set, or not, or whether there even *is* such a thing as Truth. Rather, BMI is conditional inference: we merely condition on Truth being in the model set, without any philosophical baggage. It is completely legitimate to say "I don't believe Truth is in that model set, but if you ask me to pretend it is, I can assign such and such probability to Truth being *this* particular model, or *that* particular model."[3] Model probabilities are never unconditional, but always conditional on the Model Set. So though we may write $\pi_i = \text{Pr}(\textit{Model i})$, what we really mean is

$$\pi_i = \text{Pr}(\textit{Model i}|\textit{Model 1 or Model 2 or } \cdots \text{ or Model K}).$$

The bottom line? The bucket of models is *itself* a model, and no more dubious than any other model. It is merely a mathematical convenience to describe our uncertainty.

7.1.1 Example: BMI for Two Fully Specified Models

A geometric random variable has *pdf* $g(y) = p(1-p)^y$ and mean value $(1-p)/p$; a Poisson random variable has *pdf* $f(x) = e^{-\lambda}\lambda^y/y!$ and mean λ. Both take values $y = 0, 1, 2, \ldots$.

Suppose that we have a sample $Y = \{Y_1, Y_2, \ldots, Y_5\}$ of values which either come from a geometric distribution ($M = M_1$) or a Poisson distribution ($M = M_2$). Here, M is a categorical random variable describing Nature's multinomial choice of model. Then the probability of the data is either

$$\text{Pr}(Y|M_1, p) = \prod_{i=1}^{5} g(Y_i) = p^5(1-p)^{5\bar{Y}}. \tag{7.2}$$

or

$$\text{Pr}(Y|M_2, \lambda) = \prod_{i=1}^{5} f(Y_i) = \frac{\exp(-5\lambda)\lambda^{5\bar{Y}}}{\prod_{i=1}^{5} Y_i!}. \tag{7.3}$$

Strictly speaking, we should write the probabilities as conditional on $M = M_i$, as for example $\text{Pr}(Y|M = M_1, p)$, emphasizing that we are treating M as a random variable, with M_1 being an outcome; ease of notation and convention favor the simpler notation.

To make things easy, suppose that we know the population mean is 3 (i.e., either $\lambda = 3$ if the data come from a Poisson distribution, or $p = 1/4$, if the data come from a geometric distribution). The two densities are displayed in Fig. 7.2. Substituting $\lambda = 3$ and $p = 1/4$ in (7.2) and (7.3), we obtain $\text{Pr}(Y|M_i)$ for $i = 1, 2$. Note that the models are fully specified, hence we do not need to include p or λ in the conditional description.

Given prior probability $\pi = \text{Pr}(M_1)$, straightforward application of Bayes' theorem yields posterior probabilities

$$\text{Pr}(M_1|Y) = \frac{\pi \, \text{Pr}(Y|M_1)}{\pi \, \text{Pr}(Y|M_1) + (1-\pi) \, \text{Pr}(Y|M_2)}. \tag{7.4}$$

Consider the data set $Y = \{0, 1, 2, 3, 8\}$. The sample mean is 2.8, consistent with both models, but the sample variance is $s^2 = 9.7$. Given that the Poisson mean and variance are equal, the data would seem to favor the geometric model, which allows for greater variability relative to the mean (compare Fig. 7.2.) If we had no a priori reason to favor the Poisson over the geometric model, we would probably set $\pi = 0.5$ in Eq. (7.4) and obtain posterior probability

---

3. To illustrate: imagine an urn with 99,999,997 red marbles, 2 blue marbles, and 1 white marble. If I were to draw a marble at random, and report that it was not red, you might not believe me, but could still assign odds to whether it were blue or white, conditional on my claim.

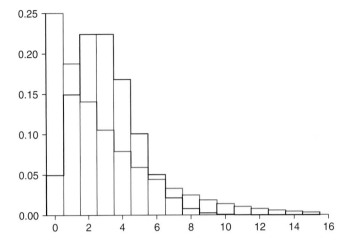

FIGURE 7.2   Poisson (blue) and geometric (red) densities, means $= 3$.

of 0.852 on the geometric model: the data have shifted the odds from being even (0.5:0.5) or (1:1) to odds of (0.852:0.148) or (5.75:1) in favor of the geometric model.

This change in odds is a very useful summary in BMI. Note that Eq. (7.4) implies

$$\Pr(M_2|Y) = \frac{(1-\pi)\Pr(Y|M_2)}{\pi\Pr(Y|M_1)+(1-\pi)\Pr(Y|M_2)}. \tag{7.5}$$

Thus, the ratio of the left-hand sides of Eqs. (7.4) and (7.5) equals the ratio of their right-hand sides, namely

$$\frac{\Pr(M_1|Y)}{\Pr(M_2|Y)} = \left(\frac{\pi}{1-\pi}\right) \times \frac{\Pr(Y|M_1)}{\Pr(Y|M_2)}. \tag{7.6}$$

Equation (7.6) relates the posterior model odds (left-hand side) to the prior model odds $\pi/(1-\pi)$. Prior model odds are scaled by the relative probabilities of the data, under the two models.

As a mathematical expression, $\Pr(Y|M)$ is a function of the data, Y, and of the model, M. Restricting our attention to a fixed set of data, $\Pr(Y|M)$ is a function of M alone, the likelihood function for the model. The ratio of model likelihoods in Eq. (7.6) is called the *Bayes factor*. We may thus put Eq. (7.6) into words:

Posterior model odds = Prior model odds × Bayes factor.

For the example considered, prior odds of 1:1 were converted to posterior odds of 5.75:1, hence the Bayes factor is 5.75.

7.1.2 Example: BMI with Unknown Parameters

Suppose that instead of comparing a geometric distribution with known $p=1/4$ to a Poisson distribution with known $\lambda=3$, the choice was between an unknown geometric distribution and an unknown Poisson distribution. To calculate the model likelihoods, we would need prior distributions for p and λ.

That is, instead of substituting specific values of p and λ in Eqs. (7.2) and (7.3), we must calculate the average probability of the data under each model, weighted by the prior distributions.

Given a prior distribution $g(p)$ for the parameter p of the geometric distribution, we would calculate

$$\Pr(\boldsymbol{Y}|M_1) = \int \Pr(\boldsymbol{Y}|M_1,p)g(p)\,dp = \int p^5(1-p)^{5\bar{Y}}g(p)\,dp; \qquad (7.7)$$

given a prior distribution $h(\lambda)$ for the mean λ of the Poisson distribution, we would calculate

$$\Pr(\boldsymbol{Y}|M_2) = \int \Pr(\boldsymbol{Y}|M_2,\lambda)h(\lambda)\,d\lambda = \int \frac{\exp(-5\lambda)\lambda^{5\bar{Y}}}{\prod_{i=1}^5 Y_i!}h(\lambda)\,d\lambda. \qquad (7.8)$$

Choice of prior distributions for parameters in BMI is a ticklish business, which we discuss subsequently. Suppose that we choose a $U(0,T)$ prior for the mean λ of the Poisson distribution. It seems reasonable then to choose a prior on p for the geometric distribution such that the mean $(1-p)/p$ also has a $U(0,T)$ distribution. It can be shown using the change of variables theorem (2.2.4) that we require a prior

$$g(p) = \frac{1}{Tp^2},$$

for $1/(T+1) < p < 1$.

For the data set $\boldsymbol{Y} = \{0, 1, 2, 3, 8\}$, it can be shown that as $T \to \infty$, the resulting Bayes factors in favor of the geometric model approach 13.84. It is natural to ask why the evidence in favor of the geometric model appears so much stronger when the mean is unknown, than in our previous analysis when the mean was known (mean $= 3$ implies BF favoring geometric model is 5.75).

Some intuition is gained by considering Fig. 7.3. It turns out that if the mean value is known, the evidence favoring the geometric model over the Poisson model is minimized when the

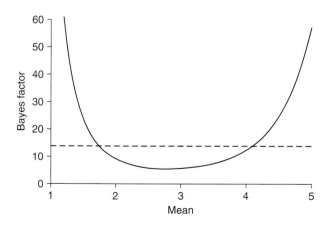

FIGURE 7.3 Bayes factors favoring geometric model over Poisson data as a function of known mean, for data $\boldsymbol{Y} = \{0,1,2,3,8\}$. Dashed line at 13.84 is Bayes factor for analysis with vague prior on mean.

true mean and the sample mean coincide, at $\bar{Y}=2.80$. There, BF $= 5.61$, favoring the geometric model. If the mean were known to be a very small or very large value, the data would provide very strong evidence favoring the geometric model over the Poisson. Incorporating prior uncertainty about the mean essentially averages the evidence favoring the geometric model over values of the mean consistent with the observed data.

7.2 BAYES FACTORS

The Bayes factor provides a way of comparing pairs of competing models. The models need not be nested; neither need be a special case of the other, as seen in the example of Section 7.1.2. In this section, we describe some features of Bayes factors.

7.2.1 Bayes Factors and Likelihood Ratio Statistics

The Bayes factor is a likelihood ratio for models. Suppose that we have models M_k with unknown parameters θ_k, for $i=1,2,\ldots,K$. Using the bracket notation described in Section 2.2.1, we have data distributions $[Y|M_k,\theta_k]$ for data and prior distributions $[\theta_k|M_k]$ for parameters. For a fixed data set Y, $[Y|M,\theta]$ is a joint likelihood for model and parameter. We obtain a marginal likelihood for M by integrating the joint likelihood against the prior for the parameter. That is,

$$[Y|M]=\int [Y,\theta|M]d\theta=\int [Y|\theta,M][\theta|M]d\theta. \tag{7.9}$$

The Bayes factor for comparing M_i to M_j is calculated as the ratio of marginal likelihoods, averaged across the parameters:

$$\mathrm{BF}_{i,j}=\frac{[Y|M_i]}{[Y|M_j]}.$$

As such, the Bayes factor is a Bayesian analog of the frequentist likelihood ratio test statistic. The difference is that the likelihood ratio test statistic is the ratio of maximum likelihoods. That is, instead of averaging against a distribution for the unknown parameter [as in Eq. (7.9)], one substitutes a specific value, the MLE $\hat{\theta}_M$:

$$\mathrm{LR}_{ij}=\frac{[Y|\hat{\theta}_i,M_i]}{[Y|\hat{\theta}_j,M_j]}.$$

If the models are nested (i.e., model M_i is a special case of model M_j), and if M_i is true, -2 times the natural logarithm of the likelihood ratio statistic can often be treated as having an asymptotic chi-squared distribution.

7.2.2 Bayes Factors are Multipliers of Odds

Given a set of K models, prior model probabilities π_k, $k=1,2,\ldots,K$, and data Y, Bayes' theorem yields

$$\Pr(M_i|Y) = \frac{[Y|M_i]\,\pi_i}{\sum_{k=1}^{K}[Y|M_k]\,\pi_k} \tag{7.10}$$

from which it follows that

$$\frac{\Pr(M_i|Y)}{\Pr(M_j|Y)} = \frac{[Y|M_i]}{[Y|M_j]} \times \left(\frac{\pi_i}{\pi_j}\right)$$

$$= \mathrm{BF}_{i,j} \times \left(\frac{\pi_i}{\pi_j}\right),$$

generalizing Eq. (7.6). That is, the Bayes factor is simply a multiplier for changing prior odds into posterior odds. In the special case of prior odds equal to 1, the Bayes factor itself is the posterior odds.

Bayes factors do not depend on the prior model probabilities; they also do not depend on the model set, being nothing more than pairwise comparisons of models. There is, however, a relation among Bayes factors within a model set. It is clear from their definition that

$$\mathrm{BF}_{1,3} = \mathrm{BF}_{1,2}\mathrm{BF}_{2,3}. \tag{7.11}$$

Similarly, if we know the Bayes factor for model 1 against model 2, $\mathrm{BF}_{1,2}$, the Bayes factor for model 2 against model 1 is $\mathrm{BF}_{2,1} = 1/\mathrm{BF}_{1,2}$.

Equation (7.10) is sometimes rewritten in terms of Bayes factors. Dividing the numerator and denominator of the right-hand side by $[Y|M_1]$, we obtain

$$\Pr(M_i|Y) = \frac{\mathrm{BF}_{i,1}\,\pi_i}{\sum_{k=1}^{K}\mathrm{BF}_{k,1}\,\pi_k}. \tag{7.12}$$

7.2.3 Updating Bayes Factors

A very appealing feature of Bayes factors is that they update naturally as more data are collected. Let $\mathrm{BF}_{i,j}(Y_1)$ denote the Bayes factor based on data set Y_1, and let $\mathrm{BF}_{i,j}(Y_2|Y_1)$ denote the Bayes factor based on Y_2 having used Y_1 to inform priors on unknown parameters. Thus,

$$\mathrm{BF}_{i,j}(Y_1) = \frac{[Y_1|M_i]}{[Y_1|M_j]} = \frac{\int [Y_1|M_i,\theta]\,[\theta|M_i]\,d\theta}{\int [Y_1|M_j,\theta]\,[\theta|M_j]\,d\theta},$$

and

$$\mathrm{BF}_{i,j}(Y_2|Y_1) = \frac{[Y_2|Y_1,M_i]}{[Y_2|Y_1,M_j]} = \frac{\int [Y_2|Y_1,M_i,\theta]\,[\theta|Y_1,M_i]\,d\theta}{\int [Y_2|Y_1,M_j,\theta]\,[\theta|Y_1,M_j]\,d\theta}.$$

Note that the priors on θ change from the first to the second of these calculations: in the second case, they have been informed by the data Y_1.

Consequently

$$BF_{i,j}(Y_1,Y_2) \equiv \frac{[Y_2,Y_1|M_i]}{[Y_2,Y_1|M_j]} = \frac{[Y_2|Y_1,M_i]}{[Y_2|Y_1,M_j]} \times \frac{[Y_1|M_i]}{[Y_1|M_j]}$$
$$= BF_{i,j}(Y_2|Y_1) \times BF_{i,j}(Y_1).$$

It makes sense that there should be a simple mechanism for describing the accumulation of evidence in favor of one model over another, as new data are obtained. This feature is conspicuously absent from sequences of hypothesis tests under the frequentist paradigm, though various ad hoc mechanisms have been proposed.

7.2.4 Bayes Factors as Measures of Relative Support

The larger the value of $BF_{i,j}$, the greater the support provided by the data to Model M_i relative to Model M_j. But how ought we to interpret the numbers? What does $BF_{i,j} = 5$ mean? On what scale are we operating?

We can get some idea of how to interpret Bayes factors by evaluating their effect on prior model weights in producing posterior model weights. Suppose that the model set consists of two models. From Eq. (7.4) it follows that

$$Pr(M_1|Y) = \frac{BF_{1,2}\,\pi}{BF_{1,2}\,\pi + (1-\pi)}, \tag{7.13}$$

and consequently that

$$Pr(M_1|Y) \geq p_0 \quad \text{if and only if} \quad \pi \geq \frac{p_0}{p_0 + BF_{1,2}(1-p_0)}.$$

Thus for example, if BF = 50, any prior probability $\pi \geq 0.16$ will produce a posterior model weight of at least 90%.

Equation (7.13) is plotted for various values of BF in Fig. 7.4. The larger the Bayes factor, the closer the posterior model probability is to 1; if the Bayes factor is large enough, the posterior probability must be nearly 1 unless the prior probability is nearly zero.

Bayes factors are treated as quantitative measures of the strength of evidence in favor of one model relative to another. In this regard, they stand in marked contrast to frequentist hypothesis tests, which measure only the evidence *against* a model. Harold Jeffreys (Jeffreys, 1961) suggested that the strength of evidence can be categorized according to the classification in Table 7.1. Alternative but similar classifications have been proposed by various authors (for example, Kass and Raftery, 1995). We tend to evaluate Bayes factors using Eq. (7.13), with $\pi = 0.50$. Given that $M = M_1$ or $M = M_2$, the posterior probability of model M_1 is

$$Pr(M_1|M_1 \text{ or } M_2, Y) = \frac{BF_{1,2}}{1 + BF_{1,2}};$$

this probability, or the description as odds multiplier [Eq. (7.11)] is better than an arbitrary cut-off in the spirit of frequentist α levels. As Kass and Raftery point out, "the interpretation may depend on the context." "Overwhelming evidence" for the superiority of one laundry detergent over another might not suffice in a criminal trial.

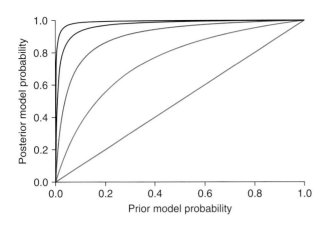

FIGURE 7.4 Posterior model probabilities (y-axis) as a function of prior model probabilities (x-axis) for Bayes factors 1, 5, 25, 125, and 625 (green, red, purple, blue, black).

TABLE 7.1   Bayes factors as weights of evidence.

Result	Interpretation
$1 < B_{1,2} < 3$	There is little support for M_1 over M_2
$3 < B_{1,2} < 12$	There is some support for M_1 over M_2
$12 < B_{1,2} < 150$	M_1 is strongly supported over M_2
$B_{1,2} > 150$	The support for M_1 is overwhelming

7.2.5 Problems with Vague Priors on Parameters

Bayesian analyses often use vague priors on parameters to let the data speak for themselves. For example, if $[X|\mu] = N(\mu, 1)$, the prior distribution $[\mu] = N(\mu_0, \sigma^2)$ results in posterior distribution

$$[\mu|X] = N\left(\frac{1}{1+\sigma^2}\mu_0 + \frac{\sigma^2}{1+\sigma^2}X, \frac{\sigma^2}{1+\sigma^2}\right),$$

which approximates the likelihood function for μ as $\sigma \to \infty$. Thus if we use a large enough value of σ in the prior, representing very limited prior knowledge about μ, inference will be based essentially on the likelihood alone, on the information provided by data rather than prior.

The "prior" with $\sigma = \infty$ is improper: it is not a true distribution function. However, we can use proper priors based on large but finite values of σ, increasingly vague in their specification of prior knowledge, and with diminishing effects on estimation. For instance, suppose $X = 1$; setting $\mu_0 = 0$, inference based on priors with $\sigma = 100$ and $\sigma = 10^6$ will yield nearly identical results, and both approximate the result based on $\sigma = \infty$.

Thus it is often the case that for the purpose of single model inference (estimation), the choice among vague priors and even of improper priors is of little consequence. Unfortunately this boon does not extend to multimodel inference, particularly when the number of parameters varies among models. Continuing with the same example, suppose that we wish to compare two models under which X is a sample from a normal distribution with variance equal to 1. Under Model 1, $\mu = 0$; under Model 2, μ is unknown. Model 1 has no unknown parameters, so the marginal distribution is $[X|M=1]=N(0,1)$. For Model 2 we choose the prior $[\mu]=N(\mu_0,\sigma^2)$, obtaining marginal distribution $[X|M=2]=N(\mu_0,1+\sigma^2)$. Thus the Bayes factor is

$$
\mathrm{BF}_{1,2} = \frac{\frac{1}{\sqrt{2\pi}}\exp\left(\frac{-1}{2}X^2\right)}{\frac{1}{\sqrt{2\pi(1+\sigma^2)}}\exp\left(\frac{-1}{2(1+\sigma^2)}(X-\mu_0)^2\right)}
$$

$$
= \sqrt{1+\sigma^2}\,\exp\left(-\frac{1}{2}\left\{\frac{X^2\sigma^2+2X\mu_0-\mu_0^2}{1+\sigma^2}\right\}\right).
$$

For any fixed X and μ_0, $\mathrm{BF}_{1,2}$ acts like $\sigma\exp\left(-\frac{1}{2}X^2\right)$ for large values of the prior variance σ^2; that is, $\mathrm{BF}_{1,2}\to\infty$ as $\sigma\to\infty$, regardless of the value X. The vaguer the prior, the greater the prejudice in favor of the simpler model; with the improper prior ($\sigma=\infty$) on μ, we need not even collect data, as the matter will be decided in advance, in favor of Model 1.

Aitkin (1991) suggested comparing models using the ratio of posterior mean likelihoods, a quantity which he named the Posterior Bayes factor (PBF). Recall that the marginal distribution used in computing Bayes factors is the average value of the likelihood against the prior $[\theta_M|M]$, viz.,

$$
[Y|M]=\mathrm{E}_{[\theta_M|M]}([Y|M,\theta_M])=\int[Y|M,\theta_M][\theta_M|M]d\theta_M;
$$

Aitkin's proposal was to replace the prior $[\theta_M|M]$ by the posterior $[\theta_M|M,Y]$ in the calculation, using ratios of

$$
\mathrm{E}_{[\theta_M|M,Y]}([Y|M,\theta_M])=\int[Y|M,\theta_M][\theta_M|M,Y]d\theta_M.
$$

The PBF thus avoids problems with vague priors on parameters by using an informative prior, one that has been informed by the data.[4]

The PBF has been roundly criticized by Bayesian statisticians on the grounds that it amounts to double dipping, using the data twice, once to estimate parameters (i.e., to obtain the posterior distribution of the parameters) and then again to compute model weights (see Discussion following (Aitkin, 1991; Berger and Pericchi, 1996). Using the posterior, the argument goes, is to overstate the fit of the model to the data, by suggesting that unknown parameters take values consistent with the data. Although the essence of the criticism is legitimate, the PBF may be a useful tool for comparing models, developed in the spirit of BMI, but avoiding problems associated with vague priors.[5] Aitkin ably addresses the criticisms of the PBF both in the concluding comments of his paper, and after the subsequent discussion.

---

4. The BIC, discussed subsequently in Section 7.3.3, makes implicit use of a similar default prior, though one that is intended to be minimally informative.
5. The PBF can be understood and justified as a measure of fit based on the posterior predictive distribution; see Section 5.1.2.

Another suggestion, offered in the discussion of Aitkin's paper, is to split off some of the data into a training sample, and to use the posterior distributions arising from the training samples as informative priors for analysis of the remaining data; Berger and Pericchi (1996) develop the idea further, describing an "intrinsic Bayes factor" based on priors trained by the smallest sample sizes needed for estimation.

We conclude with this summary: the choice of priors on parameters matters in multimodel inference. There is no easy or automatic choice available. Our view is that priors on parameters should be chosen with the goal of avoiding a priori preference for one model over another; that such preference (e.g., for parsimonious modeling) should be reflected in the prior weights on models. We illustrate this perspective in our multimodel analysis of return rates for tagged trout, Section 7.4.3.

7.3 MULTIMODEL COMPUTATION

BMI can present serious computational challenges. In addition to the usual problem of computing posterior distributions for quantities of interest under each model, we must also be able to compute Bayes factors for comparing models and computing model weights.

Ideally, we would compute marginal likelihoods $[Y|M]$ for each model M, and use these to compute Bayes factors directly. Unfortunately, this approach is rarely feasible.

In this section we review two alternative approaches. The first approach is to use Gibbs sampling with *Model* treated as an unobserved quantity. Bayes factors are then computed as the ratio of (observed) posterior model odds to (specified) prior model odds. We illustrate this in Section 7.3.1 with two examples using program BUGS. In Section 7.3.2, we describe a special implementation of MCMC designed for multimodel inference, reversible jump MCMC (RJMCMC). It is possible, in some cases, to implement RJMCMC in BUGS. Programming and tuning RJMCMC can be challenging, but the basic ideas are fairly straightforward.

The second approach is to approximate $[Y|M]$ using the BIC, which we review in Section 7.3.3.

7.3.1 Multimodel Inference in BUGS

BMI can sometimes be implemented using program BUGS, with Model treated as a categorical random variable. Considerable care is required; analysts should be on the lookout for long autocorrelations in Model. We illustrate the general approach with two examples.

An Example with Nonnested Models

In Section 7.1.2, we compared the fit of geometric and Poisson models to a data set consisting of five observations. There, we calculated the integrals in (7.7) and (7.8) exactly, obtaining $BF_{1,2} = 13.84$ for the given data set.

The analysis can be carried out using Gibbs sampling, with the BUGS code given in Panel 7.1. Several features of this code require comment. First, note that Model is a categorical random variable, taking the values 1 or 2, for the geometric and Poisson models, respectively. We have assigned prior probabilities of 0.10 and 0.90 to the two models, and compute the Bayes factor as the ratio of posterior odds to prior odds. We generated a chain of

PANEL 7.1   BUGS code for comparison of geometric and Poisson models, data of Section 7.1.2.

```
Data: list(Y=c(0,1,2,3,8),pi=c(0.10,0.90))
Inits: list(Model=2)
model{
  Model ~ dcat(pi[1:2])
  mu ~ dunif(0,1000)
  lambda <- mu
  p <- 1/(1+mu)

  for (i in 1:5){
    GeomProb[i] <- p*pow(1-p,Y[i])
    PoisProb[i] <- exp(-lambda+Y[i]*log(lambda)-logfact(Y[i]))

    b[i] <- pow(GeomProb[i],equals(Model,1))
        + pow(PoisProb[i],equals(Model,2)) - 1

    B[i] <- 1
    B[i] ~ dbern(b[i])
  }
}
```

length 1 million, observing Model$=1,2$ in the proportions 605858:394142. Thus, the Bayes factor in favor of Model 1 is approximated as

$$\text{BF}_{1,2} \approx \left(\frac{605858}{394142}\right) \Big/ \left(\frac{0.1}{0.9}\right) = 13.83.$$

It can be shown that the Bayes factor is most precisely estimated when the posterior model probabilities are nearly equal. It is therefore useful to run a Markov chain sampler several times, adjusting the prior model weights so as to achieve nearly equal posterior model weights. This is not cheating since the Bayes factor does not depend on the prior model weights.

We have used what is known as the "1's" trick in this code: we introduced Bernoulli random variables B_i with success probability b_i, and supposed that all $B_i = 1$. The observation $B_i = 1$ makes the same contribution to the likelihood as an observation of a random variable $Y_i = y_i$ with $\Pr(Y_i = y_i) = b_i$. The "1's" trick is useful when we wish to model a probability distribution not included in the BUGS suite. In the present case, our observations are either from a geometric distribution (if Model = 1) or a Poisson distribution (if Model = 2). Our code is written so that

$$b_i = \Pr(Y_i = y_i | \text{Model}) = \begin{cases} p(1-p)^{y_i}, & \text{if Model} = 1 \\ e^{-\lambda}\lambda^{y_i}/y_i!, & \text{if Model} = 2 \end{cases}.$$

Our analysis links the prior distributions for the parameters of the two models. We specify a uniform prior distribution for the mean $\mu = \text{E}(Y|M)$, which we assume to be independent of the model. The Poisson parameter λ is its mean, so we set $\lambda = \mu$; the geometric distribution's parameter p is related to its mean by $\mu = (1-p)/p$, hence $p = 1/(1+\mu)$.

The mean value of our Markov chain sample of values μ was 4.22. This Markov chain is actually a 0.606:0.394 mix of samples from two posterior distributions, the first for the

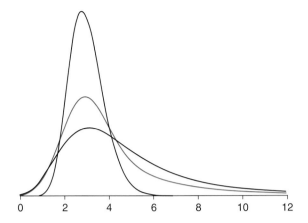

FIGURE 7.5 Posterior distributions of $\mu = E(Y)$ under geometric model (black), Poisson model (blue), and their 0.606:0.394 mixture (red).

geometric model, the second for the Poisson model. The two posterior distributions of μ are quite different (Fig. 7.5). In particular, the posterior means are 5.01 and 3.00, respectively; the overall mean of 4.22 is a weighted average:

$$4.22 = 0.606 \times 5.0 + 0.394 \times 3.0.$$

Thus 4.22 is a model averaged estimate of the mean, assuming prior model weights of 0.1 and 0.9. Credible intervals taken from the mixed posterior average over the two models. 95% CI's for Models 1 and 2 are (1.41, 14.7) and (1.68, 4.70); the 95% Model averaged CI is (1.51, 12.4).

Suppose that we wish to assign equal prior model weights to the two models. The Bayes factor is invariant to the choice of prior weights; we know it is 13.83. Thus posterior odds in favor of model 1 are (BF $\times$ prior odds) = 13.83, from which it follows that the posterior probability of model 1 is 0.933 = 13.83/(13.83 + 1). We can compute the model averaged estimate of μ as

$$4.88 = 0.933 \times 5.0 + 0.067 \times 3.0.$$

To compute credible intervals based on MCMC output, we need samples of the two posterior distributions, in the ratio 0.933:0.067. One option is to generate two new Markov chain samples, one under each model. Chains of length 933,000 and 67,000 could be combined to produce a sample from the mixed distribution. An easier option is based on the chain we have already produced: we had 605,858 samples from Model 1, and 394,142 from Model 2. If we discard all but 43,507 of the samples from Model 2, the remaining 649,365 = 605,858 + 43,507 values will stand in the desired proportions: (605,858/649,365) = 0.933.[6]

An Example with Nested Models

Another example of BMI in BUGS is given in Panel 7.2. Here, the models are band recovery models (described in more detail in Section 11.2) with three years of data. We thus have X_{ij}

6. This procedure can be formally viewed as an example of rejection sampling as described in Section 4.2.2.

PANEL 7.2   BUGS code for comparison of Brownie models.

```
Data: list(
    X = structure(.Data = c(
              50,22,14,314,
              NA,28,21,351,
              NA,NA,25,375),.Dim = c(3,4)),
    pi=c(0.25,0.25,0.25,0.25))
Inits: list(Model=1,xS=c(0.5,0.5),xf=c(0.05,0.05,0.05))

model{
    for (i in 1:3){
        xf[i] ~ dunif(0,1)
        f[i] <- pow(xf[1],constF)*pow(xf[i],1-constF)
        p[i,4] <- 1-sum(p[i,i:3])
        X[i,i:4] ~ dmulti(p[i,i:4],400)
    }
    for (i in 1:2){
        xS[i] ~ dunif(0,1)
        S[i] <- pow(xS[1],constS)*pow(xS[i],1-constS)
    }
    Model ~ dcat(pi[1:4])
    constF <- equals(Model,2)+equals(Model,4)
    constS <- equals(Model,3)+equals(Model,4)
    p[1,1]<-f[1]
    p[1,2]<-S[1]*f[2]
    p[1,3]<-S[1]*S[2]*f[3]
    p[2,2]<-f[2]
    p[2,3]<-S[2]*f[3]
    p[3,3]<-f[3]
}
```

recoveries in year j of bands released in year i, for $i=1,2,3$ and $j=i,i+1,\ldots,3$; we let $X_{i,4}$ denote the number of bands released in year i but never recovered.

The data are independent multinomial vectors, $X_i = (X_{i,i}, X_{i,i+1}, \ldots, X_{i,4})'$, $i=1,2,3$. These have cell probabilities governed by recovery rates $f_i, i=1,2,3$, and survival rates S_i, $i=1,2$.

We consider four models. Model 1 places no constraints on the five parameters. Model 2 assumes no variation in recovery rates through time and Model 3 assumes no variation in survival rates through time. Model 4 incorporates both assumptions, having constant recovery and survival rates.

Coding this analysis in BUGS is easy (Panel 7.2). Model is treated as a categorical variable. Indicator (0/1) variables constF and constS determine whether modeled rates are allowed to vary through time. For example, the modeled rate $f_i \equiv$ xf[1] if constF=1; otherwise $f_i =$ xf[i]. It is important to note that the code in Panel 7.2 sets up model-specific priors as follows:

Model 1: $f_i \sim U(0,1)$ for $i=1,2,3$ and $S_i \sim U(0,1)$ for $i=1,2$
Model 2: $f_1 = f_2 = f_3 \equiv f. \sim U(0,1)$ and $S_i \sim U(0,1)$ for $i=1,2$

Model 3: $f_i \sim U(0,1)$ for $i=1,2,3$ and $S_1 = S_2 = S_3 \equiv S_. \sim U(0,1)$
Model 4: $f_1 = f_2 = f_3 \equiv f_. \sim U(0,1)$ and $S_1 = S_2 = S_3 \equiv S_. \sim U(0,1)$.

This example (Panel 7.2) differs from the previous (Panel 7.1) in two important ways. First of all, models 2, 3, and 4 are "nested" within Model 1: each is a special case of Model 1. This suggests a generic approach to model description, with a set of derived parameters corresponding to the fullest model (the f and S in Panel 7.2), calculated from a palette of parameters (xf[i]'s and xS[i]'s in Panel 7.2) subject to the constraints imposed by the categorical variable Model. The notion of a "palette of parameters" is key to understanding RJMCMC, and will be discussed more subsequently.

The second important difference is that the number of parameters varies among models. Varying numbers of parameters among models can cause no end of frustration in BMI computation. Suppose that parameter θ_j is in model M_j, but not in Model M_i, for $i \neq j$. Parameter θ_j must be sampled when its turn comes up in the Gibbs sampling algorithm, even if the current value of Model is M_i, for some $i \neq j$. Model M_i provides no information about θ_j, so θ_j winds up being sampled from its prior. If the prior for θ_j is vague and has a large range, the Markov chain for θ_j can go wandering off into regions of low likelihood. Then, the next time the Gibbs sampler considers switching models, model M_j looks very unappealing; the chain for Model will not visit M_j. The phenomenon is illustrated for node f[2] in Fig. 7.6. A history plot of length 25,000 for f[2] has black points plotted when Model = 2 or 4 (models which do not include f_2) and red points plotted when Model = 1 or 3 (models which include f_2). Note the long sequence of observations, indices from 7522 to 11,618, where Models 1 and 3 were not visited; during this period f[2] was sampled from the uniform prior, and only rarely would a value consistent with Models 2 or 4 be sampled. The result is rather poor mixing,[7] and the necessity of a long MCMC run.

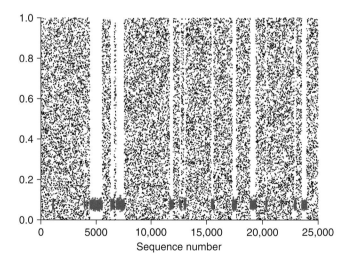

FIGURE 7.6 History plot for node f[2]. Red dots indicate samples when Model = 2 or 4, black dots indicate Model = 1 or 3.

---

7. In MCMC, Markov chains are described as mixing well if they produce representative samples of the stationary distribution, without excessive autocorrelation.

In a chain of length 10 million, the transitions between models were summarized by the matrix

$$T = \begin{bmatrix} 744,865 & 1442 & 431,151 & 1819 \\ 1472 & 1,347,519 & 1618 & 866,074 \\ 431,101 & 1653 & 1,689,340 & 3984 \\ 1838 & 866,069 & 3970 & 3,606,084 \end{bmatrix}.$$

For example $T_{1,4} = 1819$ means that there were 1819 transitions from Model 1 to Model 4. All told, there was only a 26% chance of moving from one model to another. Inspecting matrix T also shows that most transitions were between Models 1 and 3, or 2 and 4. The posterior model probabilities were $(0.12, 0.22, 0.21, 0.45)$[8]; if Model were sampled independently according to these probabilities, we would have expected T to look like

$$\begin{bmatrix} 144,000 & 264,000 & 252,000 & 540,000 \\ 264,000 & 484,000 & 462,000 & 990,000 \\ 252,000 & 462,000 & 441,000 & 945,000 \\ 540,000 & 990,000 & 945,000 & 2,025,000 \end{bmatrix}.$$

In particular, there would have been a 69% chance of moving from one model to another, rather than the 26% realized.

Thus having parameters specific to individual models, with vague priors, leads to poor mixing for the node Model. Multiply these problems by a large number of parameters specific to individual models, throw in a large model set, and you have a recipe for frustration: the Markov chain will mix very poorly, and require huge run times.

In such cases, a special McMC algorithm is required, which we now describe.

7.3.2 Reversible Jump Markov Chain Monte Carlo

Reversible jump Markov chain Monte Carlo (RJMCMC) Green (1995) is an implementation of MCMC designed for efficient multimodel analysis. RJMCMC is nothing more than Gibbs sampling with alternating updates of model M and a "palette of parameters" ψ. The distinctive feature of RJMCMC is its use of the palette to simultaneously govern the parameters of all the models in the model set.

The palette ψ is a vector of dimension d greater than or equal to the dimension of the most complex model in the model set. It is defined in terms of mappings $g_M(\psi) = (\theta^{(M)}, u^{(M)}) \equiv \Theta^{(M)}$. Thus applying $g_M(\cdot)$ to ψ, we obtain the parameter $\theta^{(M)}$ of model M, and supplemental variables $u^{(M)}$. We might consider $u^{(M)}$ as flotsam and jetsam; its sole purpose is to match dimensions of ψ and $g_M(\psi)$; both are of dimension d. This matching is necessary in order that the mappings can be bijections, mapping each value in the domain to a value in the range, with each value in the range reached from a unique value in the domain. In particular, the functions $g_M(\cdot)$ are invertible, meaning that given values of $\theta^{(M)}$ and $u^{(M)}$, we can determine the unique corresponding value of ψ by $g_M^{-1}(\Theta^{(M)}) = \psi$.

---

8. Despite the length of the chain, rather poor mixing limits the accuracy of these summaries to two significant digits.

This all seems rather mysterious until one looks at a specific example. The BUGS code of the previous section (Panel 7.2) is an implementation of RJMCMC. There, the palette was (xf[1], xf[2], xf[3], xS[1], xS[2]) and the bijections were defined by mapping xf[i] to f[i] in models with nonconstant f_i, and mapping xf[1] to all of the f[i] in models with $f_1=f_2=f_3$; xS[i] were handled similarly. Thus model 2 (constant f_i and nonconstant S_i) has parameter vector $\boldsymbol{\theta}^{(2)}=(f.,S_1,S_2)'$ and $\boldsymbol{u}^{(2)}=(u_1,u_2)'$ defined by

$$g_2(\boldsymbol{\psi})=g_2\begin{pmatrix}\text{xf[1]}\\\text{xf[2]}\\\text{xf[3]}\\\text{xS[1]}\\\text{xS[2]}\end{pmatrix}=\boldsymbol{\Theta}^{(2)}=\begin{pmatrix}\boldsymbol{\theta}^{(2)}\\\boldsymbol{u}^{(2)}\end{pmatrix}=\begin{pmatrix}f.\\S_1\\S_2\\u_1\\u_2\end{pmatrix}=\begin{pmatrix}\text{xf[1]}\\\text{xS[1]}\\\text{xS[2]}\\\text{xf[2]}\\\text{xf[3]}\end{pmatrix}.$$

Similarly, model 4 (constant f_i and constant S_i) has

$$g_4(\boldsymbol{\psi})=g_4\begin{pmatrix}\text{xf[1]}\\\text{xf[2]}\\\text{xf[3]}\\\text{xS[1]}\\\text{xS[2]}\end{pmatrix}=\boldsymbol{\Theta}^{(4)}=\begin{pmatrix}\boldsymbol{\theta}^{(4)}\\\boldsymbol{u}^{(4)}\end{pmatrix}=\begin{pmatrix}f.\\S.\\u_1\\u_2\\u_3\end{pmatrix}=\begin{pmatrix}\text{xf[1]}\\\text{xS[1]}\\\text{xf[2]}\\\text{xf[3]}\\\text{xS[2]}\end{pmatrix}.$$

It is useful to think of f_i, S_i, $f.$, $S.$, and u_i as derived parameters: the BMI model is described in terms of a global parameter vector, the palette $\boldsymbol{\psi}$.

Why go to all of this trouble? There are two reasons. First, the palette can be seen as a device for reducing the dimension of the problem. At worst, we could have started out by specifying completely distinct parameters under the four models, resulting in a $5+3+4+2=14$ dimensional prior. Regardless of which model was currently being sampled, there would be at least 9 $(=14-5)$ parameters being sampled from their priors. The problems of poor mixing indicated in Fig. 7.6 would be very much in evidence. Given the nature of the models under consideration we might not have started with 14 dimensions; a more natural choice would have included priors for three f_i, two S_i, $f.$, and $S.$, for a total of seven dimensions. The palette reduced the dimension to the lowest possible, namely 5. The second reason for using the palette is that exploiting similarities among parameters *of itself* improves the mixing of the Markov chain simulation. Given that the Markov chain is currently in a constant recovery model, the chance of moving to a nonconstant recovery model is enhanced by associating $f.$ and f_1 through xf[1].

The astute reader will no doubt ask why we did not associate $f.$ with the mean $\bar{f}=(f_1+f_2+f_3)/3$. Reasoning that $f.$ in Models 2 and 4 should be well-approximated by $\bar{f}$ in the others, we might anticipate better mixing will result from changing the line

$$\text{f[i] < -pow(xf[1],constF)*pow(xf[i],1-constF)}$$

in Panel 7.2 to

$$\text{f[i] < -pow((xf[1]+xf[2]+xf[3])/3,constF)*pow(xf[i],1-constF)}.$$

This change could indeed be made in defining bijections if we were coding our own RJMCMC. However, in the BUGS code this change induces a prior on $f.$ in Models 2 and 4 that is different from the uniform $U(0,1)$ prior we intended. Instead, the prior on $f.$ is the distribution that results from averaging three independent $U(0,1)$ random variables (which is closely approximated by a normal distribution with mean $1/2$ and standard deviation of $1/6$, hence not at all uniform).

Therein lies the challenge of RJMCMC. The problem is to specify model-specific priors on the palette, $[\psi|M]$, in accord with the priors we desire for the model-specific parameters $[\theta^{(M)}|M]$. This is accomplished by specifying a joint distribution $f_M(\Theta^{(M)}) = [\theta^{(M)}, u^{(M)}|M]$ and using the multivariate version of the change of variables theorem (Section 2.2.4), calculate

$$[\psi|M] = f_M\left(g_M(\psi)\right)\left|J_{g_M}(\psi)\right|;$$

here, $J_{g_M}(\psi)$ is the Jacobian determinant of the transformation $g_M(\psi)$. The simplest choice for $[\theta^{(M)}, u^{(M)}|M]$ is also the most natural: given that $u^{(M)}$ is irrelevant to model M, we may as well assume prior independence, $[\theta^{(M)}, u^{(M)}|M] = [\theta^{(M)}|M] \times [u^{(M)}|M]$. The choice of prior on $u^{(M)}$ under model M will have no bearing on inference.

Gibbs Sampling

It is possible to become overwhelmed with the welter of details, so let us review the fundamentals. We have model set $\mathcal{M}$ and model weights $[M]$; we have model-specific data distributions $[Y|M, \theta^{(M)}]$. We have a palette of parameters ψ and bijections $g_M(\cdot)$ from which we can obtain the model-specific parameters $\theta^{(M)}$; thus we may write the data distributions as $[Y|M, \theta^{(M)}] = [Y|M, \psi]$. Having specified priors $[\theta^{(M)}|M]$ and $[u^{(M)}|M]$, we obtain model-specific priors on the palette, $[\psi|M]$.

Gibbs sampling will consist of cyclical sampling of full conditional distributions, alternating between $[\psi|M, Y]$ and $[M|\psi, Y]$. These full conditional distributions are proportional to

$$[Y, \psi, M] = [Y|\psi, M][\psi|M][M].$$

For a fixed model, we update ψ by drawing from the full-conditional

$$[\psi|\cdot] \propto [Y|\psi, M][\psi|M].$$

This may be more easily done by drawing a sample for $\theta^{(M)}$ from the full-conditional

$$[\theta^{(M)}|\cdot] \propto [Y|\theta^{(M)}, M][\theta^{(M)}|M],$$

and a sample $u^{(M)}$ from $[u^{(M)}|M]$, then using the inverse-transformation $\psi = g_M^{-1}\left((\theta^{(M)}, u^{(M)})'\right)$ to obtain a sample for ψ. The two approaches are exactly equivalent.

Sampling the full conditional for M is done using the Metropolis–Hastings algorithm. Given the current value is M, a candidate value M^* is selected at random, with probability $J(M^*|M)$. Our sample from the full conditional $[M|\cdot]$ is either M^* or the current value M, on the basis of a Bernoulli trial. The candidate value is selected with probability $\min\{r, 1\}$, where

$$r = \frac{[Y|\psi, M^*][\psi|M^*]}{[Y|\psi, M][\psi|M]} \times \frac{[M^*]J(M|M^*)}{[M]J(M^*|M)}. \tag{7.14}$$

RJMCMC for Comparing Two Binomial Models

It is best to illustrate the process with an example. Suppose that we have independent binomial random variables $Y_i \sim B(N_i, p_i)$, $i = 1, 2$; we write $Y = (Y_1, Y_2)'$. Under Model 1 we place no restrictions on the values p_i; under Model 2 we assume $p_1 = p_2 \equiv \pi$.

The unknown quantities in our multimodel inference are a categorical variable M (for model) and a two-dimensional parameter ψ which determines the values of p_1, p_2, and π under the two models. For the example at hand, the two bijections are the identity mapping $g_1(\psi) = \psi = (p_1, p_2)'$, and

$$g_2(\psi) = \left(\frac{\psi_1 + \psi_2}{2}, \psi_2 \right)' = (\pi, u)'.$$

Note that u has no role in the likelihood when $M = 2$; it is merely there as a placeholder, to match the dimensions of $g_2(\psi)$ and ψ.

For our sample problem, under Model 1 we choose independent $Be(a_i, b_i)$ priors for p_i, $i = 1, 2$. Under Model 2, we choose a $Be(a_\pi, b_\pi)$ prior for π. As mentioned earlier, the choice of prior for u is irrelevant for inference, though the choice may influence the performance of the RJMCMC sampler. We choose a $Be(a_u, b_u)$ prior for u, independent of the prior for π.

Consider the update of ψ for fixed M. For $M = 1$, likelihood and prior combine to produce independent beta full conditional distributions for ψ_i; we sample $\psi_i \sim Be(Y_i + a_i, N_i - Y_i + b_i)$, for $i = 1, 2$.

For $M = 2$, we sample π from $Be(Y_1 + Y_2 + a_\pi, N_1 + N_2 - Y_1 - Y_2 + b_\pi)$ and u from $Be(a_u, b_u)$. Parameter π has been informed by both Y_1 and Y_2 under Model 2; u has been informed by neither, and is sampled from its prior. The sampled $\Theta^{(2)} = (\pi, u)'$ determines a sampled value for $\psi = (\psi_1, \psi_2)'$ drawn from $[\psi | M_2]$, by $\psi = g_2^{-1}(\Theta^{(2)})$. That is, we set $\psi_1 = 2\pi - u$, and $\psi_2 = u$.

Now consider the sampling of M for fixed ψ. Equation (7.14) can be simplified by makings some judicious choices. First, if we specify equal prior model weights, the terms $[M_1]$ and $[M_2]$ cancel from numerator and denominator. Further, if we never consider the current state as a possible candidate, $J(M_1 | M_2) = J(M_2 | M_1) = 1$, so the numerator and denominator in this term cancel as well. The form of $[Y | \psi, M]$ being known from the model specification, it only remains to calculate formulas for $[\psi | M]$.

For $M = 1$, we have $g_1(\psi) = \psi = (p_1, p_2)'$. The Jacobian of transformation is the 2×2 identity matrix, with determinant equal to one, hence from Eq. (7.14) we conclude that

$$[\psi | M_1] = \prod_{i=1}^{2} \frac{\Gamma(a_i + b_i)}{\Gamma(a_i)\,\Gamma(b_i)} \, \psi_i^{a_i - 1} (1 - \psi_i)^{b_i - 1} \, \mathbf{I}(\psi_i \in (0, 1)). \tag{7.15}$$

For $M = 2$, we have $g_2(\psi) = (\pi, u)'$. The Jacobian of transformation is

$$\begin{pmatrix} 1/2 & 1/2 \\ 0 & 1 \end{pmatrix}$$

with determinant equal to one half. From Eq. (7.14) we conclude that

$$[\psi | M_1] = \frac{\Gamma(a_\pi + b_\pi)}{2\,\Gamma(a_\pi)\,\Gamma(b_\pi)} \, \bar{\psi}^{a_\pi - 1} (1 - \bar{\psi})^{b_\pi - 1} \times \mathbf{I}(\bar{\psi} \in (0, 1)) \, \mathbf{I}(\psi_2 \in (0, 1)), \tag{7.16}$$

where $\bar{\psi} = (\psi_1 + \psi_2)/2$. Note that in Eqs. (7.15) and (7.16) we have been careful to retain the normalizing constants, because these do not cancel in the acceptance ratio [Eq. (7.14)].

Let's consider a specific data set. Suppose that $Y_1 = 8, Y_2 = 16, N_1 = 20, N_2 = 30$, and all of the beta hyperparameters are ones, for uniform priors.

First, consider updates of $\boldsymbol{\psi}$. If $M = 1$, we sample $\psi_1 \sim \mathrm{Be}(9,13)$ and $\psi_2 \sim \mathrm{Be}(17,15)$; if $M = 2$, we sample $\pi \sim \mathrm{Be}(25,27)$ and $u \sim \mathrm{Be}(1,1)$, converting these to $\psi_1 = 2\pi - u$ and $\psi_2 = u$.

Given $\boldsymbol{\psi}$, the chain moves from Model 1 to Model 2 with probability min{r,1}, where

$$r = \frac{\bar{\psi}^{24} (1 - \bar{\psi})^{26} \left(\frac{1}{2}\right) \mathbf{I}(\bar{\psi} \in (0,1)) \, \mathbf{I}(\psi_2 \in (0,1))}{\psi_1^8 (1 - \psi_1)^{12} \psi_2^{16} (1 - \psi_2)^{14} \mathbf{I}(\psi_1 \in (0,1)) \, \mathbf{I}(\psi_2 \in (0,1))}. \tag{7.17}$$

The probability of moving from Model 2 to Model 1 is min{$1/r$,1}.

The indicator functions in Eq. (7.17) deserve some attention. It is possible, when in Model 2, for $\bar{\psi}$ and ψ_2 both to fall in the interval $(0,1)$, but for ψ_1 to be negative. In this case, $1/r = 0$, so it is impossible to move from Model 2 to Model 1. The situation can arise, for instance, if $\bar{\psi} = 0.2$ and $\psi_2 = 0.9$; these imply $\psi_1 = -0.5$.

The frequency with which this occurs will depend on the choice of prior for u under Model 2. Regardless of the choice, the Markov chain will eventually provide stable estimates of posterior probabilities. However, judicious choice of prior for u will reduce autocorrelation in the chain for M.

A chain of length 10^7 generated using this Gibbs sampler took approximately 3 minutes to run on a 2.6 GHz dual-core processor and approximated the posterior probability of model 2 as 65.81%. The autocorrelation tailed off rapidly, being essentially zero by lag 5. Treating the values as equivalent to $10^7/5$ independent observations the estimate has standard error of 0.03%. For this problem we can also compute an analytical solution because under model 1, the marginal distribution of the data, $[Y|M]$, is the product of two independent beta-binomial distributions and under model 2 it is a single beta-binomial distribution. The calculated value is 65.80%.

7.3.3 Bayesian Information Criterion

In Sections 7.3.1 and 7.3.2, we have considered MCMC techniques for computing Bayes factors. We shift gears now, considering an approximation to the Bayes factor based on maximum likelihood estimation.

Recall that the Bayes factor is the ratio of marginal model likelihoods $[Y|M]$, calculated as the average value of $[Y|M, \theta_M]$ against the prior $[\theta_M|M]$, viz.,

$$[Y|M] = \int [Y|M, \theta_M][\theta_M|M]d\theta_M.$$

If we could compute this integral for all of the models under consideration, then we could compute all of the Bayes factors we need, and Bayesian multimodel computation would be no problem. However, cases where this integral can be computed analytically are rare. We need to look for ways of computing or approximating $[Y|M]$.

The BIC (Schwarz, 1978) is an approximation to $-2\log([Y|M])$. Let us assume that

1. The data vector Y consists of n observations, conditionally independent given model M and its k-dimensional parameter θ_M.
2. The posterior distribution $[\theta_M|M,Y]$ is closely approximated by a multivariate normal distribution. This assumption is often reasonable for analyses based on relatively large data sets, a result known as the "Bayesian Central Limit Theorem."
3. The prior for θ_M of model M is closely approximated by a multivariate normal distribution known as the "unit information prior." This prior, discussed by Weakliem (1999) and Raftery (1999) has mean equal to the MLE $\hat{\theta}_M$, and variance matrix such that the prior "contains the same amount of information as a single, typical observation" (Raftery, 1999). It is a weakly informative prior, not strongly contradicted by the data.[9]

Then

$$\text{BIC}_M = -2\log([Y|\hat{\theta}_M, M]) + k\log(n) \approx -2\log([Y|M]). \tag{7.18}$$

We will return to discuss information criteria subsequently; here, we present BIC because it serves as the basis of a large-sample approximation to a Bayes factor. Note that we say *a* Bayes factor, rather than *the* Bayes factor, in acknowledgment that Bayes factors reflect the choice of priors on parameters. For the choice of priors on parameters implicitly assumed by the BIC approximation, we may approximate the Bayes factor $\text{BF}_{i,j}$ by

$$\tilde{\text{BF}}_{i,j} = \exp\left(-\frac{1}{2}(\text{BIC}_i - \text{BIC}_j)\right).$$

This approximation is useful in that it allows us to fit one model at a time rather than having to deal with problems arising in applying BMI to the bucket of models, whether using program BUGS as in Section 7.3.1, or RJMCMC as in 7.3.2. It should be remembered, however, that this approach involves an implicit default choice of priors on parameters, and that the resulting multimodel inference might not correspond with that which would result from our particular choice of priors, especially if our choice involves informative priors.

7.4 INDICES OF MODEL ACCEPTABILITY: AIC AND DIC

Given data and a set of models, we are faced with decisions. Which model best describes the data? for that matter, what do we mean by "best"? Will we choose one model and subsequently ignore the rest? or will we somehow combine inferences across models?

Multimodel inference often boils down to the calculation of a simple numerical summary of candidate models in light of the data, an index of model acceptability. Many exist, including the PBF described in Section 7.2.5, and the BIC (Section 7.3.3). We regard the model likelihood as the ideal index of model acceptability. BMI posits the "bucket of models" meta-model of Section 7.1, with prior model weights reflecting knowledge available in the absence of the data,

---

9. Of course, a prior distribution cannot be based on the MLE; you can't choose your priors on the basis of the data. The BIC is an approximation, most nearly justified under the conditions described.

and chosen to favor model having features we deem important. Ratios of model likelihoods are Bayes factors, which convert prior model probabilities into posterior model probabilities, which can then be used either to select or weight models.

We regard the use of the BMI model as the gold-standard of multimodel inference. We have seen however that there are difficulties associated with it, some computational, some operational. Given these difficulties, analysts might be content with alternative, approximate, simpler indices of model acceptability. Many exist, including the posterior mean likelihood used in computing the PBF (Section 7.2.5), and the BIC (Section 7.3.3), which is based on approximations of the model likelihood, and implicitly selected vague priors. These two are loosely associated with Bayes factors, and lead naturally enough to weighting schemes.

We anticipate that readers of this book may have some familiarity with AIC and the DIC which has been proposed as a Bayesian analog of AIC. These two are also used to compute ad hoc model weights, though their connection to BMI is weaker. In this section we describe AIC, noting its tendency to favor highly parameterized models. We also provide a brief introduction to the DIC. Finally, we compare AIC, DIC, and Bayes factors in an analysis of return rates from a study of brown trout (*Salmo trutta*) spawning in a tributary of Lake Brunner, located in the West Coast region, South Island, New Zealand.

7.4.1 Akaike's Information Criterion

AIC has been popularized in the ecological literature by David Anderson and Ken Burnham (Burnham and Anderson, 2002). AIC estimates an index to the Kullback–Leibler (K-L) divergence of a candidate model from the true model. The K-L divergence of a function $g(y)$ from a function $f(y)$ is

$$K(f,g) = \int f(y) \log \left\{ \frac{f(y)}{g(y)} \right\} dy.^{10}$$

Let f and g be distribution functions, f determined by the unknown true model, and g which we consider as an approximation to f. The K-L divergence is the difference between the expected values of the true loglikelihood and the approximating loglikelihood, the expectation being taken over f:

$$K(f,g) = E_f \left\{ \log f(y) - \log g(y) \right\}$$

$$= E_f \left\{ \log f(y) \right\} - E_f \left\{ \log g(y) \right\}. \tag{7.19}$$

The first term in (7.19) is a constant, in comparisons of candidate models; it cancels out in calculating $K(f,g_1) - K(f,g_2)$. Thus for comparing candidate models, the second term in (7.19) is a suitable index for comparing candidate models. We cannot compute $E_f \left\{ \log g(y) \right\}$, not knowing f. However, we do have a sample from f, the data, and can use this to estimate an approximation to $E_f \left\{ \log g(y) \right\}$.

---

10. $K(f,g)$ is sometimes referred to as "K-L distance" but the term is incorrect, because $K(f,g) \neq K(g,f)$. Rather $K(f,g)$ should be considered as a discrepancy of g from f, measured by f.

Suppose that we have a model M described by an unknown k-dimensional parameter θ and distribution functions $g(y|\theta_M)$. Letting $\hat{\theta}_M$ denote the MLE based on data Y, Akaike derived $-2\log g_M(Y|\hat{\theta}_M)+2k$ as an estimator of the minimum value of the index to K-L divergence consistent with model M. Expressed in bracket notation,

$$\text{AIC} = -2\log([Y|\hat{\theta}_M M])+2k.$$

An alternative, preferred for small sample sizes, is

$$\text{AIC}_c = \text{AIC} + \frac{2k(k+1)}{n-k-1}.$$

Small values of AIC or AIC_c are taken as evidence of small K-L divergences, hence good model fit.

These indices, the BIC, the PBF, and Bayes factors are all based on likelihoods $\mathcal{L}(M,\theta_M) = [Y|M,\theta_M]$, which depend on the models M and parameters θ_M. We wish to compare models M; the difficulty is that values θ_M must be specified to make the comparison. Which value should be used? Bayes factors and the PBF use *all* possible values of the parameter: model comparisons are based on the average value of $\mathcal{L}(M,\theta_M)$, against distributions representing our uncertainty about the parameters. The BIC can be interpreted similarly, with the prior having been chosen implicitly.

Alternatively, BIC (and AIC, and AIC_c) can be thought of as based on penalized maximum likelihoods. Instead of averaging $\mathcal{L}(M,\theta_M)$ over values of the parameter, we simply choose the best value of θ_M for each model (the MLE, $\hat{\theta}_M$) and compute maximum values $\mathcal{L}(M,\hat{\theta}_M)$. Recognizing the tendency for maximum likelihoods to increase with model complexity, we compensate by comparing scaled versions of the maximum likelihood, the scaling being heaviest for models with larger numbers of parameters. On the log scale, this translates to creating indices of the form

$$\mathcal{C}(M) = \log\left(\mathcal{L}(M,\hat{\theta}_M)\right) - g(k_M),$$

where $g(k)$ is an increasing function of k, and k_M is the number of parameters in model M. AIC and BIC are of the form $-2\mathcal{C}(M)$, with $g(k)=k$ and $(k/2)\log(n)$, respectively. Multiplying by -2 (as compare the likelihood ratio statistic, Section 7.2.1) has no effect on the use of the indices, except to reverse the ordering: the best models are the ones with smallest AIC or BIC values.

Observe that the complexity penalty for BIC depends on the sample size as well as the number of parameters in the model. This seems sensible, as relating to the tendency for $\mathcal{L}(M,\hat{\theta}_M)$ to increase with model complexity: the phenomenon is most evident with larger sample sizes. The penalty for model complexity based on BIC exceeds that for AIC if $(k/2)\log(n) > k$, i.e., if $n \geq 8$. Thus AIC will tend to favor more highly parameterized models than BIC.

Given prior model weights π_i we can make use of BIC values and Eqs. (7.12) and (7.18) to compute approximate model weights, provided the dimension of the model is known, as

$$[M_i|Y] \approx \frac{\exp\left(-\frac{1}{2}\text{BIC}_i\right)\pi_i}{\sum_j \exp\left(-\frac{1}{2}\text{BIC}_j\right)\pi_j}. \tag{7.20}$$

Setting

$$\pi_i = \frac{\exp\bigl(k_i \ln(n)/2 - k_i\bigr)}{\sum_j \exp\bigl(k_j \ln(n)/2 - k_j\bigr)},$$

(7.21)

and substituting this into the right-hand side of (7.20), one obtains

$$w_i = \frac{\exp\bigl(-\tfrac{1}{2} \mathrm{AIC}_i\bigr)}{\sum_j \exp\bigl(-\tfrac{1}{2} \mathrm{AIC}_j\bigr)}.$$

These AIC weights (Burnham and Anderson, 2002) are widely used. Burnham and Anderson (2004) suggest that (7.20) and (7.21) provide a Bayesian justification for AIC weights. We question this observation on two grounds: first, that it requires the BIC approximation, which is itself not strictly Bayesian; second, and more importantly, it involves what is at best an unconventional set of prior model weights [Eq. (7.21), which Burnham and Anderson (2004) refer to as a "savvy prior"]. Priors describe what is known in the absence of data; they can be specified before data are collected. Equation (7.21) says that our prior knowledge depends on the sample size we will collect.

Nevertheless, Eqs. (7.20) and (7.21) provide a useful basis for evaluating the performance of AIC or similarly defined AIC_c weights.[11] For instance, if model $k_i > k_j$ for all $j \neq i$, $\pi_i \to 1$ as $n \to \infty$; all of the prior mass is on model i. In Fig. 7.7, we display AIC_c-based "savvy" prior model weights for a set of five models having 1, 2, 2, 3, and 4 parameters, based on sample sizes ranging from 10 to 200. These model dimensions correspond to the trout return data to be analyzed in Section 7.4.3; note that with $n = 200$, more than 93% of the prior mass is placed

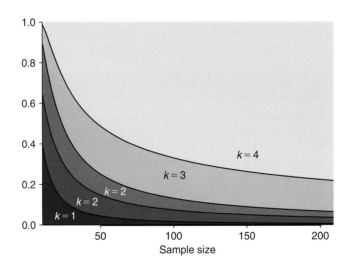

FIGURE 7.7 AIC_c-based "savvy" prior weights on a set of five models.

11. AIC_c results from using $\pi_i \propto \exp[k_i \ln(n)/2 - nk_i/(n - k_i - 1)]$ in Eq. (7.20).

on the two most complex models. The trout return data actually consist of 1961 records, for which the "savvy" prior weights corresponding to AIC_c are 0.0002, 0.0036, 0.0036, 0.0575, and 0.9351; if these were our prior convictions, there would be little reason to even consider the three simplest models.

7.4.2 Deviance Information Criteria

One criticism of the AIC is that it is unclear how it should be calculated for mixed-effects models. In particular, how should one calculate the effective number of parameters k? A solution that has been proposed for a linear mixed-effects model with just one random effect and a large sample size by Burnham and Anderson (2002) corrects the parameter count using linear model theory. Conditional AIC has also been recently proposed for mixed-effects models (Vaida and Blanchard, 2005) and leads to a calculation similar to the proposal of Burnham and Anderson (2002) for the case of a model with one random effect. However, as pointed out by Burnham and Anderson (2002): "A general approach to K-L model selection when the models include random effects remains elusive."

The DIC is an information criterion for Bayesian modeling introduced by Spiegelhalter *et al.* (2002). The development of DIC was motivated by a desire for an information criterion useful in a Bayesian setting where prior information was not negligible (cf. BIC), where the model was not fitted by maximum-likelihood (cf. AIC), and where models are nonnested and data not independent, such as under hierarchical or time-series models. We have used -2 times the loglikelihood in describing AIC and BIC; call this the "deviance," $D(\theta) = -2\log[Y|M,\theta_M]$. The DIC is defined as the posterior mean value of the deviance, plus a measure p_D of model complexity:

$$\text{DIC} = \bar{D}(\theta_M) + p_D,$$

where

$$p_D = \bar{D}(\theta_M) - D(\bar{\theta}_M).$$

Here $\bar{D}(\theta_M)$ is the average value of the deviance, averaged across the posterior distribution of the parameters, and $D(\bar{\theta}_M)$ is the deviance computed at the posterior means of the parameters for the model.[12]

7.4.3 Example: Trout Return Rates

We conclude this section by evaluating the performance of AIC, BIC, and DIC with a fully Bayesian analysis of a sample data set (Link and Barker, 2006). The data are return rates for brown trout (*Salmo trutta*) in a spawning tributary of Lake Brunner, on the West Coast of the South Island, New Zealand. In 1987, 1961 trout were tagged and released. The following year,

---

12. Although $D(\bar{\theta}_M)$ is usually computed using the posterior mean, it is sometimes computed using another measure of central tendency for the posterior distribution such as the posterior mode or median. The quantity p_D is usually greater than zero and is interpreted as the effective number of parameters in the model (Spiegelhalter *et al.*, 2002). Given vague priors, Gelman *et al.* (2004) suggest replacing p_D with $p_V = \text{Var}(D(\theta)|Y)/2$, i.e., one half the posterior variance of the deviance. The latter is much easier to compute, is always nonnegative, is invariant to model reparameterization, and does not require an arbitrary choice among mean, median and other measures of central tendency.

TABLE 7.2　Candidate models for trout return data.

Model 1:	$\text{logit}(\pi_i) = \beta_1$
Model 2:	$\text{logit}(\pi_i) = \beta_1 + \beta_2 S_i$
Model 3:	$\text{logit}(\pi_i) = \beta_1 + \beta_3 L_i$
Model 4:	$\text{logit}(\pi_i) = \beta_1 + \beta_2 S_i + \beta_3 L_i$
Model 5:	$\text{logit}(\pi_i) = \beta_1 + \beta_2 S_i + \beta_3 L_i + \beta_4 S_i L_i$

94 of the fish were recaptured. Investigators were interested in whether the return probability was sex- or length-dependent.

A natural model for the analysis is that the return indicator Y_i is a Bernoulli random variable with probability π_i. Letting S_i be an indicator of the event that the ith fish is female, and L_i be its length, we consider five candidate models (Table 7.2) for the return rate data: We standardize the values S_i, L_i, and $S_i L_i$, subtracting their means and dividing by their standard deviations.

Recall that the logit link function maps the interval $(0, 1)$ to the entire real line via the transformation $\text{logit}(\pi) = \log(\pi/(1 - \pi))$. We thus chose mean-zero normal priors for the parameters β_k. The precision of these priors we governed through specification of a global parameter T. Letting K_j denote the number of β_k's in Model j, we set

$$\text{Precision}(\beta_k | \text{Model } j) = \begin{cases} K_j T, & \beta_k \text{ in Model } j \\ T, & \text{otherwise} \end{cases} . \qquad (7.22)$$

This choice requires some explanation!

First, we note the sensitivity of Bayes factors to differing levels of variability in marginal distributions (cf. Section 7.2.5). To assign equal prior variances to all parameters, independent of model, would be to prejudice our analysis in favor of less complex models. Some writers take this as a virtue of BMI, an "automatic" penalty against model complexity. We are skeptical of automatic procedures, the effects of which may not be readily quantified or consistent across applications. Penalties for model complexity should be chosen by the analyst and explicitly incorporated in the choice of prior model weights. Bayes factors, which convert the prior model weights to posterior model weights, should be as free as possible from prejudices.

Next, note that Eq. (7.22) implies that conditional on Model j, the variances of β_k in Model j are identical, and sum to $1/T$, and that this total level of prior variability is constant among models. Using standardized regressors and with prior mean zero for all β_k, we may think of the prior distribution of a typical value of $\text{logit}(\pi_i)$ as approximately normal with mean zero and precision T, independent of the model. For an objective analysis, it seems reasonable to choose $T = 0.40$, or (perhaps better) $T \sim Ga(3.29, 7.80)$, as these produce prior distributions for $\text{logit}(\pi_i)$ such that π_i is approximately uniform on $(0, 1)$.

Finally, note that Eq. (7.22) defines prior distributions for β_k even when β_k is not in Model j. This specification has no influence on inference, but is needed for MCMC analyses, and can have an operational effect. For Gibbs sampling, if β_k is not in the present model, it must be sampled from its prior; our choice here was intended to encourage reasonably high transition probabilities between models.

We fit this model by maximum likelihood to calculate AIC and BIC values. Next, we fit the models, one at a time, using MCMC to compute DIC statistics[13] and posterior mean values of the likelihood used in computing Aitkin's PBF; chains were of length 100,000 after trimming 10,000.[14] Finally we conducted BMI using BUGS reproduced in Panel 7.3; once again, chains were of length 100,000 after trimming 10,000.

Several features of the BUGS code in Panel 7.3 require comment. First, note that *Ind.Arg* is an indicator variable based on the categorical variable `Model`. Thus for instance `Ind.Sex = 1` if and only if `Model = 2, 4, or 5`. The specification of logit(π) thus includes or excludes (zeros out) parameter values as appropriate for the present model.

PANEL 7.3   BUGS code for BMI of Trout return data.

```
model{
  for (i in 1:1961){
    Returned[i] ~ dbern(pi[i])
    logit(pi[i]) <- constant + Ind.Sex*sex*S[i]
      +Ind.Length*length*L[i]
      +Ind.SexbyLength*sex.length*SL[i]
  }
  ####################################
  T ~ dgamma(3.29,7.80)
  tau.constant <- T*num.par[1,Model]
  tau.sex <- T*num.par[2,Model]
  tau.length <- T*num.par[3,Model]
  tau.sex.length <- T*num.par[4,Model]
  constant ~ dnorm(0,tau.constant)
  sex ~ dnorm(0,tau.sex)
  length ~ dnorm(0,tau.length)
  sex.length ~ dnorm(0,tau.sex.length)
  ####################################
  Model ~ dcat(p.model[1:5])
  for (i in 1:5){
    Ind.Model[i] <- equals(Model,i)
  }
  Ind.Sex <- equals(Model,2)+equals(Model,4)+equals(Model,5)
  Ind.Length <- equals(Model,3)+equals(Model,4)+equals(Model,5)
  Ind.SexbyLength <- equals(Model,5)
}
```

13. We computed DIC using p_D and Gelman *et al.* p_V; there was very little difference in the results, though (as previously noted) p_V is much more easily computed.

14. The likelihood is proportional to $\prod_{i=1}^{1961} \pi_i^{y_i}(1-\pi_i)^{1-y_i}$, the product of 1961 numbers $0 < \pi_i < 1$, hence a very small number. To avoid round off and domain errors, we multiply by $\left(\frac{1961}{94}\right) = \exp(374.11)$; we chose this multiplier because there were 94 returns among the 1961 fish. Thus the likelihood was calculated as $\exp\left(374.11 + \sum y_i \log(\pi_i) + (1-y_i)\log(1-\pi_i)\right)$.

A 4×5 matrix of values num.par is referenced in the BUGS code of Panel 7.3. This matrix, entered in the BUGS data statement, is

$$\text{num.par} = \begin{bmatrix} 1 & 2 & 2 & 3 & 4 \\ 1 & 2 & 1 & 3 & 4 \\ 1 & 1 & 2 & 3 & 4 \\ 1 & 1 & 1 & 1 & 4 \end{bmatrix}.$$

Its rows correspond to the four β_i, its columns to the five models. Thus num.par[1, 3] = num.par[3, 3] = 2, because the constant term (parameter β_1, row 1) and the length effect (β_3, row 3) are the two parameters of Model 3. This matrix encodes the prior specification given at Eq. (7.22), which is conditional on Model.

Results of our analyses are summarized in Table 7.3. PBF, AIC, and DIC have implicitly chosen prior model weights. The PBF, BIC, and BF weights are based on uniform prior model weights of 20% on each of the five models.

Note that Bayes factors are estimated from MCMC output as

$$\widehat{\text{BF}}_{i,j} = \left(\frac{\text{MCMC Frequency } Model=i}{\text{MCMC Frequency } Model=j} \right) \Big/ \left(\frac{\text{Prior Probability } Model=i}{\text{Prior Probability } Model=j} \right).$$

As noted previously, the precision of this estimate is greatest when the prior model probabilities have been specified so as to nearly equalize the sampled frequencies for the various models. One can accomplish this, at least approximately, by several preliminary MCMC runs, decreasing prior probabilities on models visited too frequently, and increasing prior probabilities on models visited too rarely. We used prior probabilities {0.0015, 0.0653, 0.0267, 0.4226, 0.4839}, and observed *Model* frequencies {20,088, 19,903, 21,285, 21,739, 16,985} in a Markov chain of length 100,000. Thus

$$\widehat{\text{BF}}_{1,3} = \left(\frac{20,088}{21,285} \right) \Big/ \left(\frac{0.0015}{0.0267} \right) = 16.80.$$

This is the same as the ratio of BF weights (0.9185/0.0547) from Table 7.3, because those weights were based on uniform prior model weights.

TABLE 7.3   Model weights for Lake Brunner trout return-rate data, calculated based on Aitkin's PBF, AIC, DIC, and BIC weights, and Bayes factors obtained by multimodel Markov chain simulation (BF).

Model	PBF	AIC	DIC	BIC	BF
			Model weights		
1. Constant	0.07	0.25	0.25	0.9197	0.9185
2. Sex	0.05	0.10	0.09	0.0217	0.0209
3. Length	0.14	0.26	0.26	0.0571	0.0547
4. Sex + length	0.10	0.09	0.09	0.0013	0.0035
5. Sex × length	0.62	0.30	0.31	0.0002	0.0024

TABLE 7.4   Model (j), $BF_{1,j}$ = Bayes factor for comparing models 1 and j, four prior model weights (KL corresponding to AIC, Comp = "complexity" weights $\propto \exp(k)$ favoring complex models, Const = uniform prior model weights, Ock = "Ockham" with weights $\propto \exp(-k)$ favoring parsimony), and resulting posteriors (priors weighted by $BF_{1,j}$).

j	$BF_{1,j}$	Prior weights				Posterior weights			
		KL	Comp	Const	Ock	KL	Comp	Const	Ock
1	1.00	0.000	0.030	0.200	0.521	0.068	0.767	0.918	0.970
2	43.94	0.004	0.080	0.200	0.191	0.025	0.047	0.021	0.008
3	16.80	0.004	0.080	0.200	0.191	0.066	0.124	0.055	0.021
4	260.34	0.057	0.218	0.200	0.071	0.069	0.022	0.004	0.001
5	381.54	0.935	0.592	0.200	0.026	0.772	0.040	0.002	0.000

Several features of Table 7.3 are noteworthy. First, that AIC and DIC weights nearly coincide, as do BIC and BF weights. Thus if maximum likelihood estimation is difficult, DIC provides an MCMC-based alternative that produces similar results. On the other hand, if maximum likelihood estimation is easily implemented, and the priors on parameters are vague, use of the BIC provides a simple alternative to the computational challenges associated with Bayes factors, and similar results. The difference between AIC and DIC, on the one hand, and BIC and BF on the other, can be explained in terms of the implicit K-L priors [Eq. (7.21)] described in Section 7.4.1: AIC favors more highly parameterized models, and more emphatically so as sample size increases. The PBF favors more highly parameterized models, but for a different reason, namely its use of a highly informative "prior" on parameters.

Our view is that if model complexity is to be taken into account, it should be done explicitly in the selection of prior model weights. In Link and Barker (2006), we suggested prior model weights proportional to $\exp(k)$ as favoring complex models and $\exp(-k)$ as favoring parsimony. We compare these and the K-L weights $\propto \exp(k\log(n)/2 - k)$ in Table 7.4, using Bayes factors to compute posterior model weights. The Bayes factors are such that, even with the Complexity weight prior, Model 1 has the greatest posterior probability. It is only with the extremely prejudicial K-L prior that the most highly parameterized model is favored.

7.5 AFTERWORD

MMI is Difficult

When we first set out to write this book, we were inclined to duck the issue of multimodel inference entirely, thinking it to be an intellectual morass, a tar-baby we were not inclined to embrace. We are not the only ones to feel some reluctance on the topic. Consulting the first edition of the text by Gelman *et al.* (1995) on Bayesian methods, one finds in the index "model selection: why we do not do it." But that position is not sustainable. Consulting the second edition (Gelman *et al.*, 2004), one finds "model selection, why we avoid it".

Multimodel inference is difficult, no matter whether one is Bayesian or frequentist. And why shouldn't it be? Choosing among models is central to science and it would be naive to think that the process could be automated, that all we need to do is collect data, gobs of it, and let our information criteria sort through it.

Use and Abuse

Model selection techniques and model averaging are well-suited to cases where there are a small number of distinct and carefully thought out model choices available. We regard model selection as appropriate for identification of parsimonious models when there are large numbers of possible nuisance parameters (e.g., in mark-recapture modeling) and less appropriate for making inference about large numbers of alternative biological processes.

Model selection and model averaging techniques are not well-suited to cases where there may be hundreds or thousands of models to choose from. Model selection from a very large class of potential models has an obvious flaw, the heavy dilution of data. The more parameters there are to estimate, the more data that is needed, period.

An important issue in multimodel inference is the need to account correctly for model uncertainty. Where a model has been selected as part of data analysis, estimation and predictions should be less precise, since part of the evidence is "spent" specifying the model (Leamer, 1978 and Chatfield, 1995). If one selects a single model, it is important to at least note when inference is conditional on that model; it is better to use model averaging to account for this uncertainty, and mathematically cleanest to do so in the BMI framework.

Repeat Analyses

It is important to distinguish the evaluation of hypotheses from the generation of hypotheses. Given a treasured data set, it is tempting to milk it for all it's worth. Although there is nothing virtuous in *not* having a good look for patterns in our data, we really must distinguish evaluation of patterns posited a priori from patterns that seem to emerge after the fact. To do otherwise is to risk reporting ephemera as facts.

Good science requires the replication of studies in time and space. When we see data that are in accord with predictions we gain confidence in our models. Newtonian gravitational theory is widely accepted not because of its beautiful calculus, but because its predictions are consistent with everyday observation; high school physics students around the world daily corroborate its usefulness.[15] Unfortunately, many ecological studies involving model selection are never repeated to see whether predictions are borne out. If a study is not worth repeating it might not be worth doing.

Model Averaging

A good historical treatment of model averaging, and Bayesian model averaging in particular, is given in Hoeting *et al.* (1999). BMI provides a formal justification for the sort of model averaging attempted by Buckland *et al.* (1997) [Eq. (7.1)] without the need to resort to adhockeries in evaluating the precision of such estimates. Where there is a parameter (or, better,

---

15. On the other hand, cosmologists routinely account for inadequacies in Newton's theory.

a prediction) Ψ common to all models, one may simply compute its marginal distribution across models, as

$$[\Psi|Y] = \sum_{k=1}^{K}[\Psi|M_k, Y][M_k|Y],$$

where $[\Psi|M_k, Y]$ is the posterior distribution of Ψ under model k, and $[M_k|Y]$ is the posterior model weight for model M_k. Thus, inferences are averaged across the model set so as to ensure that each model contributes in proportion to its relative support by the data. As a case in point, we refer to the calculations of posterior means under the Poisson/geometric analysis in Section 7.3.1.

When considering model averaging, it is important to decide whether averaging actually makes sense for the quantities of interest. For example, if a model set includes an unbounded exponential growth model, and a Gompertz model with a finite carrying capacity, and these have posterior probabilities of 0.001 and 0.999, it makes no sense to estimate carrying capacity as

$$0.001 \times \infty + 0.999 \times \text{(whatever)}.$$

While nobody would do that (we hope), we *have* come across model averaged estimates of regression coefficients, which can have entirely different meanings under distinct models. Scale also matters: the average of $\phi_1 = 0.65$ and $\phi_2 = 0.95$ is 0.80 but the average of logit(0.65) and logit(0.95) is 1.78 which is logit(0.86).

Complexity and AIC's Prejudice

How should the varying complexity of our models influence the choice of prior model weights? This is a philosophical issue, rather than a mathematical one. Ockham's razor is usually paraphrased as "we should use the simplest model that fits the facts." But does this mean that we should favor simple models a priori? No: we cannot forget the requirement that the model fit the facts. Similarly, we should not favor complexity a priori, because the addition of irrelevant parameters to a model automatically increases complexity. Fisher (1925a) argued that we should adopt an initial position that favors simple models and only increase the complexity of a model when there is strong support from the data for increased complexity. Whatever the position one adopts on the virtue of simplicity, we believe that this position should be represented by explicit model priors. Thus, we would argue against the automatic use of AIC weights on the basis of their implicit use of the K-L prior [Eq. (7.21)]. Although the K-L prior might serve a useful purpose in guiding choice of prior model weights, the scientist ought at least to work out the values in advance, and see whether they appear reasonable. Their dependence on n and the tendency to highly favor model complexity are, to us, disconcerting features. The weighted BIC [Eq. (7.20)] is no more difficult to calculate than AIC or DIC weights, and has the benefit of forcing the scientist to specify prior model weights, rather than to accept default (and perhaps unreasonable) weights associated with AIC and DIC.

PART III

APPLICATIONS

Hidden Data Models

O U T L I N E

8.1	Complete Data Likelihood	164
8.2	Randomized Response Data	165
	8.2.1 Calculating Posterior Distributions	168
	8.2.2 Remarks	174
8.3	Occupancy Models as Hierarchical Logistic Regressions	176
	8.3.1 American Toads in Maryland	178
	8.3.2 More Complex Occupancy Models	180
8.4	Distance Sampling	181
8.5	Finite Population Sampling	186
	8.5.1 Muskrats: A Simple Sample Survey	188
	8.5.2 Stratification	192
	8.5.3 Cluster Sampling	194
	8.5.4 Auxiliary Variables	197
8.6	Afterword	199

One of the most appealing features of Bayesian analysis is the ease with which hierarchical models are handled.

Hierarchical models are richly structured, with multiple levels of stochasticity. Stochastic modeling extends beyond description of data as sampled from fixed distributions, to descriptions of the parameters governing those distributions. Thus, conventional modeling might use data to estimate a set of time-indexed parameters $\theta_1, \theta_2, \ldots, \theta_T$; hierarchical modeling allows the investigation of stochastic structure among the θ_t. We need not impose a deterministic relation among the parameters, such as $\theta_t = a + bt$. We need not estimate θ_t as though they were unrelated quantities, and then seek to uncover pattern among the estimates $\hat{\theta}_t$, struggling to distinguish between variation among the parameters θ_t and scientifically irrelevant sampling variation of their estimates $\hat{\theta}_t$. Rather, hierarchical modeling allows us to posit stochastic structure among parameters, for example, $\theta_t = a + bt + \epsilon_t$, where ϵ_t are iid $N(0, \sigma^2)$ random variables, and to investigate that structure directly from observed data.

Hierarchical models arise in a variety of contexts and have many uses. They can be used as a device for introducing parsimony when there are many nuisance parameters in a regression model; they can be used in studies with multiple levels of variation (e.g., a split-plot experimental design); they can be used when a response variable is not fully observed; they can be used to describe latent variables.

In this chapter, we investigate "hidden data models" – hierarchical models which begin with descriptions of an ideal data set, which (borrowing a phrase from Draper, 1995) consists of "the data we wish we had." We then go on to describe the processes by which those data are corrupted into the data we actually observe. Our focus is on computational and inferential benefits associated with identifying this structure. We begin by comparing complete data likelihoods (CDL's) with observed data likelihoods (ODL's) typically used in frequentist analyses. CDL's are based on the "the data we wish we had"; they are much more easily specified than ODL's but result in the same inference as the conventional ODL's. "Data augmentation" is the process of specifying a CDL consistent with a particular ODL, a process that can have substantial computational benefits. These matters are introduced in Section 8.1.

Next, we consider a randomized response data set, investigating the misdeeds of a sample of our ecological colleagues (Section 8.2). These data have several features common to ecological data and provide a simple illustration of the use of CDL's rather than ODL's.

In Sections 8.3–8.5, we describe occupancy models, distance sampling models, and finite population sample survey models. These models, typically viewed as completely distinct, will be seen to share remarkably similar structure among themselves, and with the randomized response data. Each involves corrupted or hidden data; in each case, modeling begins with a description of 'the data we wish we had'; in each case, Bayesian analysis based on a CDL is a simple matter.

8.1 COMPLETE DATA LIKELIHOOD

Generally speaking, statistical inference begins with specification of a data distribution $[Y|\theta]$ describing the probability of data Y in terms of unknown parameters θ. Viewed as a data distribution, $[Y|\theta]$ is a function of variable Y and θ is a fixed quantity. Given data, we may wish to treat $[Y|\theta]$ as a function of varying θ with Y fixed; we compare values of $[Y|\theta]$ for different θ to measure the relative support provided by the data to each θ. In this context, we write

$$\mathcal{L}(\theta|Y) = k\,[Y|\theta],$$

with k an arbitrary constant, and refer to $\mathcal{L}(\theta|Y)$ as a likelihood function (see Section 3.1 for more details).

It is often the case that the data we get to see are a subset of the data we wish we had. That is, we can partition Y as $Y = \{Y^{\text{obs}}, Y^{\text{mis}}\}$, where superscripts obs and mis denote 'observed' and 'missing,' respectively. Many classical statistical methods begin with specifying models for the complete data Y, computing the implied data distribution for the observed portion of the data, and using this as the basis of an ODL. The process involves integrating or summing over possible values of the unobserved data to obtain an observed data distribution

$$[Y^{\text{obs}}|\boldsymbol{\theta}] = \int [Y^{\text{obs}}, Y^{\text{mis}}|\boldsymbol{\theta}] \, dY^{\text{mis}}. \tag{8.1}$$

The resulting likelihood $\mathcal{L}(\boldsymbol{\theta}|Y^{\text{obs}}) = k\,[Y^{\text{obs}}|\boldsymbol{\theta}]$ is the ODL.

An alternative approach is to treat the unobserved Y^{mis} in $[Y^{\text{obs}}, Y^{\text{mis}}|\boldsymbol{\theta}]$ exactly as we treat parameters $\boldsymbol{\theta}$. That is, we can treat Y^{obs} as fixed, and compare the support for various values of Y^{mis} and $\boldsymbol{\theta}$. Thus, we define a CDL as

$$\mathcal{L}(\boldsymbol{\theta}, Y^{\text{mis}}|Y^{\text{obs}}) \propto [Y^{\text{obs}}, Y^{\text{mis}}|\boldsymbol{\theta}].$$

There are a number of benefits to using CDL's rather than ODL's. First, the ODL can be a much more complicated function of parameters than the CDL. Second, use of the CDL provides a natural framework for prediction of unobserved quantities; describing missing data in terms of CDL's also tends to clarify assumptions required for predictions. CDL's are sometimes used in frequentist analyses (e.g., the EM algorithm, Dempster *et al.*, 1977). However, they are much more naturally and easily handled under the Bayesian paradigm, in which all unobserved quantities – be they parameters, predictions, missing values, whatever – all are treated equally (Section 5.1).

Before presenting examples, we take a moment to mention *data augmentation*, a concept closely related to that of CDL's, and one which users of Bayesian methods are likely to encounter. Data augmentation consists of expanding a model for observed data in terms of unobserved structures, usually with the goal of creating computational efficiencies. The augmented model includes a data distribution for observed data and the unobserved structures; integrating over values of the augmenting variables produces the ODL, as in Eq. (8.1). An example of data augmentation is found in Section 8.4.

8.2 RANDOMIZED RESPONSE DATA

One of our favorite demonstrations of the difference between ODLs and CDLs involves data from a randomized response survey. Randomized response surveys are used to estimate population rates of stigmatized behaviors, without requiring specific information about individuals surveyed. Such surveys were first described by Warner (1965), who begins thus:

> "For reasons of modesty, fear of being thought bigoted, or merely a reluctance to confide secrets to strangers, many individuals attempt to evade certain questions put to them by interviewers … either refusing outright to be surveyed, or consenting to be surveyed but purposely providing wrong answers … The questions that people tend to evade are the questions which demand answers that are too revealing."

Warner (1965) provided a clever solution. His technique has been applied in studies of alcohol use and abuse, smoking, abortion, academic cheating, criminal recidivism, sexual habits, and regulatory compliance (ranging from payment of taxes to obedience to fishing regulations). Brewer (1981) reports on its use by the Australian Bureau of Statistics, acting on behalf of the South Australian Royal Commission, in an investigation of marijuana use in Canberra.

Teaching a workshop for some of our colleagues — details of time and place suppressed, to protect the innocent — we wondered whether any of them might have ever engaged in Behavior X.[1] So we gathered some data.

We could not directly ask the question of interest because our colleagues may have been reluctant to own up to Behavior X. Instead, participants were asked to first toss a coin, without informing us of the outcome. They were asked to respond "Yes" if they had X'ed or if their coin came up heads. Those whose coin came up tails and who had not X'ed were asked to respond "No."

In this survey design, the data are deliberately confounded with the outcome of the coin toss in order to ensure that it is impossible to tell whether a person answering "Yes," responded positively because they had X'ed or whether it was because they obtained "Heads" on the coin flip. Despite this confounding, we can still make inference about the rate of Behavior X in the population and in our study group, modeling the confounding process using a binomial distribution.

It is easy to see how the probability of Behavior X can be estimated by considering a 2×2 table for the possible outcomes. Letting p denote the probability of Behavior X, and π denote the probability associated with the randomizing mechanism, and assuming the two events are independent, we calculate probabilities of four outcomes in Table 8.1.[2] In our randomized response data, three of the four outcomes (Heads and Behavior X, Tails and Behavior X, and Heads and no Behavior X) are all subsumed into the one response "Yes." The probability that a person answers "Yes" is $\theta = \pi + (1-\pi)p$; the probability of "No" is $(1-\pi)(1-p)$.

We can solve θ for p, obtaining $p = (\theta - \pi)/(1-\pi)$. Letting x denote the number responding "Yes" in a sample of n individuals, the maximum likelihood estimator (MLE) of θ is $\hat{\theta} = x/n$. Thus, given that $\hat{\theta} \geq \pi$, it follows from the invariance property of maximum likelihood

TABLE 8.1   Cell and margin probabilities for the 2×2 table of possible outcomes in the randomized-response to the Behavior X question.

	Behavior X		
	Yes	No	
Heads	πp	$\pi(1-p)$	π
Tails	$(1-\pi)p$	$(1-\pi)(1-p)$	$1-\pi$
	p	$(1-p)$	

Parameter π is the probability that the coin comes up Heads and p is the probability of Behavior X. Shaded cells correspond to response "Yes" in randomized-response survey, and have total probability $\theta = \pi + (1-\pi)p$.

1. We can't say what it was. Readers are asked to use their imagination.
2. We shall assume the coins flipped to be fair, hence $\pi = 1/2$. Randomization schemes could use alternative mechanisms, leading to alternative values of π; the important feature is that π is known.

estimation that the MLE of p is

$$\hat{p} = \frac{\hat{\theta} - \pi}{1 - \pi} = \frac{x - n\pi}{n(1 - \pi)}.$$

If $\hat{\theta} < \pi$, that is, if there are fewer positive responses than attributable to the randomizing event, the MLE of p is 0.[3]

From a modeling point of view it is instructive to think of how we would simulate the data. First, we would need to assign each subject a Behavior X status, using an indicator variable D_i for individual i. The next step would be to simulate the outcome of the individual's coin flip, with indicator variable C_i for the outcome "Heads." Because of the randomized response both D_i and C_i are latent variables, meaning they are potentially unobservable. What we always observe, however, is a variable Y_i which equals 0 if person i obtained tails and had never X'ed. Otherwise, Y_i is 1. We can write Y_i as a function of D_i and C_i, namely

$$Y_i = 1 - (1 - D_i)(1 - C_i). \tag{8.2}$$

If our n individuals can be regarded as a random sample from a large population, a reasonable way to simulate a value for D_i is to generate a Bernoulli random variable with success parameter p. Having simulated values for D_i, we next draw values C_i from a Bernoulli distribution with success parameter π. The observed variables Y_i are then constructed from D_i and C_i as in Eq. (8.2).

We represent this process using the *directed acyclic graph* (DAG) in Fig. 8.1. DAG's are an intuitive means for depicting relationships among quantities in hierarchical models. Single arrows denote stochastic dependencies (e.g., Bernoulli trials D have probability distributions depending on success parameters p) while double arrows denote deterministic relationships (e.g., Y is exactly calculated from D and C). The independence of D and C is indicated by the absence of shared stochastic dependencies. The DAG is *directed*: we write $p \to D$ because our model will describe the relation between p and D in terms of $[D|p]$ rather than the other way around. The DAG is *acyclic* in that we avoid model specifications of the form $X \to Y \to Z \to X$ which cycle back on themselves, and might wind up describing models that do not make sense.[4]

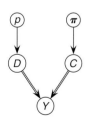

FIGURE 8.1 Directed acyclic graph (DAG) representation of the complete data likelihood for the Behavior X example.

---

3. The MLE is biased because of the truncation. Warner (1965) describes the estimator as unbiased, but the result requires the acceptance of negative estimates.
4. For example, $[X|Y] = N(Y, 1)$ and $[Y|X] = N(X, 1)$ might seem a reasonable model specification, but there is no joint probability distribution on $\{X, Y\}$, which yields these conditional distributions.

In this particular problem, the variables C and D are partially observed through Y. For cases where $Y_i = 0$, we know that $C_i = D_i = 0$. If we use the superscript obs to denote observed values and mis to denote missing values, we can write the CDL as:

$$\mathcal{L}(p, \mathbf{C}^{\text{mis}}, \mathbf{D}^{\text{mis}} | \mathbf{Y}) \propto [\mathbf{Y}, \mathbf{C}, \mathbf{D} | p]$$

$$= \prod_{i=1}^{n} I(Y_i, C_i, D_i) \times [\mathbf{C}][\mathbf{D}|p]$$

$$= \prod_{i=1}^{n} I(Y_i, C_i, D_i) \times \pi^{C_i}(1-\pi)^{1-C_i} \times p^{D_i}(1-p)^{1-D_i}.$$

Here, $I(Y_i, C_i, D_i)$ is an indicator for the defining relationship given in Eq. (8.2). Thus, $I(Y_i, C_i, D_i) = 1$ if either (1) $Y_i = 1$ and $\{C_i, D_i\} \neq \{0,0\}$ or (2) $Y_i = 0$, $C_i = 0$, and $D_i = 0$. Otherwise $I(Y_i, C_i, D_i) = 0$. The role of this indicator function is to enforce constraints that the data impose on allowable values for C_i and D_i, and in particular that we cannot have $Y_i = 1$, $C_i = 0$, and $D_i = 0$. This constraint defines the role of the observed data ($\mathbf{Y}$) in providing information about p in the CDL.

As a problem in Bayesian inference, we require the posterior distributions of all unknown quantities. This is proportional to the CDL multiplied by the joint prior on unknown parameters. If we also use a $\text{Be}(\alpha, \beta)$ prior for p, then

$$\left[p, \mathbf{C}^{\text{mis}}, \mathbf{D}^{\text{mis}} | \mathbf{Y}\right] \propto \left\{\prod_{i=1}^{n} \left\{I(Y_i, C_i, D_i)\right\} \left\{\pi^{C_i}(1-\pi)^{1-C_i}\right\} \left\{p^{D_i}(1-p)^{1-D_i}\right\}\right\} \times p^{\alpha-1}(1-p)^{\beta-1}.$$
(8.3)

8.2.1 Calculating Posterior Distributions

For inference about p, we can proceed in a number of ways. One approach is to try to find an explicit expression for the density $[p|\mathbf{Y}]$. Formally, we integrate (in this case sum) over the possible values for the latent variables C_i and D_i in Eq. (8.3). We spare the reader the ugly details, noting only that the result is the same as obtained by first computing the ODL based on Table 8.1, then multiplying by the prior for p.

We will follow that easier course. The values Y_i are exchangeable (conditionally independent) Bernoulli trials. Examining Table 8.1, we see that $\theta \equiv \text{Pr}(Y_i = 1) = \pi + (1-\pi)p$; thus, $X = \sum_{i=1}^{n} Y_i$ is a binomial random variable with index n and success rate θ. The ODL is thus $\mathcal{L}(p|\mathbf{Y}) \propto \theta^X (1-\theta)^{(n-X)}$. Multiplying by the prior for p, we have

$$[p|\mathbf{Y}] \propto (\pi + (1-\pi)p)^X ((1-\pi)(1-p))^{n-X} \times p^{\alpha-1}(1-p)^{\beta-1}.$$
(8.4)

The posterior density for p does not exist in closed-form for arbitrary α and β. While it is possible to find the required normalizing constant using software like R, an easier solution is desirable. Nevertheless, we have done so for two priors, graphing the densities in Fig. 8.2.

Analysis of the Behavior X data is made much easier through Markov chain Monte Carlo (MCMC). We present three different approaches, based on the CDL, the ODL, and a partially integrated version of the CDL.

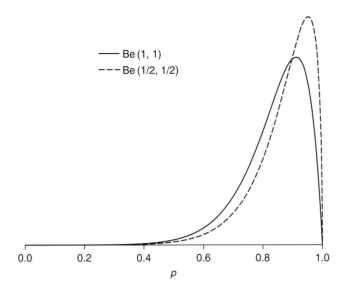

FIGURE 8.2 Posterior density for p in the Behavior X example for $X = 22$ and $n = 23$, with a Be(1,1) (solid line) and Be(1/2,1/2) prior on p.

MCMC Based on CDL Using BUGS

For posterior summary, it is easy to fit the model in BUGS, using the code in Panel 8.1. An interesting feature of the code is that we have modeled Y_i as a Bernoulli random variable with parameter `alpha[i]=1-(1-D[i])(1-C[i])`. Given that Y_i is a deterministic function of C_i and D_i (as indicated in Fig. 8.1) this coding is perhaps counterintuitive. Given also that `alpha[i]` is always either 0 or 1, so that $Y_i \equiv$ `alpha[i]`, the coding might seem unnecessarily complicated. However, specification of Y_i as stochastic in the BUGS code ensures that the likelihood is a function of the observed variable Y_i through the indicator function $I(Y_i, C_i, D_i)$.[5]

We generated a Markov chain posterior sample of size one million, using program BUGS and the code in Panel 8.1. Discarding the first 10,000 values as a burn-in, we conclude that there is a 95% chance that the parameter p lies between 0.58 and 0.98. This result was obtained using a Be(1,1) prior. Using Jeffrey's prior, the Be(1/2,1/2) distribution, the interval is (0.62, 0.99). These results agree with analytical calculations summarized in Fig. 8.2.

As is evident in Fig. 8.2, there is some sensitivity to the choice of prior. The explanation is that the observed value $X = 22$ suggests a value of p close to 1. Inference about a parameter that lies near the boundary of its allowable range is a difficult statistical problem. In frequentist inference, estimation of parameters near boundaries requires very large sample sizes for asymptotic results to apply.

For estimation of the binomial success rate $\theta = \pi + (1 - \pi)p$, it is commonly said that asymptotic results based on the standard normal distribution can be trusted provided that $n\theta = E[X]$ and $n(1 - \theta) = E[n - X]$ both exceed 5. Without endorsing this claim, we note that $n - X = 1$

---

5. To have instead written `Y[i] <- 1-(1-D[i])*(1-C[i])` would have generated an error message, stating that `Y[i]` were defined twice, that is, once in the data statement, and once as functions of independently generated `D[i]` and `C[i]`.

PANEL 8.1   BUGS code for the randomized-response model of Behavior X data, based on complete data likelihood.

```
model{
     for(i in 1:23){
          D[i] ~ dbern(p)
          C[i] ~ dbern(pi)
          alpha[i] <- 1-(1-D[i])*(1-C[i])
          Y[i] ~ dbern(alpha[i])
     }
     p ~ dbeta(1,1)
     pi <- 1/2
     Xers <- sum(D[1:23])
}

Data
list(C=c(NA,NA,NA,NA,NA,NA,NA,NA,NA,NA,NA,NA,NA,NA,
NA,NA,NA,NA,NA,NA,NA,NA,0), D=c(NA,NA,NA,NA,NA,NA,
NA,NA,NA,NA,NA,NA,NA,NA,NA,NA,NA,NA,NA,NA,NA,NA,0),
Y=c(1,1,1,1,1,1,1,1,1,1,1,1,1,1,1,1,1,1,1,1,1,1,0))
```

suggests that $n(1-\theta)$ is well below 5, so that we should not trust asymptotic results. In fact, if we calculate the usual asymptotic confidence interval (CI) $\hat{\theta} \pm z\,\mathrm{SE}(\hat{\theta})$, we obtain the 95% CI $(0.87, 1.04)$ which is not satisfactory as it extends beyond the boundary of 1. As an alternative frequentist procedure, we could compute an "exact" 95% CI obtaining $(0.78, 1.00)$ leading to a CI for p of $(0.56, 1.00)$. However, this CI has the unsatisfactory property of being conservative, in the sense that its coverage rate can be well above 95% depending on the true value for p.[6]

The Bayesian analysis of these data does not rely on asymptotic approximations. However, to conduct Bayesian inference we need to specify a prior for p, and due to the small sample size there is some sensitivity to the choice.

A useful feature of modeling using the CDL is that posterior prediction is straightforward. For example, we can predict whether a person responding "Yes" had ever X'ed by monitoring corresponding values D_i. We can similarly predict the outcome of their coin toss, although this is not of any particular interest.

Using the Be(1,1) prior for p, 909348 of the million values for D_1 were 1; the posterior probability that individual 1 had engaged in Behavior X is 0.91 under this prior. Since there are no covariates, this posterior probability is the same for any person who responded "Yes" in the randomized experiment. Similarly, 545293 of the million values for C_1 were 1, so the posterior probability that an individual flipped "Heads" on their coin toss given that they responded "Yes" is just 0.55. For Jeffrey's prior, these probabilities were only slightly changed to 0.93 and 0.54.

The parameter p describes a hypothetical infinite population from which our 23 individuals were sampled. Alternatively, we might interpret p as an a priori probability of Behavior X, the

---

6. For details on exact frequentist CI, see Section 3.2.

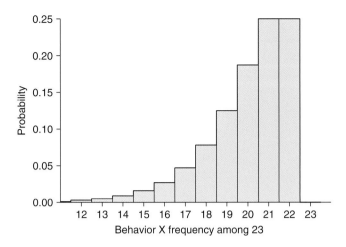

FIGURE 8.3 Posterior probabilities for $\sum_{i=1}^{23} D_i$ in Behavior X data.

value we would use to describe a 24th individual who was absent from the room, when the data were collected. As such, it is to be distinguished from the proportion of individuals among the 23 who had X'ed. This latter quantity, $\sum_{i=1}^{23} D_i/23$, could be used as an estimate of p but is not p itself. However, we might be interested in the number of individuals among our 23 that had X'ed. Use of the CDL makes inference about $\sum_{i=1}^{23} D_i$ a trivially simple matter. In our BUGS code (Panel 8.1), we simply defined a derived parameter for the sum; we obtain a sample of its posterior distribution with no additional effort. Posterior probabilities are plotted in Fig. 8.3; in particular, we conclude with 98% confidence that at least 15 of the 23 had engaged in Behavior X. However, the posterior probability on 23 is 0; our experimental design allowed us to conclude with certainty that one person had not X'ed.[7]

MCMC Based on ODL Using BUGS

The foregoing analyses in BUGS were based on the CDL. For this particular problem, the analysis could just as easily have been based on the ODL (Eq. 8.4), and coded for BUGS as in Panel 8.2.

Here the terms gamma[1] and gamma[2] are used for computing the posterior densities for the probability that a person has engaged in Behavior X, given that they responded "Yes" (γ_1) and for the probability that they obtained heads given that they reported obtaining heads (γ_2). By inspection of Table 8.1, it is clear that $\gamma_1 = p/\theta$, and that $\gamma_2 = \pi/\theta$. These are functions of p, derived parameters, hence samples of their posterior distributions are readily obtained from posterior samples for p.

---

7. That is, provided that the individual reporting "No," was honest. The odds of Behavior X are doubled by a positive response in the randomized experiment; it is conceivable that a participant might have intuited (or calculated) this change in odds and been inclined to dishonestly respond "No". However, it occurred to us subsequently that given the crowd we were sampling, there might have been a stigma associated with a *negative* response. Both responses could be randomized by asking participants to flip a coin twice, responding "Yes" if they flipped heads twice, "No" if they flipped tails twice, and honestly otherwise.

PANEL 8.2   BUGS code for the randomized-response model of Behavior X data, based on observed data likelihood.

```
model{
        X ~ dbin(theta,n)
        theta <- pi+p*(1-pi)
        p ~ dbeta(1,1)
        gamma[1] <- p/theta
        gamma[2] <- pi/theta
        pi <- 1/2
    }

Data
list(n=23, X=22)
```

Note that our analysis based on the ODL does not include nodes D_i for individual Behavior X status. However, we can make use of the posterior samples generated to calculate the posterior predictive distribution for D_i, and thus calculate $\Pr(D_i = 1 | Y_i = 1)$. We have

$$\Pr(D_i = 1 | Y_i = 1) = \int_{\gamma_1} \Pr(D_i = 1 | Y_i = 1, \gamma_1) \times [\gamma_1 | \text{Data}] d\gamma_1$$

$$= \int_{\gamma_1} \gamma_1 \times [\gamma_1 | \text{Data}] d\gamma_1$$

$$= \mathrm{E}[\gamma_1 | \text{Data}].$$

Similarly,

$$\Pr(C_i = 1 | Y_i = 1) = \mathrm{E}[\gamma_2 | \text{Data}].$$

These probabilities are calculated as the means of the Markov chains generated for γ_1 and γ_2.

MCMC Based on Partially Integrated CDL Using BUGS

Another approach is to model in terms of a partially integrated CDL in which latent variables that are a nuisance aspect to the model (in this case C_i) are integrated out. This leads to the BUGS code in Panel 8.3.

Here Y_i is modeled stochastically, as alpha[i] either takes the value 1 if the person has X'ed or $\pi = 0.5$ if they have not. Formally, we are obtaining $\Pr(\text{Report "Yes"} | D)$ by summing across the possible outcomes for the coin toss:

$$\Pr(Y_i = 1 | D_i) = \Pr(Y_i = 1 | D_i, C_i = 0) \Pr(C_i = 0)$$
$$+ \Pr(Y_i = 1 | D_i, C_i = 1) \Pr(C_i = 1),$$

which can be shown to equal π if $D = 0$, and 1 if $D = 1$, hence in general is $\pi(1 - D) + D$.

PANEL 8.3 BUGS code for the randomized-response model of Behavior X data, with D_i (but not C_i) included as a latent variable.

```
model{
        for(i in 1:23){
            D[i] ~ dbern(p)
            alpha[i] <-  D[i]+pi*(1-D[i])
            Y[i] ~ dbern(alpha[i])
        }
        pi <- 1/2
        p ~ dbeta(1,1)
    }


Data
list(D=c(NA,NA,NA,NA,NA,NA,
NA,NA,NA,NA,NA,NA,NA,NA,NA,NA,NA,NA,NA,NA,NA,NA,0),
Y=c(1,1,1,1,1,1,1,1,1,1,1,1,1,1,1,1,1,1,1,1,1,1,0))
```

Coding a Gibbs Sampler

We have, so far, described four different analyses of the Behavior X data. First, one could describe the posterior distribution for p as proportional to the product of the ODL $\mathcal{L}(p|Y)$ and the prior $[p]$, calculating the normalizing constant necessary for inference by numerical integration. Next, we suggested use of MCMC based on the CDL, using BUGS code given in Panel 8.1. Our third and fourth analyses used MCMC based on the ODL (Panel 8.2) and MCMC based on a partially integrated CDL (Panel 8.3).

In this section, we present one final approach to analysis of the data, based on explicit formulation of a Gibbs sampling scheme using the CDL. We include this section to highlight the equivalence, under the Bayesian māramatanga, of latent variables and parameters; also, to encourage our readers in understanding the Gibbs sampler. While the use of BUGS makes fitting Bayesian models easy, there is nothing like writing one's own code to build intuition for MCMC. The present analysis is a good simple example, and easily implemented in a program like R.[8]

Recall (Section 4.3.4) that Gibbs sampling is cyclical sampling of full conditional distributions. Given a set of unknowns $\boldsymbol{\theta} = (\theta_1, \theta_2, \ldots, \theta_k)'$ and data X, the full conditional for θ_i is the distribution proportional to $[X|\boldsymbol{\theta}][\boldsymbol{\theta}]$ with fixed values of θ_j for all $j \neq i$; it is denoted $[\theta_i|\cdot]$. Gibbs sampling consists of sequentially updating values θ_i by sampling from their full conditionals.

The complete data for the Behavior X study consist of three vectors of length $n = 23$, namely Y, C, and D. Components of the vectors are C_i, an indicator that the ith individual flipped "Heads"; D_i, an indicator that the individual had X'ed; and $Y_i = 1 - (1 - C_i)(1 - D_i)$, the indicator of response "Yes" in our randomized-response survey. All of the values Y_i are observed, but none of the C_i or D_i are directly observed. We can deduce $C_{23} = D_{23} = 0$ because $Y_{23} = 0$, but $Y_i = 1$ for $i \neq 23$, hence the corresponding values C_i and D_i are unknown. We must include such C_i and D_i in our Gibbs sampler.

---

8. Program R has the benefits of being free, widely used, and well-documented on the web.

The requisite full conditional distributions of C_i and D_i are readily obtained by inspection of Table 8.1. Calculation of $[C_i|\cdot]$ conditions on all of the values Y, C, and D except C_i. Because of independence assumptions, none of these matter except D_i and Y_i. We have

$$\Pr(C_i = c \,|\, D_i = 1, Y_i = 1, \pi) = \pi^c (1 - \pi)^{(1-c)}$$
$$\text{and} \quad \Pr(C_i = c \,|\, D_i = 0, Y_i = 1, \pi) = c,$$

for $c = 0, 1$. These can be summarized by saying that the full conditional distribution of C_i is Bernoulli, with rate parameter $D_i \pi + (1 - D_i)$.

Similarly, the full conditional distribution $[D_i|\cdot]$ depends only on C_i, Y_i, and p. We have

$$\Pr(D_i = d \,|\, C_i = 1, Y_i = 1, p) = p^d (1 - p)^{(1-d)}$$
$$\text{and} \quad \Pr(D_i = d \,|\, C_i = 0, Y_i = 1, p) = d,$$

for $d = 0, 1$. These are summarized by saying that the full conditional distribution of D_i is Bernoulli, with rate parameter $C_i p + (1 - C_i)$.

Finally, the full conditional distribution $[p|\cdot]$ is determined by the collection of values D_i, and the $\text{Be}(\alpha, \beta)$ prior for p. Because the values D_i are Bernoulli trials, their sum is a binomial random variable and conjugate to the beta prior. Thus the full conditional distribution $[p|\cdot]$ is $\text{Be}\left(\alpha + \sum_{i=1}^n D_i, \beta + n - \sum_{i=1}^n D_i\right)$.

We may thus describe a Gibbs sampler as follows:

Step 1: Initialize $C_i^{(1)} = 0$ and $D_i^{(1)} = 1$ for $i = 1, 2, \ldots, 22$. (These 22 are the unknown values, with $Y_i = 1$; the choice of initial values is arbitrary but must be consistent with $Y_i = 1$). Initialize $p^{(1)} \in (0, 1)$. Superscripts (j) will denote sequence number in the Markov chain of values produced.

Step 2: For $i = 1, 2, \ldots, 22$, generate $C_i^{(j)}$ as Bernoulli trials with success parameters $D_i^{(j-1)} \pi + (1 - D_i^{(j-1)})$.

Step 3: For $i = 1, 2, \ldots, 22$, generate $D_i^{(j)}$ as Bernoulli trials with success parameters $C_i^{(j)} p^{(j-1)} + (1 - C_i^{(j)})$.

Step 4: Generate $p^{(j)} \sim \text{Be}\left(\alpha + \sum_{i=1}^n D_i^{(j)}, \beta + n - \sum_{i=1}^n D_i^{(j)}\right)$.

Step 5: Repeat Steps 2 to 4 for $j = 1, 2, \ldots$

We encourage readers to try coding this Gibbs sampling algorithm in R, or some other programming language. Alternatively BUGS can be used as a simulator, and the analysis conducted using full conditional distributions, as in Panel 8.4. Having specified the full conditional distributions, there is no data statement; the observed data Y are there only implicitly, having been used in defining the full conditionals, in combination with a $\text{Be}(1, 1)$ prior for p.

8.2.2 Remarks

In this example, the confounding of responses is a deliberate data collection device. Through the coin toss, the data of interest, the D_i have been "corrupted" into the Y_i. Thus, the D_i are now latent. Our judicious choice of data corruption mechanism allows us to make inference about the probability p despite not observing the D_i. In many ways, the randomized-response survey

PANEL 8.4 BUGS code for the randomized-response model of Behavior X data, based on full conditional distributions from complete data likelihood.

```
model{
        for (i in 1:22){
            success.c[i] <- D[i]*pi+(1-D[i])
            success.d[i] <- C[i]*p+(1-C[i])
            C[i] ~ dbern(success.c[i])
            D[i] ~ dbern(success.d[i])
        }
        Xers <- sum(D[1:22])
        alpha.fc <- 1+Xers
        beta.fc <- 1+23-Xers
        p ~ dbeta(alpha.fc,beta.fc)
        pi <- 1/2
}
```

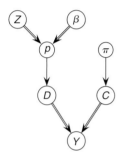

FIGURE 8.4   Directed acyclic graph (DAG) representation of the complete data likelihood for the Behavior X example where the probability of prior Behavior X is modeled using a logistic regression on covariates Z.

serves as a canonical example of a corrupted data problem in which inference is facilitated, though expressing the problem in terms of a hierarchical model.

Biological data often have a similar element of confounding. The difference, however, is that with field data the confounding mechanism has been chosen by nature rather than the researcher. As a consequence, careful thought must be given to the processes that gave rise to the data. Hierarchical models facilitate this thinking by clearly separating processes involved in generating the latent variable of interest from processes that lead to corruption of the data. Having identified a data corruption model, researchers are free to concentrate on modeling the latent variable. This approach we refer to as "modeling the data you wish you had." That is, we model in terms of latent variables of interest.

For example, suppose we had a covariate vector, Z_i, recorded for each person that we believe might be relevant in determining whether or not they have engaged in behavior X. Covariates might include age, sex, music preferences, and so forth. If we had such information it would be natural to model D_i using logistic regression on Z_i. Although the D_i are latent, it is straightforward to extend the model so that instead of D_i being modeled as a Bernoulli random variable with parameter p, we model it as Bernoulli with individual specific parameter p_i, where $\text{logit}(p_i) = Z_i'\beta$, where parameter vector β consists of regression coefficients (Fig. 8.4). The additional model structure would require only slight modifications to the BUGS code in Panel 8.1.

8.3 OCCUPANCY MODELS AS HIERARCHICAL LOGISTIC REGRESSIONS

The occupancy models considered by MacKenzie *et al.* (2002) address a class of corrupted data problems of special interest to ecologists. The basic problem is to model presence or absence of one or more species in an area, when detection methods are imperfect.

We begin by considering data for a single species. Let X_i be the indicator of the event that site i is "occupied" (the species is present), and let $Y_{ij} = 1$ be an indicator for the event that the species is detected at site i during visit j. It is assumed that there are no false detections: Y_{ij} cannot equal 1 unless $X_i = 1$.

The X_i are partially observed, in that if $Y_{ij} = 1$ for any j, then we know that $X_i = 1$. However, if we fail to detect the species during all of the visits, then either $X_i = 0$, or $X_i = 1$ and we failed to detect the species even though it was present. That is, we know we can have false negatives. The problem is really nothing more than an extension of the randomized response problem considered in Section 8.2:

- "Sites" correspond to "individuals."
- "Occupancy" corresponds to the state of never having engaged in Behavior X.
- "Not Detected" corresponds to response of "Yes" – an ambiguous response. Could have never X'ed, but might have; similarly, a nondetect might mean a site is not occupied but could occur for occupied sites. In both cases the true status is veiled by a Bernoulli trial (the coin flip, on one hand; the attempt at successful observation on the other).
- "Detected" corresponds to a response of "No" – an unambiguous response. The "No" response allows certainty of no Behavior X just as species detection allows certainty of occupancy.

The only difference with occupancy data is that we obtain multiple responses for a site, as though we had repeatedly asked about Behavior X, with independent coin flips each time, keeping track of each individual's answers. Having multiple responses allows for estimation of the detection probability; for the Behavior X problem this was not necessary, since $\pi = 1/2$ was known.

Modeling site occupancy as a Bernoulli random variable with parameter $\psi_i = $ Pr(Site i is occupied) and detection probability

$$p_{ij} = \text{Pr(Detected at site } i, \text{on sampling occasion } j \mid \text{Site } i \text{ is occupied)},$$

MacKenzie *et al.* (2002) described the model in terms of the ODL, which has contribution

$$\psi_i \prod_{j=1}^{t} p_{ij}^{Y_{ij}} (1-p_{ij})^{(1-Y_{ij})}$$

for the sites where $Y_{ij} = 1$ for at least one j, and

$$1 - \psi_i + \psi_i \prod_{j=1}^{t} (1-p_{ij})$$

for the sites where $Y_{ij} = 0$ for all j. For identifiability, constraints must be imposed on the p_{ij} and ψ_i, for example, constraining the probabilities to be the same across sites, or modeling them using observed covariates.

A useful way to think of these models is as regression models for partially observed, or corrupted, data. Let Y_i denote a vector $(Y_{i1}, Y_{i2}, \ldots, Y_{it})'$ of detection indicators on t sampling occasions at site i, with j indexing sampling occasion. Let $p_i = (p_{i1}, p_{i2}, \ldots, p_{it})'$, supposing that these are determined by a vector of covariates W_i and a vector-valued parameter γ. Furthermore, suppose that the occupancy probability, ψ_i, is determined by a vector of covariates Z_i and a parameter vector β. Finally, let Y (without subscript) denote the collection of Y_i's; define W, Z, p, ψ, and X, similarly. Then a CDL representation of the model is given by

$$[Y|X, \gamma, W][X|\beta, Z]. \tag{8.5}$$

The term $[X|\beta, Z]$ describes (for instance) a logistic regression for the partially observed random occupancy variables X_i, and the term $[Y|X, \gamma, W]$ describes data corruption mechanisms.

As a problem in Bayesian inference, the CDL representation Eq. (8.5) offers a number of advantages including:

(i) It helps focus inference on the appropriate level of interest to the biologist, namely the model $[X|\beta, Z]$.
(ii) It serves as the basis for inference about unobserved X_i, based on their posterior predictive distributions (5.1.2).
(iii) It facilitates construction of an MCMC sampler in terms of explicit full-conditional distributions.

Of these three, (i) and (ii) are by far the most important for the process of scientific inference. But these boons require our capacity to fit the model in question. So, we illustrate the third point, constructing a Gibbs sampler.

Gibbs Sampler for Occupancy Model $\{p_t, \psi\}$

We describe full-conditional distributions under the model $\{p_t, \psi\}$, that is, assuming ψ is constant across sites and detection depends on sampling occasion but not site. Set $Y_{ij} = NA$ if site i was not visited on occasion j. Then,

(i) Assigning a $Be(\alpha_j, \beta_j)$ prior for p_j, the full-conditional distribution $[p_j|\cdot]$ is $Be(a_j, b_j)$ with

$$a_j = \alpha_j + \sum_{i:Y_{ij} \neq NA} Y_{ij}$$

$$b_j = \beta_j + \sum_{i:Y_{ij} \neq NA} X_i(1 - Y_{ij}).$$

The summation used in defining a_j gives the number of sites visited on sample occasion j at which detections occurred; the summation used in defining b_j gives the corresponding number of sites without detection.

(ii) Assigning a $Be(\alpha_\psi, \beta_\psi)$ prior to ψ, the full conditional distribution $[\psi|\cdot]$ is $Be(a_\psi, b_\psi)$, with $a_\psi = \alpha_\psi + \sum_i X_i$ and $b_\psi = \beta_\psi + n - \sum_i X_i$.

(iii) For the sites with no detections registered, X_i is unknown. Implementing Gibbs sampling will require sampling from its full-conditional distribution, a Bernoulli distribution with parameter

$$\pi_i = \frac{\psi \prod_j q_{ij}}{1 - \psi + \psi \prod_j q_{ij}};$$

here $q_{ij} = (1 - p_j)^{I(Y_{ij} \neq NA)}$. Note that π_i is the probability that site i is occupied given that no detections were made and given the history of visits.

We now present a sample data set to which this Gibbs sampler can be applied.

8.3.1 American Toads in Maryland

MacKenzie *et al.* (2002) report the results from a study of American toads (*Bufo americanus*) at 29 wetland sites in Maryland, USA, on 82 sampling occasions (days) between 9 March and 30 May 2000. Table 8.2 gives a reduced version of the data, with observations summarized for five 14-day periods, and a final period of 12 days.

Analysis of these data is challenging, and results are likely to be sensitive to model choice. Maximum likelihood estimates and asymptotic standard errors (ASE) under model $\{p_t, \psi\}$ are given in Table 8.3. Note that one estimate is on the edge of the parameter space ($\hat{p}_2 = 1$) and that

TABLE 8.2   American toad data, based on MacKenzie *et al.* (2002).

1	2	3	4	5	6	7	8	9	10	11	12	13	14	15
.	0	.	0	.	.	1	0	0	0	.	.	.	0	0
0	1	0	1	0	1	1	0	0	1	.	0	0	0	0
.	.	.	0	0	.	1	0	0	.	.	0	0	0	0
0	0	0	0	0	0	1	0	.	0	1	0	.	0	0
0	.	0	0	0	.	1	0	0	0	0	0	.	0	0
0	0	0	0	.	0	1	0	0	.	0	0	0	0	0

16	17	18	19	20	21	22	23	24	25	26	27	28	29
0	0	0	.	0	0	0	.	0	0	0	.	.	.
.	0	0	0	1	1	0	0	0	1	0	0	0	1
.	0	.	.	.	.	0	0	0	0	0	0	.	1
0	0	0	0	1	0	0	0	0	0	0	0	0	1
0	0	0	.	0	0	0	0	.	0	.	.	.	0
.	.	0	0	0	0	0	0	0	0	.	0	.	0

29 columns represent sites; six rows correspond to sampling occasions. Dots are missing values. Thus site 29 was visited on all but the first occasion, and frogs were detected on the second, third and fourth occasions.

TABLE 8.3 Maximum likelihood estimates and asymptotic standard errors.

	MLE	ASE	95% CI	95% PLI
p_1	0.139	0.129	$(-0.114, 0.392)$	$(0.008, 0.487)$
p_2	1.000	–	.	$(0.737, 1.000)$
p_3	0.500	0.250	$(0.010, 0.990)$	$(0.107, 0.893)$
p_4	0.392	0.154	$(0.090, 0.694)$	$(0.142, 0.693)$
p_5	0.122	0.114	$(-0.101, 0.345)$	$(0.007, 0.439)$
p_6	0.111	0.105	$(-0.095, 0.317)$	$(0.007, 0.406)$
ψ	0.352	0.090	$(0.176, 0.528)$	$(0.194, 0.535)$

Maximum likelihood estimates, asymptotic standard errors, and 95% confidence intervals for the parameters $p_1, p_2, \ldots, p_6$ and ψ for the model $\{p_t, \psi\}$ fitted to the American toad data of MacKenzie *et al.* (2002). CI is MLE $\pm$ 1.96 ASE; PLI is a profile likelihood interval.

conventional asymptotic CI's $\hat{\theta} \pm 1.96$ ASE$(\hat{\theta})$ for p_1, p_5, and p_6 extend well below 0; the interval for p_3 comes close to covering the entire possible range of the parameter. Profile likelihood intervals given in Table 8.3 at least have the virtue of being consistent with the range of the parameters, but coverage rates for these also rely on asymptotic approximations.

One approach to dealing with the limited precision associated with detection probabilities is to fit model $\{p, \psi\}$, without temporal variation in p_t. The estimate of ψ increases to 0.43 (SE $=$ 0.12). If p_2 is estimated as 1, with $p_1 = p_3 = \cdots = p_6$, the MLE of ψ is unchanged, to three decimal place accuracy, at 0.352.

A better course of action, followed by MacKenzie *et al.* (2002), is to describe detection parameters as functions of relevant covariates. Site- and site∗sample-specific covariates were collected for the American toad study. Site-specific covariates allow examination of factors influencing site occupancy, the focus of the study. The site∗sample covariates allow more complex models to be fitted to detectability, perhaps better dealing with these nuisance parameters.

Our view is that all of these analyses are best performed in a Bayesian context, where there is no need to rely on questionable asymptotic results. Granted, there may be discussion about which priors to use on model parameters, but at least given a choice, the mathematics leading to the posterior distributions is exact. If we must use MCMC to evaluate posterior distributions, the results can be made arbitrarily precise by sampling longer chains. Furthermore, modifications to BUGS code required for extending the model are simple.

Bayesian inference for model $\{p_t, \psi\}$ can be conducted using the Gibbs sampler described at the beginning of this section. However, BUGS code for the occupancy problem is almost startlingly simple (Panel 8.5). The data statement (not included) is the matrix of values Y_{ij}; X_i need not be included because our knowledge that certain X_i equal 1 comes from observations $y_{ij} = 1$, for which the model requires $X_i = 1$. Unknown X_i are treated the same as all other unknown quantities; we wind up with samples from their posterior distributions. The same happens for detection events Y_{ij}, when sites were not visited: we could predict whether an animal would have been detected, had the visit been made.

PANEL 8.5   BUGS code for American toad data.

```
model{
     for (i in 1:29){
            x[i] ~ dbern(psi)
            for (j in 1:6){
                   pi[i,j] <- x[i]*p[j]
                   y[i,j] ~ dbern(pi[i,j])
              }
     }
     for (j in 1:6){
              p[j] ~ dbeta(0.5,0.5)
        }
     psi ~ dbeta(0.5,0.5)
}
```

Notice also that our interest may well be in the number of occupied sites *among those in our sample* rather than in a hypothetical infinite population rate ψ. The MLE under model $\{p_t, \psi\}$ was 35.2%, not terribly different from the proportion of sites at which toads were detected. A naive analyst might scoff at the effort that went into finding this estimate, noting that frogs were detected at 10 sites, and $10/29 = 34.5\%$. To our mind, one of the most appealing products of the modeling process is one that is most easily obtained under the Bayesian framework: we are able to assign probabilities of occupancy to the 29 sites, and to do so coherently, without dubious mathematical approximations. For 18 of the 19 sites of unknown status, this probability was between 0.9 and 2.7%; for site 16, it was 19.4%.[9] We are also able to conclude that the number of occupied sites without detection was 0, 1, or 2 with probability 0.671, 0.252, and 0.052, respectively; there is only a 2.5% chance that the total was 3 or larger.

8.3.2 More Complex Occupancy Models

From a biological perspective, occupancy studies focus on the term $[X|\beta, Z]$ in the CDL. Distinguishing this feature of the model from biologically irrelevant sampling features is facilitated by a graphical representation of the model as in Fig. 8.5.

So far, we have considered occupancy models in which X_i is a scalar-valued random variable. A much richer class of models can be considered. For example, we may suppose that X_i is a two-dimensional array with one dimension being species and the other dimension being time (or season). The models we have considered so far assume that occupancy status does not change during the sampling period; we refer to such X_i as describing 1 species × 1 time.

The case where X_i is a 1 species × k times array leads to the single-species multiple season model of MacKenzie *et al.* (2002) in which occupancy of site i by the study species changes between seasons. In terms of the graphical representation of the model, the only change is in the definition of the node beta, which now includes elements for modeling extinction and

---

9. Site 16 was only visited three times, and not on the second sampling occasion, when detection probabilities were high.

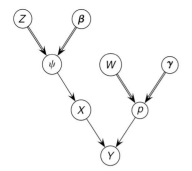

FIGURE 8.5   Directed acyclic graph (DAG) for occupancy model with detection and occupancy probabilities modeled as functions of covariates.

colonization events, and X, which now is a summary of vector-valued observations at each site. Another extension of the model is to let X_i be an s species $\times$ 1 time array, describing multiple-species single season models, with β including elements for describing species interactions among sites. The most general case is when X_i is an s species $\times$ k times array leading to multiple species-multiple season models.

In a multiple species and multiple season model, X_i can be represented as a $2 \times s \times k$ cross-classification, with potential for a large number of parameters especially if covariates influence the cross-classification. In addition, modeling the data corruption process can lead to a very large number of models and parameters for $[Y|X, \gamma, W]$, where γ represents nuisance parameters. The challenge is identification of a set of models that capture the essential features of interest from the point of view of research hypotheses without the nuisance components leading to an intractably large set of models.

The situation with the occupancy models mirrors that of the mark-recapture models where many of the parameters and parameter sets describe nuisance aspects of sampling. For the occupancy models, we see potential for the application of loglinear modeling methods to the analysis of cross-classified data in which constraints on between-species and between-times interactions can reduce the number of parameters needed in the model to represent relatively complex relationships. We advocate the use of loglinear models for describing occupancy dynamics and relationships among species, and to use hierarchical models for nuisance parameters as a strategy for dimension reduction on the nuisance aspects of the model.

8.4   DISTANCE SAMPLING

In the occupancy model, inference about detection probabilities is made possible through repeated sampling of sites; in this regard, occupancy models are closely related to the mark-recapture models discussed in Chapter 9. We now turn our attention to distance sampling, in which detection probabilities are estimated by entirely different means.

Distance sampling models emphasize the importance of one very specific covariate: the distance between the objects searched for and a reference location. For line transects, distances are measured from the transect centerline; for circular plot sampling and trapping webs, distances are measured from the plot center. By making assumptions about the functional relationship between detection probability and distance, and assumptions about the spatial distribution of objects, these models allow inference about the numbers of objects missed.

Enshrined in the name "distance sampling" is the perspective that the basic data to be modeled are a set of distances, x. A collection of n distances is regarded as a sample from a known probability distribution $\pi(x)$, typically (but not necessarily) a uniform distribution. The value n is unknown, and not all of the distances are observed. The goal is to make inference about n, or about a population of size N, of which the n are a subset.

Inference is made possible by assuming that associated with distance x_i is a detection probability $g_\theta(x_i)$, where $g_\theta(\cdot)$ is a known function, except for unknown parameters θ. The vector x is partitioned as $(x^{obs}, x^{mis})'$; conventional modeling is based on a calculated probability distribution for the observed portion x^{obs}. The first step, then, is to compute a distribution for the observed distances, namely

$$f(x|\text{Object is detected}) = \frac{\text{Pr}(\text{Object is detected}|\text{Distance}=x)\,\pi(x)}{\text{Pr}(\text{Object is detected})}$$

$$= \frac{g_\theta(x)\,\pi(x)}{\int g_\theta(x)\,\pi(x)\,dx}. \tag{8.6}$$

Inference is based on the ODL

$$\mathcal{L}(\theta; x^{obs}) \propto \prod f(x_i^{obs}|\text{Object is detected}).$$

Equation 8.6 should look familiar: it is the computation of $[x|D=1]$ via Bayes theorem, relating distances x to detections D, based on "prior" $\pi(x)$ on distances, and Bernoulli data distribution

$$\text{Pr}(D|x) = g_\theta(x)^D \left(1 - g_\theta(x)\right)^{(1-D)}.$$

We believe that a more natural approach is to consider a hierarchical model for line-transect sampling, treating the *detections* as the data to be modeled and the distances as covariates. Distance sampling is really object sampling; the distances are of no intrinsic biological interest. The sole function of the distances is to help deal with a nuisance aspect of the model, one which describes our inability to detect all the objects. As well as being in the spirit of "modeling the data you wish you had," the object-centered approach puts distance sampling into the same hierarchical framework as closed-population mark-recapture modeling and the occupancy models of Section 8.3.

A Line-Transect Survey

In this section, we consider line-transect sampling, noting that the ideas discussed here readily generalize to other forms of distance sampling. We consider a single line transect of width $2w$ and length L, randomly placed in a region of size A. The area covered by the line transect (the "quadrat") is thus of size $a = 2wL$. We will refer to the objects as "animals" although line-transect sampling can be applied to a variety of subjects.[10] We consider a population of unknown size N.

10. Methodological evaluations have involved searches for wooden stakes, bricks in Lake Huron, and polystyrene tortoises. Also beer cans, an ingenious incentive for student participation in field work.

There are two ways an animal can be missed by the survey. These define two parts of a hierarchical model. First, the animal might not be located within the sampled quadrat. We will let n denote the number of individuals in the quadrat; $N - n$ animals are missed simply because they are not in our limited sampling area. Second, it is almost inevitably the case that not all of the n animals within the quadrat will be observed. Our goal, is to describe the survey in terms of "the data we wish we had," using a CDL that relates n to the unknown N and relates the data obtained within the quadrat to the unknown n.

If the probability of finding an animal at any one point is the same for all points in the population area, and if the presence of one individual does not influence the presence of another, then the number n of individuals that appear in a quadrat of size a from a total population of size N in an area of size A can be modeled as a binomial random variable (Seber, 1982, 22):

$$[n|N] = \binom{N}{n} p_c^n (1 - p_c)^{N-n} \tag{8.7}$$

where $p_c = \frac{a}{A}$.

Conditional on presence within the area defined by the quadrat, detections of animals are independent Bernoulli random variables, with success parameters determined by their distance from the centerline. For animal i at distance x_i, this Bernoulli trial is labeled D_i; its success parameter is

$$g_\theta(x_i) = \Pr(\text{Detection}|\text{Distance} = x_i);$$

here θ represents unknown parameters governing the detection process. Let $\mathbf{D} = (D_1, D_2, \ldots, D_n)'$, and let $\mathbf{x} = (x_1, x_2, \ldots, x_n)'$.

To make inference about population size we must assume that $g_\theta(0)$ is known. It is usually assumed that every animal is detected on the centerline, that is, that $g_\theta(0) = 1$.

We model the data for animals within the quadrat as

$$[\mathbf{D}, \mathbf{x}|n, \theta, \gamma] = [\mathbf{D}|\mathbf{x}, n, \theta]\,[\mathbf{x}|n, \gamma], \tag{8.8}$$

where

$$[\mathbf{D}|\mathbf{x}, n, \theta] = \binom{n}{m} \prod_{i=1}^{n} g_\theta(x_i)^{D_i} (1 - g_\theta(x_i))^{1-D_i}.$$

Here, m is the number of animals detected, and $[\mathbf{x}|n, \gamma] = \prod_{i=1}^{n}[x_i|\gamma]$ describes a model for the joint distribution of animal locations. The model for $\mathbf{x}$ is required because distances are not observed for the animals that we have missed within the quadrat. As noted above, the usual assumption is that $[x_i|n, \gamma]$ is a uniform $U(0, w)$ density function, although any other model could be considered.

Putting the two components together, we obtain the CDL for the hierarchical line-transect model as

$$[\mathbf{D}, \mathbf{x}, n|N, \theta, \gamma] = [\mathbf{D}|\mathbf{x}, \theta, n] \times [\mathbf{x}|n, \gamma] \times [n|N], \tag{8.9}$$

with DAG representation in Fig. 8.6. The portion of the DAG to the right of center, corresponding to $[n|N]$, is the model for the data we wish we had, that is, the number of animals that were located within our quadrat(s). If we have multiple surveys, this portion of the model can be

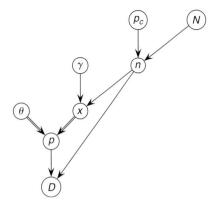

FIGURE 8.6   Directed acyclic graph (DAG) representation of the complete data likelihood for the line-transect model.

extended to allow for higher-level modeling, for example, to smooth abundance over time or space. The other components of Fig. 8.6 model nuisance aspects of the model, not of interest per se, but essential for reliable inference.

The CDL is readily used in Bayesian analyses. Frequentist analyses use an ODL, obtained from the CDL by integrating out the missing components, x^{mis}. Substituting Eqs. (8.7) and (8.8) in Eq. (8.9) with $\pi(x)$ representing the density of distances, and integrating over x^{mis}, we obtain

$$\binom{N}{m}(1-p_cE_x[g_\theta(x)])^{N-m}\prod_{i=1}^{m}g_\theta\left(x_i^{obs}\right)\pi\left(x_i^{obs}\right).$$

the likelihood given by Borchers *et al.* (2002).

Example

Buckland *et al.* (1993) (Section 8.3) discuss the analysis of data from a line-transect study of a population of 150 wooden stakes in which one transect through the study area was repeatedly sampled by 11 observers. Here, we analyze the data for the first observer, using a simple half-normal model for the detection function $p(x,\sigma^2)=\exp(-x^2/2\sigma^2)$, for $0<x<w=20$, the width of the transect. Full-conditional distributions for N and σ^2 can be obtained easily enough, to code one's own software. Fitting the CDL is straightforward using software BUGS and the code provided in Panel 8.6 or 8.7.

We begin, in Panel 8.6, with prediction of the number n of stakes occurring within the transect. An interesting feature of the BUGS code in Panel 8.6 is the use of data augmentation to model n. Population size n is unknown, thus must be treated as random. The number of unobserved quantities is, thus, also a random variable. For instance, if $n=m=72$, there are no unobserved distances x_i, but if $n=172$, there are 100 unobserved x_i for which posterior distributions must be sampled. The situation is similar to that of implementing multimodel inference using reversible jump MCMC: we require a palette of parameters of dimension at least equal to the highest dimension of the models in our model set (Section 7.3.2).

We have chosen a discrete uniform $DU(\{1,2,\ldots,200\})$ prior for n, which we implement by modeling $[n|\psi]=B(200,\psi)$, with $[\psi]=U(0,1)$. It is easily shown that the result is a discrete

PANEL 8.6   BUGS code for the line-transect model, using half-normal detection function.

```
list(w=20,M=200,

x=c(0.13,0.26,0.39,0.53,0.66,0.79,0.92,1.06,1.19,1.46,1.59,
1.72,1.85,1.98,2.12,2.25,2.38,2.52,2.65,2.78,2.91,3.05,3.18,3.31,
3.58,3.71,3.84,3.97,4.24,4.77,4.90,5.03,5.17,5.43,5.56,5.82,5.96,
6.22,6.36,6.49,6.62,6.75,7.15,7.41,7.68,7.95,8.08,8.34,8.61,8.87,
9.27,10.20,10.33,10.46,10.73,10.86,11.92,12.05,12.98,13.51,13.77,
14.03,14.17,14.96,15.90,16.02,16.29,16.95,17.21,18.14,18.67,19.34,
NA,NA,NA,NA,NA,NA,NA,NA,NA,NA,NA,NA,NA,NA,NA,NA,NA,NA,NA,NA,NA,
NA,NA,NA,NA,NA,NA,NA,NA,NA,NA,NA,NA,NA,NA,NA,NA,NA,NA,NA,NA,NA,
NA,NA,NA,NA,NA,NA,NA,NA,NA,NA,NA,NA,NA,NA,NA,NA,NA,NA,NA,NA,NA,
NA,NA,NA,NA,NA,NA,NA,NA,NA,NA,NA,NA,NA,NA,NA,NA,NA,NA,NA,NA,NA,
NA,NA,NA,NA,NA,NA,NA,NA,NA,NA,NA,NA,NA,NA,NA,NA,NA,NA,NA,NA,NA,
NA,NA,NA,NA,NA,NA,NA,NA,NA,NA,NA,NA,NA,NA,NA,NA,NA,NA,NA,NA),

D=c(1,1,1,1,1,1,1,1,1,1,1,1,1,1,1,1,1,1,1,1,1,1,1,1,1,1,1,1,1,1,
1,1,1,1,1,1,1,1,1,1,1,1,1,1,1,1,1,1,1,1,1,1,1,1,1,1,1,1,1,1,1,1,
1,1,1,1,1,1,1,1,0,0,0,0,0,0,0,0,0,0,0,0,0,0,0,0,0,0,0,0,0,0,0,0,
0,0,0,0,0,0,0,0,0,0,0,0,0,0,0,0,0,0,0,0,0,0,0,0,0,0,0,0,0,0,0,0,
0,0,0,0,0,0,0,0,0,0,0,0,0,0,0,0,0,0,0,0,0,0,0,0,0,0,0,0,0,0,0,0,
0,0,0,0,0,0,0,0,0,0,0,0,0,0,0,0,0,0,0,0,0,0,0,0,0,0,0,0,0,0,0,0,
0,0,0,0),

I=c(1,1,1,1,1,1,1,1,1,1,1,1,1,1,1,1,1,1,1,1,1,1,1,1,1,1,1,1,1,1,
1,1,1,1,1,1,1,1,1,1,1,1,1,1,1,1,1,1,1,1,1,1,1,1,1,1,1,1,1,1,1,1,
1,1,1,1,1,1,1,1,NA,NA,NA,NA,NA,NA,NA,NA,NA,NA,NA,NA,NA,NA,NA,NA,NA,
NA,NA,NA,NA,NA,NA,NA,NA,NA,NA,NA,NA,NA,NA,NA,NA,NA,NA,NA,NA,NA,NA,
NA,NA,NA,NA,NA,NA,NA,NA,NA,NA,NA,NA,NA,NA,NA,NA,NA,NA,NA,NA,NA,NA,
NA,NA,NA,NA,NA,NA,NA,NA,NA,NA,NA,NA,NA,NA,NA,NA,NA,NA,NA,NA,NA,NA,
NA,NA,NA,NA,NA,NA,NA,NA,NA,NA,NA,NA,NA,NA,NA,NA,NA,NA,NA,NA,NA,NA,
NA,NA,NA,NA,NA,NA,NA,NA,NA,NA,NA,NA,NA,NA,NA,NA,NA,NA,NA,NA,NA,NA,
NA))

model{
    # Likelihood
    for(i in 1:M){
        I[i] ~ dbern(psi)
        x[i] ~ dunif(0,w)
        p[i] <-exp(-tau/2*pow(x[i],2))
        pi[i] <- I[i]*p[i]
        D[i] ~ dbern(pi[i])
    }
    n <- sum(I[])

    # Priors
    tau ~ dunif(0,10)
    psi ~ dbeta(1,1)
}
```

uniform prior for n on $\{0,1,\ldots,200\}$. Data statements in Panel 8.6 include 128 ($=200-m$) missing values for nodes I[i] and x[i], and 128 zeros for corresponding D_i.

One can think of the analysis as describing 200 wooden stakes in the quadrat, some of them real, others imaginary, and that I[i] is an indicator for the event that the ith modeled stake is real. We have I[1] = I[2] = ... = I[72] = 1; these correspond to the observed stakes, which we know are real. Note that D[73] = D[74] = ... = D[200] = 0; some of these correspond to real stakes which were not detected, while others to imaginary stakes which could not be detected. The node n <- sum(I[1:200]) is the total of "real stakes" in the quadrat, the quantity we desire to estimate.

If n could be much larger than m, it will be necessary to include many "imaginary" nodes in the BUGS code, and the run time will increase substantially. A good course of action is to try various upper bounds M on n, seeing how large a value could possibly be supported by the data. We originally set $M = 1500$ in the present case, but found that in a Markov chain of length 100,000, the maximum observed value of n was 198, and that less than one half of 1% of sampled values exceeded 167. Changing the prior on n from $DU(\{1,2,\ldots,1500\})$ to $DU(\{1,2,\ldots,200\})$, therefore, seemed reasonable; the choice has no effect on inference, but reduced run time by approximately 90%.

The BUGS code of Panel 8.6, like that for the Behavior X data (Panel 8.1), and the frog occupancy data (Panel 8.5), is easily modified for alternative models. For instance, inclusion of covariates explaining detection requires only simple modifications of the definition of p[i], though note that models for distributions of covariates become necessary because of unobserved values. The BUGS code is also easily modified for data from multiple surveys, or to predict population size N for a known region extending beyond the quadrat.

Panel 8.7 incorporates two changes to Panel 8.6. First, we distinguish nodes by extensions .obs and .mis. This allows specification of data values within loops for observed values, and avoids the need for all those NA's in the data statement. The other change is the specification of two indicator variables, I.N for inclusion in the population of size N, and I.n for inclusion in the quadrat. For this analysis, we assume that the quadrat covered 80% of the region in which the stakes were located, so that $p_c = 0.80$. The important thing is that although x's, I's, and D's are distinguished in the two "for" loops, the parameters governing their distributions are not: it is the same psi and tau in both parts. Note also that the observed D's are all 1, the unobserved D's are all 0, and the observed I's are 1; unobserved I's could be 0 or 1.

8.5 FINITE POPULATION SAMPLING

One of the most basic sources of data is the sample survey, in which data are collected for part of a population, with the goal of making inference about the entire population. While it is always better to sample more data than less, the gains associated with larger sample sizes are even greater when sampling from a finite population. For example, a simple random sample of size m from an infinite population with mean μ and variance σ^2 produces an estimator $\bar{x}$ with standard error $\sigma/\sqrt{m}$, but if the population is finite, of size M, the standard error is

$$\text{SE}(\bar{x}) = \frac{\sigma}{\sqrt{m}} \times \sqrt{\frac{M-m}{M-1}}. \tag{8.10}$$

PANEL 8.7   Alternative version of BUGS code from Panel 8.6.

```
list(w=20, pc=0.80,

x.obs=c(0.13,0.26,0.39,0.53,0.66,0.79,0.92,1.06,1.19,1.46,1.59,
1.72,1.85,1.98,2.12,2.25,2.38,2.52,2.65,2.78,2.91,3.05,3.18,3.31,
3.58,3.71,3.84,3.97,4.24,4.77,4.90,5.03,5.17,5.43,5.56,5.82,5.96,
6.22,6.36,6.49,6.62,6.75,7.15,7.41,7.68,7.95,8.08,8.34,8.61,8.87,
9.27,10.20,10.33,10.46,10.73,10.86,11.92,12.05,12.98,13.51,13.77,
14.03,14.17,14.96,15.90,16.02,16.29,16.95,17.21,18.14,18.67,19.34))

        model{
            # Likelihood
            for(i in 1:72){
                I.N.obs[i] <- 1
                I.N.obs[i] ~ dbern(psi)
                I.n.obs[i] <- 1
                I.n.obs[i] ~ dbern(pc)
                x.obs[i] ~ dunif(0,w)
                p[i] <- exp(-tau/2*pow(x.obs[i],2))
                pi[i] <- I.N.obs[i]*I.n.obs[i]*p[i]
                D.obs[i] <- 1
                D.obs[i] ~ dbern(pi[i])
            }
            for(i in 73:300){
                I.N.mis[i] ~ dbern(psi)
                I.n.mis[i] ~ dbern(pc)
                x.mis[i] ~ dunif(0,w)
                p[i] <- exp(-tau/2*pow(x.mis[i],2))
                pi[i] <- I.N.mis[i]*I.n.mis[i]*p[i]
                D.mis[i] <- 0
                D.mis[i] ~ dbern(pi[i])
            }
            N <- sum(I.N.obs[1:72])+sum(I.N.mis[73:300])
            n <- inprod(I.N.obs[1:72], I.n.obs[1:72])
                    +inprod(I.N.mis[73:300], I.n.mis[73:300])
            # Priors
            tau ~ dunif(0,10)
            psi ~ dbeta(1,1)
        }
```

The second term on the right-hand side of Eq. (8.10) is referred to as a finite population correction; for sample size $m > 1$, it is always less than 1.

An appealing feature of Bayesian inference is the straightforward way in which finite population inference is carried out; there is an automatic adjustment of uncertainty for the fraction of the population sampled. This benefit extends to hierarchical sampling designs, for which the adjustments are a challenging problem under frequentist analyses.

The Bayesian approach to finite population modeling, pioneered by Rubin (1976), is to treat the problem as one of missing data (for a detailed discussion, see Gelman *et al.*, 2004). Thus, we conclude the present chapter on hidden data with several examples of Bayesian finite population sampling. Once again, use of the CDL will be seen to greatly simplify the modeling process.

8.5.1 Muskrats: A Simple Sample Survey

Williams *et al.* (2002) describe a study in which the abundance of muskrat *Ondatra zibethi-cus* houses in a 100-ha marsh was estimated using air boat searches. The study marsh was divided into 50 2-ha square plots, of which 10 were selected at random for a complete search (Table 8.4).

Williams *et al.* (2002) provide the standard frequentist solution to estimate the number of muskrat houses denoted N. Letting M denote the number of plots and m the number of samples, N is estimated by

$$\hat{N} = M\bar{y} = 50 \times 12.1 = 605,$$

with estimated standard error

$$\hat{\text{SE}}(\hat{N}) = \sqrt{\hat{\text{Var}}(\hat{N})} = \sqrt{M^2 \frac{s_y^2}{m}\left(1 - \frac{m}{M}\right)} = 48.28. \tag{8.11}$$

Note that the infinite population estimator for the variance has been scaled down by a factor equal to the sampling fraction m/M.

CDL, DAG, and Bayesian Model for Muskrat Data

We describe finite population sampling using a set $y = \{y_1, y_2, \dots, y_M\}$ of population values, and a set $I = \{I_1, I_2, \dots, I_M\}$ of inclusion variables; I_j is an indicator of the event that y_j is in our sample of size m. We, thus, define $y^{\text{obs}} = \{y_j | I_j = 1\}$ and $y^{\text{mis}} = \{y_j | I_j = 0\}$. We wish to make inference about some quantity $f(y)$ based on observation of I and y^{obs}. For the muskrat house survey, $f(y) = \sum_{j=1}^{M} y_j \equiv N$.

A probability model for finite population sampling requires a joint distribution for y and I. The DAG in Fig. 8.7 presents a general formulation, with parameter vector θ governing a probability distribution of which y is regarded as a sample, and parameter vector ϕ governing the sampling mechanisms that determine I and consequently y^{obs}. Note that we have included a single arrow from y to I, acknowledging the possibility, at least, that the probability a sample unit is included has something to do with the value of the unit. Double arrows indicate deterministic relations: given y, $f(y)$ is fixed; given y and I, y^{obs} is also known.

A CDL for $y^{\text{mis}}, \theta, \phi$ is thus of the form

$$\mathcal{L}(y^{\text{mis}}, \theta, \phi | y^{\text{obs}}, I) \propto [I, y | \theta, \phi]$$
$$= [I | y, \phi] [y | \theta].$$

TABLE 8.4   Numbers of muskrat houses counted in a simple random sample of 10 fixed-area plots (from Williams *et al.*, 2002).

Plot i	1	2	3	4	5	6	7	8	9	10	$\bar{y}$	s_y
Count y_i	13	18	10	6	16	13	12	13	9	11	12.10	3.41

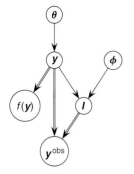

FIGURE 8.7   Directed acyclic graph (DAG) for the complete data in a sample survey. In the muskrat house survey, $f(y) = \sum y_i = N$.

Under simple random sampling, $[I|y, \phi]$ does not depend on y (as it does, say, in length-biased sampling). What is more, there are no parameters of this sampling distribution to be estimated; $[I] = c$ is a uniform distribution over the collections of sets $\{I_1, I_2, \ldots, I_M\}$ having all $I_j = 0$ or 1, and $\sum_{j=1}^{M} I_j = m$, the fixed sample size. In this design, the missing data are said to be missing "completely at random" (Gelman *et al.*, 2004). The result is that the CDL is proportional to the population data distribution $[y|\theta]$.

To complete the model, we must specify the distribution $[y|\theta]$. Since the counts are discrete, natural candidates are the Poisson or negative binomial distributions. Given that the sample mean and variance are nearly identical, a Poisson distribution seems reasonable. In the analysis of Williams *et al.* (2002), no distribution is specified for y and inference is based on an assumption of asymptotic normality. While not having to specify a distribution may seem appealing, this approach has no basis for probability-based inference, and must rely on asymptotic frequency calculations; we comment further on the matter of distributional assumptions at the end of this section.

Bayesian Analysis of the Muskrat House Data

Bayesian analysis is based on the CDL and a prior for θ. BUGS code for this problem is given in Panel 8.8 for a Poisson data model with a vague gamma prior on the Poisson mean μ.

The posterior distribution for N suggests that there are about 603 muskrat houses[11] with a 95% CI of [513, 705]. This compares with the estimate of 605 by Williams *et al.* (2002) and an approximate 95% CI of [510, 700]. From Panel 8.8, it is easy to see how Bayesian inference automatically accounts for finite population sampling. In predicting N, the total for the sampled plots `sum(y.obs[1:m])` is known. The only uncertainty in N is due to the unknown numbers on the plots that were not sampled. If there are relatively few plots that were not sampled, `sum(y.obs[1:m])` is a large fraction of N and the uncertainty in the prediction of N is small.

Finite- versus Infinite-Population Inference

Often researchers are interested in the finite population mean $\bar{Y} = \frac{1}{M} \sum_{i=1}^{M} y_i = \frac{N}{M}$.[12] In the muskrat sample, this is the average number of muskrat houses in the 50 plots. From a posterior

---

11. Posterior median $= 603$, posterior mean $= 605$.
12. We write $\bar{Y} = \frac{1}{M} \sum_{i=1}^{M} y_i$ to distinguish this quantity from $\bar{y} = \frac{1}{m} \sum_{i=1}^{m} y_i$, the mean for the observed sites.

PANEL 8.8   BUGS code for estimating the number of muskrat houses using simple random sampling.

```
list(m=10,M=50,y.obs=c(13,18,10,6,16,13,12,13,9,11))

model{
    for(i in 1:m){
        y.obs[i] ~ dpois(mu)
    }
    for(i in m+1:M){
        y.mis[i] ~ dpois(mu)
    }
    mu ~ dgamma(0.001,0.001)
    N <- sum(y.obs[1:m])+sum(y.mis[m+1:M])
}
```

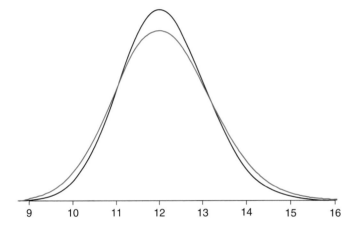

FIGURE 8.8   Posterior distribution of finite population mean $\bar{Y} = \frac{1}{M}\sum_{i=1}^{M} y_i$ (black) and infinite-population mean μ (red), for muskrat house data of Williams *et al.* (2002).

sample for N, we simply divide each value by M to obtain a posterior sample for $\bar{Y}$. The posterior distribution of $\bar{Y}$ for the muskrat data is given in Fig. 8.8 along with the posterior density for μ.

The distinction between $\bar{Y}$ and μ is important. The parameter μ represents the mean number of muskrat houses in a theoretical infinite population of 2-ha fixed-area plots, but $\bar{Y}$ represents the realized value of the mean number of muskrat houses treating the 50 plots as the population of interest. Given that we know exactly how many houses there were in 10 of these plots, it makes sense that we should have better information about $\bar{Y}$ than about μ. This is reflected in the two posterior samples summarized in Fig. 8.8; the means are nearly identical, but the standard deviation for $\bar{Y}$ is approximately 10% smaller. This is in agreement with the finite population correction factor of $\sqrt{1 - m/M} = \sqrt{0.8}$.

Another context in which a similar issue of finite- versus infinite-population inference arises is in the occupancy model discussed in Section 8.3. In this context, the parameter ψ, which is the probability that a site is occupied, is an infinite-population summary. Researchers may well be interested in a particular set of sites and wish to know how many of these are occupied, or equivalently, what fraction are occupied. Adopting a CDL approach, as outlined in Section 8.3, finite-population inference about the actual number of sites that were occupied can be accomplished using MCMC simply by summing up each sample value of the vector X that has element i equal to 1 if site i is occupied and 0 otherwise (for more information and an example see MacKenzie *et al.*, 2006).

Asymptotic Normality

Looking at the code of Panel 8.8, we note that the only use of the unobserved values `y.mis[]` is through their total. Simply as a matter of efficiency, we might choose to model their sum, rather than the individual values. The sum of independent Poisson random variables has a Poisson distribution, with mean equal to the sum of the means, so we can replace the second "for" loop with two lines,

```
Mu.mis.total <- (M-m)*mu
y.mis.total ~ dpois(Mu.mis.total)
```

Generating a single value for the missing total of observations, rather than $M - m = 40$ values leads to a 40% reduction in run time.

This gain in computational efficiency depends on our knowledge of the distribution of the sum of observations. Although we do not have such knowledge for every possible choice of distribution, we can sometimes rely on the Central Limit Theorem for an asymptotic approximation. Given that the number of unobserved values is reasonably large, and their variance is not too large, the distribution of their sum can be approximated as normal with mean $(M - m)\mu$, and variance $(M - m)\sigma^2$, where μ and σ^2 are the population mean and variance.

If the population distribution is itself normal, the $N\left((M - m)\mu, (M - m)\sigma^2\right)$ distribution for the sum of the missing values is not approximate, but exact. In this case, we can also summarize the observed data by sufficient statistics $\bar{y}$ and s^2. We do so in Panel 8.9, in analysis of counts

PANEL 8.9   BUGS code for estimating the number of cottontail rabbits using simple random sampling and assuming a normal distribution for the numbers of rabbits counted on each 1-ha plot.

```
list(M=1000,m=100,ybar=16,ss=40)

model{
    ybar ~ dnorm(mu,tau.ybar)
    tau.ybar <- m*tau
    ss ~ dgamma(a,b)
    a <- (m-1)/2
    b <- a*tau

    y.mis.total ~ dnorm(Mu.mis.total,Tau.mis.total)
    Mu.mis.total <-(M-m)*mu
    Tau.mis.total <- tau/(M-m)
    N <- m*ybar+y.mis.total

    mu ~ dnorm(0,1.0E-6)
    tau ~ dgamma(0.001,0.001)
}
```

of cottontail rabbits from a simple random sample of $m = 100$ 1-ha plots sampled from a 1000-ha study area (Williams *et al.*, 2002); the sample mean $\bar{y}$ was 16, and the sample variance was $s^2 = 40$. The BUGS code makes use of the fact that $\bar{y}$ and s^2 are independent with $\bar{y} \sim N(\mu, \sigma^2/m)$ and $(m-1)s^2/\sigma^2 \sim \chi^2_{m-1}$.[13]

Using the BUGS code in Panel 8.9, we obtain a posterior 95% prediction interval of $(14810, 17190)$, in close agreement with the 95% CI of $(14824, 17176)$ reported by Williams *et al.* (2002). We do not commend the frequentist analysis on its own merits, but as an illustration of a general feature of frequentist finite sample population methods relying on asymptotic normality. These tend to agree with fully model-based Bayesian methods when m and $M - m$ are large, or when the population distribution is assumed to be normal. More typically, finite population sampling data involve smaller sample sizes, and asymptotic approximations should be avoided.

8.5.2 Stratification

Two common alternatives to simple random sampling are stratified sampling and cluster sampling. Bayesian analysis of data obtained using these approaches is a reasonably straightforward extension of the simple random sampling problem. In the case of stratified sampling, the study population is divided into subpopulations (strata), all of which are sampled, although possibly with varying intensity. Inference is conducted at the stratum level and summarized to the population level. Parameters can be defined at the stratum level, shared across strata, or treated as random effects across strata. Stratification is often used as a strategy for efficient estimation, when the population can be divided into homogeneous subgroups.

To illustrate Bayesian inference for a stratified sampling design, we use an example from Siniff and Skoog (1964) who used a stratified random sample and aerial counts to estimate the size of an Alaskan caribou population. Sampling effort was allocated proportionally to anticipated population density. It is a simple matter to extend the code of Panel 8.9 to account for the stratification in the caribou data; we simply change the statistics and parameters to vectors with one element for each stratum (see Panel 8.10).

The overall total number of caribou can be found by summing up the predicted values for each stratum. Doing this, we obtain a 95% posterior predictive interval of $(42840, 66150)$, with median of 54500, in close agreement with the estimate of $(43050, 65943)$ found using the standard estimator as in Williams *et al.* (2002).

Note that in the BUGS code of Panel 8.10 we have specified priors of the form

$$[\mu_i, \tau_i] = [\mu_i|\tau_i][\tau_i]$$

where $[\tau_i]$ is $Ga(a, b)$ and $[\mu_i|\tau_i]$ is $N(\psi, \kappa\tau)$. This is a conjugate prior for the vector parameter $\boldsymbol{\theta} = (\mu, \tau)'$. We have set $\psi = 0$ and chosen small values for a, b, and κ, in the interest of having a noninformative prior. Given that we are using a small value for κ, it might seem a matter of little consequence whether we relate the prior precision of μ to τ, or not. The effect of the choice is more substantial than one might think: for one thing, the scaled inverse chi-squared

---

13. In Panel 8.9, the distribution of s^2 is written as

$$s^2 \sim Ga\left(\frac{m-1}{2}, \frac{m-1}{2\sigma^2}\right);$$

this alternative representation is based on the definition of chi-squared random variables as gamma random variables, and on properties of gamma random variables. For details, see Appendix B.9.

Stratum	M	m	$\bar{y}$	s^2
A	400	98	24.1	5575
B	30	10	25.6	4064
C	61	37	267.6	347,560
D	18	6	179.0	22,798
E	70	39	293.7	123,580
F	120	21	33.2	9795
Total	699	211		

PANEL 8.10 BUGS code for estimating the number of caribou from Alaskan aerial surveys using stratified random sampling.

```
list(Strata=6,M=c(400,30,61,18,70,120),m=c(98,10,37,6,39,21),
    ybar=c(24.1,25.6,267.6,179,293.7,33.2),
    ss=c(5575,4064,347556,22798,123578,9795))

model{
  for(i in 1:Strata){
      ybar[i] ~ dnorm(mu[i],tau.ybar[i])
      tau.ybar[i] <- m[i]*tau[i]
      ss[i] ~ dgamma(a[i],b[i])
      a[i] <- (m[i]-1)/2
      b[i] <- a[i]*tau[i]

      y.mis.total[i] ~ dnorm(Mu.mis.total[i],Tau.mis.total[i])
      Mu.mis.total[i] <-(M[i]-m[i])*mu[i]
      Tau.mis.total[i] <- tau[i]/(M[i]-m[i])
      N[i] <- m[i]*ybar[i]+y.mis.total[i]

      mu[i] ~ dnorm(0,tau.mu[i])
      tau.mu[i]<-1.0E-6*tau[i]
      tau[i] ~ dgamma(0.001,0.001)
  }
  ybar_st <- Total/sum(M[])
  Total <- sum(N[])
}
```

full conditional for τ gains a degree of freedom. In practical terms, as discussed by Gelman *et al.* (2004), having the prior variance for the means tied to the prior precision ensures that the prior for the means is calibrated to the scale of measurement of the observations. Because of the very large sample variances in strata C and E, prior precisions even as small as 10^{-6} would still result in mildly informative priors on μ for these strata.

Given the large sample sizes, the asymptotic approximations used in these analyses seem reasonable. Bayesian and frequentist results roughly coincide. We note, however, that 95% predictive intervals for strata B and F extend below 0,[14] suggesting some inadequacy in the

14. These are $(-275, 1811)$ and $(-791, 8764)$, with predictions of 768 and 3983, respectively.

approximations. A model-based analysis, such as used for the muskrat house data in the previous section would seem preferable. However, given that the variances so greatly exceed the means, the Poisson model (Panel 8.8) is surely inadequate. A better course of action would be to specify a model such as the negative binomial, which is consistent with the data as nonnegative and highly variable counts, and to analyze the raw data rather than summary statistics.

Siniff and Skoog (1964) report that the caribou census required "135 man-days of effort, exclusive of pre-census planning ... costs, excluding salaries, were approximately $8600" [roughly $60,000 in 2009 dollars]. Given the expense associated with gathering good data, and given the computational capacity presently available, it seems irresponsible to apply 1950's technologies and approximations to such data today.

8.5.3 Cluster Sampling

Finite population studies often involve multistage samples. The simplest example is cluster sampling, in which data are collected in two stages. First, a collection of subgroups (clusters) is sampled; next, data are collected for individuals within clusters. The data reflect variation between clusters and variation within clusters; both components of variation must be accounted for in making inference about the entire population. Cluster sampling models are naturally hierarchical, with models required for each level of the hierarchy.

To illustrate Bayesian inference of cluster sampling data, we consider a survey that was used to determine the number of breeding shearwater pairs on a small (15-ha) island. Fifty 10 m × 10 m quadrats were placed using simple random sampling, and the number of shearwater burrows were counted. Each burrow was inspected for evidence of current occupancy by breeding birds using a burrowscope. The goal of the study was to estimate the total number of occupied burrows and hence, by implication, the number of breeding pairs.

The survey data reflect two levels of variation. First, there is variation associated with random sampling of $m = 50$ quadrats (the clusters) from a finite population of $M = 1500$ quadrats. The response variable at this level is the number of burrows. We let y_i denote the number of burrows in quadrat i. Second, there is variation associated with the data within quadrats. The occupancy status of each burrow is a Bernoulli random variable. Assuming that the probability of occupancy is the same for each burrow and that burrows are independent (at least conditional on the quadrat), the number of occupied burrows x_i in quadrat i is modeled as a binomial random variable with index y_i and success rate p_i; p_i is the probability of occupancy for burrows in quadrat i.

We consider two models. The first assumes that the occupancy probability is the same for each quadrat. We fitted this model using the BUGS code in Panel 8.11. Notice that the hierarchical nature of the model means that prediction of the total number of occupied burrows also occurs in two stages. In the first stage, we predict the numbers of burrows in the unsampled quadrats, and at the second stage, we predict the the numbers of these burrows that are actually occupied.[15]

---

15. Some versions of BUGS do not allow an index of 0 in the binomial distribution; an error message is generated when `x ~ dbin(p,n)` if n = 0. A workaround is to define `a <- equals(n,0)`, `pp <- (1-a)*p`, `nn <- n+a`, and to generate `x ~ dbin(pp,nn)`. If n = 0, x is a single Bernoulli trial with success rate 0, hence x = 0. If n > 0, then nn = n, and pp = p.

PANEL 8.11   BUGS code for estimating the number of occupied shearwater burrows using a simple random sample of quadrats and counting the proportion of occupied burrows.

```
list(M=1500,m=50,burrows=c(3,3,7,2,5,3,1,3,6,7,8,4,
      4,4,3,0,5,2,3,4,5,3,3,3,2,4,4,7,5,7,4,3,2,4,1,
      6,4,4,3,5,1,1,1,7,3,10,2,4,3,4),occupied=c(2,3,1,2,
      2,3,1,0,3,7,2,3,3,3,3,0,2,2,3,3,0,2,1,1,1,4,2,6,4,
      7,3,2,2,2,1,5,3,1,3,5,1,1,1,7,0,3,1,3,1,3))

model{
    for(i in 1:m){
      burrows[i]  ~ dpois(lambda)
      occupied[i]  ~ dbin(p,burrows[i])
    }
    for(i in 1:M-m){
      burrows_mis[i]  ~ dpois(lambda)
      occupied_mis[i]  ~ dbin(p,burrows_mis[i])
    }

    N.Occupied <- sum(occupied[])+sum(occupied_mis[])
    p  ~ dbeta(1,1)
    lambda  ~ dgamma(0.001,0.001)
    }
```

Under this model, a 95% prediction interval indicates that the number of occupied burrows is in the range [3109, 4371]; the posterior median is 3704. To assess the assumptions of the model, we carried out two-stage posterior predictive assessment as outlined in Section 5.6. The first stage assessed the assumption that the occupancy probability was the same for each burrow. We carried this out by posterior predictive assessment conditional on the numbers of burrows in the quadrats, using as a test statistic

$$T^{\text{obs}} = \sum_{i=1}^{50} \frac{(x_i - y_i p)^2}{y_i p}.$$

A value of T^{obs} was computed for each iteration of the Markov chain and compared to the value computed for data generated under the model:

$$T^{\text{rep}} = \sum_{i=1}^{50} \frac{(x_i^{\text{rep}} - y_i p)^2}{y_i p}$$

where $x_i^{\text{rep}} \sim B(y_i, p)$ represents the replicated data. To test the Poisson assumption for the numbers of burrows, we computed similar sets of test statistics, this time computed as:

$$T^{\text{obs}} = \sum_{i=1}^{50} \frac{(y_i - \lambda)^2}{\lambda}$$

III. APPLICATIONS

and

$$T^{\text{rep}} = \sum_{i=1}^{50} \frac{(y_i^{\text{rep}} - \lambda)^2}{\lambda}$$

for $y_i^{\text{rep}} | \lambda \sim P(\lambda)$.

Plots of observed versus replicated test statistics indicate a lack of fit of the binomial part of the model but good fit for the Poisson part (Fig. 8.9).[16]

A likely reason for violation of the binomial assumption within the quadrats is that some quadrats are located in habitat that is currently preferred by shearwaters, and hence has a higher occupancy probability. Ideally, covariates would be available with which to model habitat preferences. We could allow the occupancy probabilities to be distinct among the quadrats, but this would give us no basis for predicting the numbers of occupied quadrats in the plots not sampled. Another approach would be to model the differences between quadrats using a random effects model.

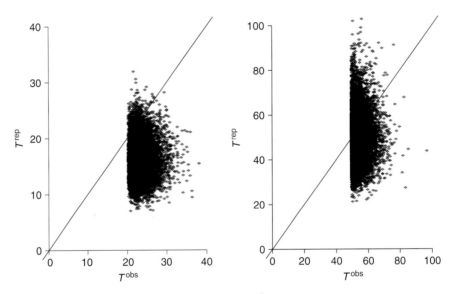

FIGURE 8.9   Scatterplot of predictive T^{rep} versus observed T^{obs} values of the discrepancy statistic for the shearwater burrow occupancy model. The plot on the left assesses the assumed binomial distribution for the numbers of occupied burrows conditional on the number present in each quadrat. The Bayesian p-value is the proportion of points above the 45° line and has a value of 0.038. The plot on the right assesses the assumed Poisson distribution for the numbers of burrows in each quadrat, and has a Bayesian p-value of 0.34.

16. In Figure 8.9, values of T^{obs} appear to cut off at a minimum value. The explanation of this phenomenon is that, for a fixed set of y_i's, T^{obs} is a chi-squared statistic, minimized by a specific choice of parameter value, known as the minimum chi-squared estimator (MCE). The MCE is sometimes considered as an alternative to the maximum likelihood estimator. For the Poisson parameter λ, the MCE is the root mean square value of the data (4.34 in this example); the MLE is the sample mean (3.84).

In our second analysis, we refitted the model but this time allowed the occupancy probability to vary among quadrats, including those not sampled, but assumed that the logits of the occupancy probabilities are normal random variables that are exchangeable among the quadrats. That is,

$$\left[\text{logit}(\boldsymbol{p})|\mu_p,\sigma_p^2\right]=\prod_{i=1}^{1500}\phi\left(\text{logit}(p_i);\mu_p,\sigma_p^2\right),$$

where μ_p is the mean occupancy probability on the logit-scale, and σ_p^2 is the variance. These changes are incorporated in the BUGS code of Panel 8.11 by replacing p's with p[i]'s defined within loops as

```
logit(p[i]) <- logitp[i]
logitp[i] ~ dnorm(mu.p,tau.p),
```

with priors

```
mu.p ~ dnorm(0,1.0E-6)
 sd.p ~ dunif(0,100)
tau.p <- 1/pow(sd.p,2).
```

Under this model, a 95% posterior prediction interval suggests that there are between 2625 and 4508 occupied burrows; the posterior median is 3539. The posterior median is lower than in the original analysis (3711), and the prediction interval is almost 50% longer. However, the fit of the binomial part of the model is much improved (Fig. 8.10).

8.5.4 Auxiliary Variables

In the shearwaters example, random effects were used to fit a model that has the flexibility of plot-specific parameters but that still allows prediction of missing values on the plots that were not sampled. An obvious improvement to this approach would be to include plot-level covariates that provide information on the missing values. There are two possible cases: (1) where covariates are measured only on the plots sampled and (2) where covariates are available for the sampled plots as well as other plots, possibly all of them.

If covariates are only available on the sampled plots then the random-effects approach used in the second shearwater analysis can be extended by adding a model for the plot covariates. Provided the plots can be regarded as exchangeable, then the model for the covariates can be used to predict the values for the missing covariates. A variation on this arises when summary information is available such as the sum of the covariate across all the plots, or its average value. This information can be exploited using methods such as ratio or regression estimation (Scheaffer et al., 1996).

Returning to the muskrat house problem discussed in Section 8.5.1, Williams et al. (2002) reported results from a second survey in which the plot sizes differed. Now, each plot has its own Poisson mean μ_i, with

$$\mu_i=\lambda a_i$$

where a_i is the area of plot i. Although the exact area of the plots is known only for the plots that were sampled, it is known that the total area covered by all the plots was 100-ha. This

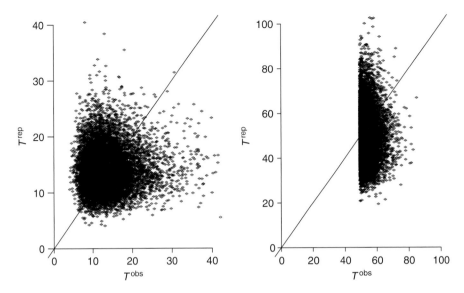

FIGURE 8.10   Scatterplot of predictive T^{rep} versus observed T^{obs} values of the discrepancy statistic for the shearwater burrow occupancy model. The plot on the left assesses the assumed binomial distribution for the numbers of occupied burrows conditional on the number present in each quadrat in which the logits of the occupancy probabilities are modeled as exchangeable $N\left(\mu_p, \sigma_p^2\right)$ random variables. The Bayesian p-value is the proportion of points above the 45° line and has a value of 0.58. The plot on the right assesses the assumed Poisson distribution for the numbers of burrows in each quadrat, and has a Bayesian p-value of 0.34.

allows us to model the observed number of muskrat houses in terms of the plot area, and then make use of the Poisson assumption and the total area of the missing plots to predict the number of muskrat houses on the plots that were not sampled.

Fitting this model using the BUGS code in Panel 8.12 leads to the posterior distribution for N displayed in Fig. 8.11, which includes a posterior predictive assessment of the model. Based on this analysis, there are around 533 muskrats with a 95% credible interval of (430, 654). The model appears to fit well. Williams *et al.* (2002) used a ratio estimator and relied on asymptotic normality to obtain a point estimate of 536 muskrats with a 95% CI of (458, 614). In this study, it appears that asymptotic normality approximations lead to a considerable overstatement of precision.

Ideally, covariates are measured on all the plots, including the ones that were excluded from the sample. Even if it is impracticable to measure the covariates on all the plots, there are still advantages if the covariates can be measured on the sampled plots and others. Double-sampling is a special case where a single covariate is selected to provide auxiliary information and chosen because it is relatively easy to measure. A sample of plots is drawn on which the covariate is measured, and a subsample of these is drawn on which the variable of interest is measured. This approach is commonly used in aerial surveys, where counts from aircraft on plots are supplemented with complete counts or abundance estimates made on a subsample of plots. When estimation is used on the subunits instead of complete censusing, the predictions must also account for uncertainty of estimation on the subunits (Barker, 2008).

PANEL 8.12   BUGS code for estimating the number of muskrat houses, making use of the area of each plot sampled and the fact that 14-ha were sampled from the 100-ha available.

```
model{
    for(i in 1:m){
            y[i] ~ dpois(mu[i])
            mu[i] <- lambda*a[i]
    }

    lambda ~ dgamma(0.001,0.001)

    lam_mis <- lambda*Area_mis
    sum_ymis ~ dpois(lam_mis)
    N <- sum(y[])+sum_ymis
}

list(m=10, y=c(15,8,6,8,7,3,3,3,9,13),
     a=c(2,1,1,1,2,1,1,1,2,2), Area_mis=86)
```

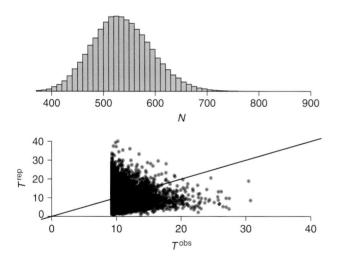

FIGURE 8.11   Posterior distribution for N, the number of muskrat houses in the variable-area plot example from Williams *et al.* (2002), analyzed using plot area as a covariate of the expected count.

8.6 AFTERWORD

In Chapter 1, we described Bayesian inference as "attractive because it is useful" and suggested that its usefulness "derives in large measure from its simplicity." The examples in this chapter surely support this perspective. In each case modeling begins with a description of "the data we wish we had," the latent data of Table 8.5. The latent data are treated as a sample from an infinite population; inference can be made either about the infinite population, or about the sampled values.

In each case, observed data are a corrupted or incomplete version of the data we wish we had. Modeling requires a statement of data corruption mechanisms, in terms of conditional

TABLE 8.5   Summary of examples, Chapter 8.

	Latent data	Data corruption
Behavior X	D_i = Use status	Randomized response
	$i = 1, 2, \ldots, M$ (known)	$Y_i = 1 - (1 - C_i)(1 - D_i)$
Occupancy	X_i = Occupancy status	Incomplete detection
(frogs)	$i = 1, 2, \ldots, M$ (known)	$[Y_{ij}|X_i] = B(X_i, p_j)$
Line transect	D_i = Detection status	Missing data
(stakes)	$i = 1, 2, \ldots, N$ (unknown)	($D_i = 0$ not observed)
Finite populations	Y_i = Count on plot	Missing data
(muskrat houses)	$i = 1, 2, \ldots, M$ (known)	(SRS of m observed)

distributions [Observed Data|Latent Data]. Frequentist analyses are usually based on the ODL, obtained by integrating these against prior distributions for the latent data. The ODL is generally much more difficult to work with than the CDL, which is based on the joint distribution of latent and observed data. Bayesian methods are readily based on CDLs. Under the Bayesian māramatanga all unknown quantities, be they latent variables, parameters, or missing values, all are treated equally. Thus, the analyses presented in this chapter reconstruct latent data – the data we wish we had – by sampling from posterior distributions.

Closed-Population Mark-Recapture Models

O U T L I N E

9.1 Introduction 201
9.2 Mark-Recapture Models and
Missing Data 202
9.2.1 Completing the Data 202

9.3 Time and Behavior Models 204
9.3.1 A Gibbs Sampler for
Model M_t 204
9.3.2 Example: Adult Female
Meadow Voles 206

9.4 Individual Heterogeneity Models 209
9.4.1 Constrained Capture Probability Models, Including
Covariates 211
9.4.2 Issues with Model M_h 214
9.4.3 Loglinear Representation 215
9.4.4 The Rasch Model and M_h 217

9.5 Example: Koalas 218
9.6 Afterword 223

9.1 INTRODUCTION

Mark-recapture, occupancy and distance sampling all involve models for missing data. In each the mechanism by which data are missing is presumed to depend on the values of unknown parameters, such as the population size or the catchability of the animals.[1] For problems of this nature, the mechanism by which the data went missing must play a dominant role in any model if reasonable inference is to result.

We begin by describing mark-recapture models in the context of missing data, an idea that we pursue further in Chapter 11 for open population models. Mark-recapture modeling has a

---

1. Such missing-data mechanisms are said to be nonignorable (Gelman *et al.*, 2004).

vast literature including Seber (1982) and subsequent reviews, and is also covered in detail in Williams *et al.* (2002). It is not our intention to show the Bayesian equivalent for each analysis but rather to emphasize some of the possibilities that result from the use of hierarchical models and Bayesian inference.

In this chapter, we highlight the usefulness of the complete data likelihood (CDL), discussed in Section 8.1, for mark-recapture modeling. We also show the use of data augmentation for abundance and hierarchical modeling for multimodel inference. In both cases, we can implement the required models in BUGS without the need for customized code based on reversible jump Markov chain Monte Carlo, despite the potential difficulties due to varying model dimensions.

9.2 MARK-RECAPTURE MODELS AND MISSING DATA

The usual way to summarize mark-recapture data is through a matrix of indicators $X_{i,j}$, with $X_{i,j} = 1$ if animal i was caught in sample j and 0 otherwise. We will refer to this matrix as X^{obs}, the observed capture history matrix. This matrix has $u.$ rows and k columns, where $u.$ is the total number of animals caught at least once during the study, and k is the number of samples. The notation $u.$ is derived from $u. = u_1 + \cdots + u_k$, where u_j is the number of unmarked animals caught in sample j.

We also refer to a matrix of missing capture histories, X^{mis}, which has $(N - u.)$ rows and k columns. The population size N denotes the total number of animals that were ever available for capture during the study. All the elements of the matrix X^{mis} are 0; the only thing that is unknown is the number of rows. Thus, we can write

$$X = \left(\begin{array}{c} X^{\text{obs}} \\ X^{\text{mis}} \end{array} \right) = \left(\begin{array}{c} X^{\text{obs}} \\ 0 \end{array} \right),$$

where 0 represents an $(N - u.) \times t$ matrix of zeros.

In carrying out a Bayesian analysis of mark-recapture data, the main issues are the same as for frequentist inference and relate to the choice of appropriate models for describing the data collection process in terms of demographic parameters. Once these models have been selected, issues surrounding choice of priors and method of describing or sketching posterior distributions are the same as for any Bayesian analysis.

9.2.1 Completing the Data

As discussed in Chapter 8, there are advantages to working with the CDL. In a closed population mark-recapture study, the complete data can be represented by the set $\{X^{\text{mis}}, X^{\text{obs}}, I\}$, where I is an N-dimension inclusion vector with ith element 1 if animal i was ever caught during the study and 0 otherwise $(i = 1, \ldots N)$. Notice that I is a deterministic function of X^{obs} and X^{mis}. Each animal that appears in X^{obs} has a 1 in the corresponding row of I and each animal that appears in X^{mis} has a 0 in the corresponding row of I. The inclusion vector I in the mark-recapture model fulfills the same function as the inclusion vector I in the simple random sample survey model of Section 8.5.1 and the vector I_x in the CDL for the line-transect model in Section 8.4.

In the CDL, the dimension N of the inclusion vector I, X^{obs}, and X^{mis} is a parameter. Conditional on N we can then treat I (and hence X^{mis}) as though it were fully observed. Once N and $u.$ are specified there is no further information in I or X^{mis}. Thus, the complete data likelihood can be written as

$$[X^{\text{obs}}, X^{\text{mis}}, I | \boldsymbol{\theta}, N] \propto [X^{\text{obs}}, u. | \boldsymbol{\theta}, N]$$
$$= [X^{\text{obs}} | u., \boldsymbol{\theta}^{\text{obs}}][u. | \boldsymbol{\theta}, N] \tag{9.1}$$

where $\boldsymbol{\theta}$ is a matrix of parameters which is partitioned into a set $\boldsymbol{\theta}^{\text{obs}}$ corresponding to the animals in X^{obs} and a set $\boldsymbol{\theta}^{\text{mis}}$ corresponding to the animals in $\boldsymbol{\theta}^{\text{mis}}$.

Normally to go from the CDL to the observed data likelihood (ODL), we need to integrate out the missing data (Section 8.1). Conditioning on N, the matrix X^{mis} becomes a matrix of zeros of known dimension, and is omitted from the likelihood. Thus, the ODL is also proportional to the right-hand side of Eq. (9.1).

The formulation of the model as in Eq. (9.1) is completely general. In practice, constraints will need to be made on $\boldsymbol{\theta}$ in order for models to be identifiable. Different models, such as those in specific closed-population families, can be constructed by specifying different forms for the terms $[X^{\text{obs}} | u., \boldsymbol{\theta}^{\text{obs}}]$ and $[u. | \boldsymbol{\theta}, N]$, and by the constraints on the elements of $\boldsymbol{\theta}$.

In closed-population studies, a series of samples is drawn from a population over a period of time during which the population is closed to additions (births and/or immigration) and deletions (deaths and/or emigration). The general parameter matrix is given by $\boldsymbol{\theta} = \{p^{\text{obs}}, p^{\text{mis}}\}$, and comprises the set of capture probabilities $p_{ij}, i = 1, \ldots, N; j = 1, \ldots, k$. Given $\boldsymbol{\theta}$, we assume that capture events are independent,[2] and a convenient form of the CDL that we will use for subsequent developments is:

$$[X | N, p^{\text{obs}}, p^{\text{mis}}] \propto \binom{N}{u.} \prod_{i=1}^{N} \prod_{j=1}^{k} p_{ij}^{x_{ij}} (1 - p_{ij})^{1 - x_{ij}}. \tag{9.2}$$

In almost all closed-population mark-recapture studies, the focus of interest is on N, and $\boldsymbol{\theta}$ can usually be regarded as a nuisance parameter: of little interest in its own right but necessary for reliable inference about N.

We distinguish two classes of models. The first class, *time and behavior models*, comprises models in which groups of animals have parameters in common. All animals, whether caught or not, can be assigned to these groups without ambiguity. Parameters may vary among sampling occasions and may also change in relation to the capture history (behavior dependency) and may depend on observed covariates. In the second class of models, the *heterogeneity models*, parameters also differ among individuals, but there is insufficient information either to unambiguously allocate them to known classes in which capture probabilities are the same for members of the class, or to express capture probabilities as functions of observed covariates.

---

2. Here, dependencies between individual captures caused by behavioral changes are expressed through p.

9.3 TIME AND BEHAVIOR MODELS

In the time and behavior models, the observed and missing animals are assumed to share parameters. That is, we assume that elements of p^{mis} equal elements of p^{obs} in a way that is completely determined. The capture probabilities in p^{obs} can vary with the sample (time) and can also depend on whether or not the animals have yet appeared in a sample (behavior). Using combinations of time or behavior constraints, we can derive the standard closed-population mark-recapture models, such as M_0, M_t, M_b, and M_{tb} of Otis *et al.* (1978).

Model M_0 specifies that $p_{ij} = p_{kl} \equiv p$ for all i, j, k, and l. That is, the capture probability is the same for all individuals and for all samples. Model M_t specifies the constraints $p_{1j} = \ldots = p_{Nj} \equiv p_j$. Thus, there are distinct capture probabilities corresponding to each sample that are the same for all individuals.

Model M_b posits detection probabilities that are constant among individuals and through time, up until the time of first capture. The model description includes a vector y, the ith element of which is the index for the sample in which the ith captured animal was first caught. Vector y can be obtained by inspection of X^{obs}. The matrix p^{obs} is constrained so that each individual has probability p of capture up until the time of first capture and probability c for all subsequent samples. We can then factor $[X^{\mathrm{obs}}|u_{\cdot}, p^{\mathrm{obs}}]$ as:

$$[X^{\mathrm{obs}}|u_{\cdot}, p^{\mathrm{obs}}] = [X^{\mathrm{obs}}|y, u_{\cdot}, c][y|u_{\cdot}, p].$$

Under M_b, if the initial capture probability is greater than the recapture probability $(p > c)$, animals are said to become trap-shy. If $p < c$, animals are said to become trap-happy. Model M_{tb}, in which the behavior effect can depend on the sample, is constructed similarly except that the single value of p is replaced by the set $\{p_1, \ldots p_k\}$ and the single value of c is replaced by the set $\{c_2, \ldots c_k\}$, where p_j and c_j are indexed by the sample j.

9.3.1 A Gibbs Sampler for Model M_t

It is instructive to consider how we can build a Gibbs sampler for model M_t. Starting with the CDL (Eq. 9.2) and the constraint $p_{1j} = \ldots = p_{Nj} \equiv p_j$, we can group terms with common p_j to obtain

$$[X|N, \theta] \propto \binom{N}{u_{\cdot}} \prod_{j=1}^{k} p_j^{n_j} (1 - p_j)^{N - n_j}, \tag{9.3}$$

where $\theta = (p_1, \ldots, p_k)'$, and n_j is the number of animals caught in sample j.

Useful priors for p_j are independent $Be(\alpha_p, \beta_p)$ distributions. Provided the prior for N does not depend on any of the p_j's, this leads to the full-conditional distribution

$$[p_j|\cdot] \propto p_j^{n_j + \alpha_p - 1} (1 - p_j)^{N - n_j + \beta_p - 1}$$

and so the full-conditional $[p_j|\cdot]$ is a $Be(n_j + \alpha_p, N - n_j + \beta_p)$ distribution.

A natural prior for N is a negative-binomial $NB(\alpha_N, \beta_N)$ distribution, arising as a Poisson distribution with its mean drawn from a $Ga(\alpha_N, \beta_N)$ distribution (Appendix B.12). As a prior for N, the Poisson distribution might be considered too restrictive because of the property that its mean and variance are the same. The negative-binomial can be thought of as a generalization of a Poisson distribution allowing the variance to take any value greater than the mean.

To derive the full-conditional distribution for N, we start with the model (9.3) and multiply it by our $NB(\alpha, \beta)$ prior to obtain,

$$[N|\cdot] \propto [X|N, \boldsymbol{\theta}][N|\alpha, \beta]$$

$$\propto \binom{N}{u.} \prod_{j=1}^{k} p_j^{n_j} (1-p_j)^{N-n_j} \times \frac{\Gamma(N+\alpha)}{\Gamma(N+1)\Gamma(\alpha)} \left(\frac{\beta}{1+\beta}\right)^{\alpha} \left(\frac{1}{1+\beta}\right)^{N}.$$

Now, we can make the substitution $N = u. + U$, where U is the number of unmarked animals in the population at the end of the experiment. By the change of variables theorem (Section 2.2.4), and ignoring terms that do not involve U (remembering that in the full-conditional for N, p_j, α, and β are all known), we obtain the full-conditional distribution for $[U|\cdot]$:

$$[U|\cdot] \propto \frac{(u.+U)!}{U!} \prod_{j=1}^{k} (1-p_j)^{u.+U} \times \frac{\Gamma(u.+U+\alpha)}{\Gamma(u.+U+1)} \left(\frac{1}{1+\beta}\right)^{u.+U}$$

$$\propto \frac{\Gamma(u.+U+\alpha)}{\Gamma(U+1)} \left(\frac{\pi_0}{1+\beta}\right)^{U}, \tag{9.4}$$

where

$$\pi_0 = \prod_{j=1}^{k} (1-p_j).$$

With a little more effort, we can show that Eq. (9.4) is the kernel of a $NB(a,b)$ distribution for U with

$$a = u. + \alpha_N$$

and

$$b = \frac{1+\beta_N - \pi_0}{\pi_0}.$$

Therefore, we can simulate a draw from the full-conditional $[N|\cdot]$ by drawing a variate U from a $NB(a,b)$ distribution, and then transforming it to N by $N = u. + U$.

Gibbs sampling now proceeds as follows:

Step 1: Beginning with a set of starting values, $\{p_j^{(0)}\}$ (anything in the range $(0,1)$ will do) draw a value $U^{(1)}$ from a $NB(a,b)$ distribution with $a = u. + \alpha_N$ and $b = \frac{1+\beta_N - \pi_0}{\pi_0}$. Set $N^{(1)} = u. + U^{(1)}$. Note that we can do this in two parts: first draw $\lambda \sim Ga(a,b)$, followed by $U \sim P(\lambda)$.

Step 2: Draw values for $p_j^{(1)}$ by sampling from $Be(n_j + \alpha_p, N^{(1)} - n_j + \beta_p)$ distributions.

Step 3: Repeat steps 1 and 2 a large number of times drawing values for $N^{(h)}$ conditional on $\boldsymbol{p}^{(h-1)}$ and values for $\boldsymbol{p}^{(h)}$ conditional on $N^{(h)}$, where h indexes the elements of the Markov chain.

To implement this Gibbs sampler we need to decide on values for α_p, β_p, α_N, and β_N that characterize our prior knowledge about $\boldsymbol{p}$ and N, respectively. With an objective analysis in mind, we can set $\alpha_p = \beta_p = 1$ (uniform prior) or $\alpha_p = \beta_p = 1/2$ (Jeffreys prior, if all other parameters are fixed). Jeffreys prior for N, when all other parameters are fixed, is given by $[N] \propto 1/N$ and is obtained by setting $\alpha_N = \beta_N = 0$ and the uniform prior by setting $\alpha_N = 1$, $\beta_N = 0$. Although both priors are improper, they lead to proper posterior distributions and so are good candidates for noninformative prior distributions. We can motivate the uniform prior by considering a proper discrete uniform distribution in which $f(N) = 1/\nu$ for $N = 1, \ldots, \nu$. For ν, large relative to N, the full conditional for N is closely approximated by the above negative binomial distribution with $\alpha = 1$ and $\beta = 0$. We can make this approximation as close as we like simply by increasing ν.

9.3.2 Example: Adult Female Meadow Voles

Williams *et al.* (2002) describe the analysis of data from a study on meadow voles (*Microtus pennsylvanicus*) at the Patuxent Wildlife Research Center. In this study, $u. = 52$ different animals were caught during five consecutive days of trapping with capture vector $n = (27, 23, 26, 22, 23)'$ (the full set of data is given as Table 14.9 of Williams *et al.*, 2002). The Gibbs sampling algorithm in Section 9.3.1 was used to generate Markov chains of 100,000 samples with two sets of starting values: $p_j^{(0)} = 0.9$ for each j and $p_j^{(0)} = 0.3$ for each j. For priors, we used the improper Jeffreys prior for N, with $\alpha_N = \beta_N = 0$, and independent $Be(0.5, 0.5)$ priors for each p_j. The resulting chains appear well-mixed, even with as few as 100 iterations (Fig. 9.1).

As a burn-in, we have discarded the first 1000 values to generate the posterior distributions summarized in Figs. 9.2 and 9.3. Posterior summaries indicate that there were around 55 females in the population with a little under half caught in each sample. There appears to be close agreement with standard frequentist results (Figs. 9.2 and 9.3).

Goodness of Fit

Because of the sensitivity of inference to model assumptions, it is important to assess how well the mark-recapture model fits the data. Assuming multinomial sampling, a frequentist test is given by Seber (1982) which uses an asymptotic chi-square distribution for the discrepancy statistic

$$T = \sum_{\omega \in \Omega} \frac{(x_\omega - \hat{e}_\omega)^2}{\hat{e}_\omega},$$

where x_ω is the number of animals with history ω, $\hat{e}_\omega$ is the expected number of animals with that history computed at the parameter estimates, and Ω denotes the complete set of histories excluding the null history $00\ldots0$.

A common problem with tests of this nature is that expected counts e_ω, estimated by $\hat{e}$, need to be reasonably large for the asymptotic approximation to be valid. A simple solution to this problem is to use posterior predictive assessment, discussed in Section 5.6. In this approach, the expected value e_ω is computed for each set of parameters drawn in the Markov chain. If we use $e_\omega^{(h)}$ to denote the hth such draw ($h = 1, \ldots, M$) for a chain of length M, the values $T_1^{obs}, \ldots, T_M^{obs}$, that are obtained using

$$T_h^{obs} = \sum_{\omega \in \Omega} \frac{(x_\omega - e_\omega^{(h)})^2}{e_\omega^{(h)}}$$

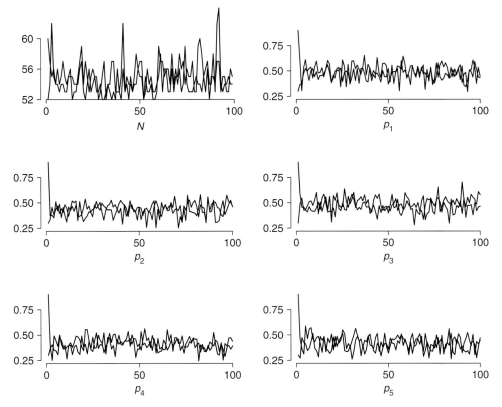

FIGURE 9.1 The first 100 values of the Markov chains generated using Gibbs sampling for the meadow vole example starting each chain at $p_j = 0.9$ for all j or at $p_j = 0.3$ for all j.

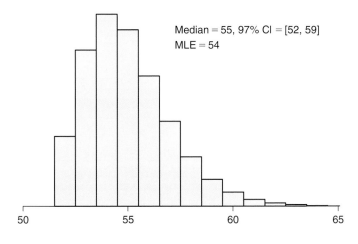

Median = 55, 97% CI = [52, 59]
MLE = 54

FIGURE 9.2 Posterior distribution for N in the meadow vole example with $Be(1/2, 1/2)$ priors for each p_j and an improper negative binomial $NB(0,0)$ prior on N.

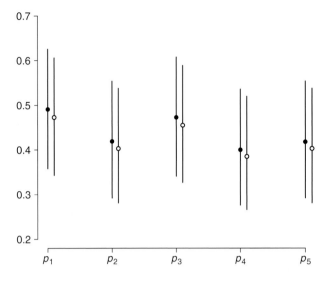

FIGURE 9.3 Posterior summaries for the capture probabilities p_j in the meadow vole example with $Be(1/2, 1/2)$ priors for each p_j and an improper negative binomial $NB(0,0)$ prior on N. The closed circles are the posterior medians and the lines denote 95% credible intervals. The open circles denote the MLE's and the associated lines asymptotic 95% confidence intervals.

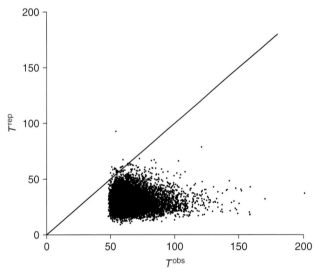

FIGURE 9.4 Scatterplot of predictive T^{rep} versus observed T^{obs} values of the discrepancy statistic for the female meadow vole data fitted to model M_t. The Bayesian p-value is estimated by the proportion of points above the 45° line and has a value of 0.002.

are compared to a similar set $T_h^{\text{rep}}, \ldots$ obtained using

$$T_h^{\text{rep}} = \sum_{\omega \in \Omega} \frac{(x_{h\omega}^{(\text{rep})} - e_\omega^{(h)})^2}{e_\omega^{(h)}},$$

where $x_{h\omega}^{(\text{rep})}$ is the hth draw from the posterior predictive distribution for x_ω. In this way, the observed data can be compared to data generated under the model.

To implement a posterior predictive goodness of fit assessment, we generate replicated data for each set of p generated by MCMC. For the female meadow vole data a plot of T^{rep} versus

T^{obs} indicates that the model fits poorly (Fig. 9.4), with a Bayesian p-value of 0.002 indicating that the observed data are considerably more dispersed than would be expected under the model, after averaging across parameters.

9.4 INDIVIDUAL HETEROGENEITY MODELS

In heterogeneity models, individuals may have distinct capture probabilities, and these may vary among samples. The CDL (Eq. 9.2) describes a model with time- and individual-specific heterogeneity, but it has more parameters (capture probabilities and N) than observations and so cannot be used for inference about N without adding constraints. To make inference possible, the various heterogeneity models introduce assumptions about how the capture probabilities of the individuals that we did not catch are related to the ones that we did catch. These assumptions are expressed through a model $[p^{obs}, p^{mis}|\theta, N]$ or $[p^{obs}, p^{mis}|\theta, N, z]$, where z denotes covariates. Because covariates are (usually) unobserved for the animals that are not captured, complete specification of the CDL also requires a model $[z|\gamma]$ for covariates if these are to be included (Fig. 9.5).

Model M_h of Otis *et al.* (1978) introduces the constraint $p_{i1} = p_{i2} = \ldots = p_{ik} \equiv p_i$. That is, each individual has its own capture probability assumed to be constant throughout the experiment. Even with this constraint, there are still too many parameters to allow useful inference for N. Huggins (2002), and also Dorazio and Royle (2003), considered a random effects parameterization of M_h, in which the p_i's are modeled as exchangeable beta random variables. Huggins (2002) developed empirical Bayes estimation procedures, in which estimates of the beta parameters were obtained by method of moments and population size estimated using a Horwitz–Thompson-type estimator. For a model such as this, frequentist inference can also be carried out using an ODL of the form: $\mathcal{L}(N, \alpha, \beta|u., X^{obs})$, obtained by integrating out the p_i's, now random effects, from the CDL represented graphically in Fig. 9.5. This is the approach taken by Dorazio and Royle (2003).

Instead of integrating out random effects to obtain an ODL, we can make use of the CDL (Eq. 9.2) to fit the model in BUGS. Fitting model M_h in BUGS is straightforward but for one minor exception: because N is updated each iteration, the dimension of p varies, but BUGS

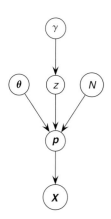

FIGURE 9.5 Directed acyclic graph (DAG) representation of the complete data likelihood for closed population mark-recapture with heterogeneous capture probabilities where the relationship between p^{obs} and p^{mis} expressed through a model $[p|\theta, N, z]$ where $p = (p^{obs}, p^{mis})'$ and z denotes covariates.

does not allow a loop index to be stochastic. A way around this is to use data augmentation in which N is modeled as the sum of M Bernoulli random variables each with probability ψ, as suggested by Royle *et al.* (2007). Adopting a uniform prior for ψ induces a discrete uniform $DU(\{0,1,\ldots,M\})$ prior on N. In this approach, we explicitly model the inclusion indicator I of Eq. (9.1). Also, each element of the vector p, which is of dimension M, is updated regardless of whether or not the animal is included in the population in the current iteration of the Markov chain.

The data augmentation algorithm suggested by Royle *et al.* (2007) is equivalent to a reversible jump algorithm, as described in Section 7.3.2, with a discrete uniform prior for N. Here, the 'palette' of parameters corresponds to the full set of probabilities in the node p [ ] and defined for all M animals in the augmented population. The bijection matching parameters on the palette to capture probabilities in the population is defined in terms of the indicator variables in the node w, which corresponds to the inclusion variable I in Eq. (9.1).

Using the BUGS code in Panel 9.1, we fit the beta-binomial model that was considered by Huggins (2002) and Dorazio and Royle (2003) to the snowshoe hare data in Agresti (1994). Specification of the data in BUGS requires the following:

- A value for M, which is data, and represents the upper limit of the discrete-uniform prior on N.
- A value for k (the number of sampling occasions); in this case $k = 6$.
- A data vector x of length M, with ith element, the number of times that individual i was caught during the study ($i = 1,\ldots,M$). This is set to 0 for the $N - u.$ animals that were not caught. For the snowshoe hare data x can be constructed from the model M_h sufficient statistics $f = (25,22,13,5,1,2)'$, where f_j is the number of animals caught exactly j times ($j = 1,\ldots,k$).
- A data vector w with element i is assigned the value 1 if individual i was caught and NA otherwise ($i = 1,\ldots,M$).

PANEL 9.1   BUGS code for model M_h fitted using data augmentation for N, modeled as a latent random variable sampled from a binomial $B(M,\phi)$.

```
model{
      mu~dunif(0,1)
      t~dt(0,1,2)
      theta<-abs(t)
      alpha<-mu*theta
      beta<-(1-mu)*theta
      psi~dunif(0,1)
      for(i in 1:M){
          w[i]~dbern(psi)
          p[i]~dbeta(alpha,beta)
          wp[i]<-p[i]*w[i]
          x[i]~dbin(wp[i],k)
      }
      N<-sum(w[])
}
```

The prior used in our BUGS code warrants some explanation. The beta model is sometimes better parameterized using $\mu = \alpha/(\alpha+\beta)$ $(0 < \mu < 1)$ and $\theta = \alpha+\beta$ $(\theta > 0)$ (Appendix B.10). These parameters are more easily interpretable and naturally assigned independent priors. Also, their use can lead to Markov chain Monte Carlo (MCMC) with better mixing[3] of the chains than if we assign independent priors to α and β.

A natural reference prior to use for μ is a $U(0,1)$ distribution. Identifying an appropriate reference prior to use for θ is more difficult, as variance-type parameters in hierarchical models need to be chosen carefully to ensure that the posterior distribution is proper. This issue has been discussed by Gelman (2006) and Gelman and Hill (2007) who recommend the use of the folded t-distribution. Note that if X has a t-distribution then $Y = |X|$, the absolute value of X, is said to have a folded t-distribution.

In the BUGS code in Panel 9.1, we have used a folded t-distribution with two degrees of freedom[4] as a prior for θ that gives moderate support for values near zero without the impropriety issues identified by Gelman (2006) and Gelman and Hill (2007) associated with the $Ga(\varepsilon,\varepsilon)$ prior, when ε is small. With the back-transformation $\alpha = \mu\theta$ and $\beta = (1-\mu)\theta$, the prior distribution that is induced for α is the same as the prior induced for β and has a mean of 0.71 and SD of approximately 2.5, with 69% of probability mass on both α and β below 1 (U-shaped beta distributions), 88% mass on both less than 2, and 98% mass on both less than 5.

To assess prior sensitivity, we also considered a folded t-distribution with scale parameter equal to 5, obtained by replacing the line specifying θ in the above code by `theta<-abs(t*5)`. The prior distribution that is induced for α and β has a mean of 3.5, an SD of approximately 3.5, with 19% of mass on both α and β below 1, 36% mass on both less than 2, and 69% mass on both less than 5. The results from fitting these two priors suggests there is quite a bit of sensitivity to the choice of prior (Figs. 9.6 and 9.7). This sensitivity reflects the limited information in a data set consisting of only 6 observations.

9.4.1 Constrained Capture Probability Models, Including Covariates

In addition to model M_h, we can also use BUGS to fit models M_t and M_b and M_0 by applying the appropriate constraints on $\boldsymbol{p}$. In these cases, the code can also be simplified, if we wish, by removing the data augmentation step for N because the dimension of $\boldsymbol{p}$ does not change.

It is also easy to fit models with individual covariates, including the special case of model M_h considered by Huggins (1989, 1991). A difficulty with this version of M_h is that the covariates are unobserved for the $N - u.$ animals that were never caught. However, provided a suitable model for these covariates exists, we can still include the covariates in the model.

Here we consider the deer mice (*Peromyscus maniculatus*) data of V. Reid (Otis *et al.*, 1978) modeled by Huggins (1991), where it is assumed that the detection probabilities are a deterministic function of covariates describing sex, age, and weight of each animal. To fit this model, we need to predict the missing sex, age, and weight of the $N - u.$ uncaught ani-

---

3. In MCMC, Markov chains are described as mixing well if they produce representative samples of the stationary distribution, without excessive autocorrelation.

4. BUGS does not allow draws from a t-distribution with 1 degree of freedom; however, we could have induced such a draw by creating a random variable $Y = Z_1/Z_2$, where Z_1 and Z_2 are independent standard normal variates.

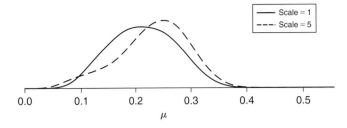

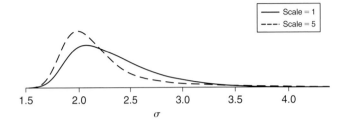

FIGURE 9.6   Posterior density for μ and θ in the snowshoe hare example using a $U(0,1)$ prior for μ and folded t-distributions for θ with scale of 1 or 5.

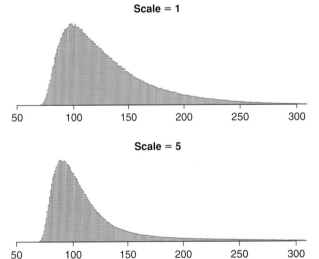

FIGURE 9.7   Posterior density for N in the snowshoe hare example using a $U(0,1)$ prior for μ and folded t-distributions for θ with scale of 1 or 5.

mals. If we use z to denote covariates, we do this by adding a term $[z|\gamma]$ to the model, with z partitioned into observed and missing components. For this example, we model standardized weight as a normal random variable with different means for each distinct age and sex class. To model age and sex, we treat them as exchangeable Bernoulli random variables. BUGS code for fitting this model is given in Panel 9.2 with a representation of the data in Panel 9.3.

The posterior distribution for N is given in Fig. 9.8. A reasonable point estimate for N is 39 (the posterior mode; note also that $Pr(N \leq 39|Data) = 0.48$ and $Pr(N \leq 40|Data) = 0.69$).

PANEL 9.2   BUGS code for model M_h including covariates for the the probability of detection. The model is fitted using data augmentation for N, modeled as a latent random variable sampled from a $B(M, \psi)$ distribution.

```
model{
      for(i in 1:M){
            w[i]~dbern(psi)
            logit(p[i])<-b[1]+b[2]*sex[i]+b[3]*age[i]+b[4]*weight[i]
            mu[i]<-g[1]+g[2]*sex[i]+g[3]*age[i]
            weight[i]~dnorm(mu[i],tau)
            sex[i]~dbern(pie[1])
            age[i]~dbern(pie[2])
            wp[i]<-p[i]*w[i]
            x[i]~dbin(wp[i],T)
      }
      N<-sum(w[])

      psi~dunif(0,1)
      for(i in 1:4){
            b[i]~dnorm(0,0.1)
            g[i]~dnorm(0,0.1)
      }
      tau~dgamma(0.001,0.001)
      pie[1]~dunif(0,1)
      pie[2]~dunif(0,1)
}
```

PANEL 9.3   BUGS data for model M_h including covariates for the probability of detection. M specifies the upper limit on population size and must equal the dimension of the data vectors x, w, sex, age, and weight.

```
list(M=100,T=6,
x=c(6,4,4,5,6,5,5,4,6,5,5,5,6,4,2,2,3,2,3,3,3,2,3,1,4,3,4,1,2,1,3,2,1,1,1,1,
1,1,0,...,0),
w=c(1,1,1,1,1,1,1,1,1,1,1,1,1,1,1,1,1,1,1,1,1,1,1,1,1,1,1,1,1,1,1,1,1,1,1,1,
1,1,NA,...,NA),
sex=c(0,1,0,0,0,0,0,0,0,0,0,1,0,0,1,1,1,0,0,0,1,1,0,1,1,1,1,1,1,0,1,0,1,0,
0,0,1,NA,...,NA),
age=c(0,0,0,0,0,1,0,1,0,0,0,1,0,0,0,1,0,0,1,0,0,1,0,0,1,0,0,1,1,1,1,0,0,0,1,
0,1,0,NA,...,NA)
weight=c(-0.522,0.098,0.098,0.098,-0.316,1.339,-0.729,0.098,-0.109,-0.109,
-0.316,1.545,-0.109,-0.729,-0.936,1.752,-1.556,-1.349,0.925,-0.316,-1.970,
1.132,-0.522,-1.763,1.545,-0.936,-0.109,0.925,0.925,1.132,0.305,-0.729,
-0.109,-0.729,1.959,-1.143,0.305,0.925,NA,...,NA))
```

The interval [38,43] has posterior probability of 0.95, hence is the 95% highest posterior density interval for N.[5] This is somewhat shorter than the approximate interval of (38.3, 45.6) obtained using the Horwitz–Thompson-type estimator by Huggins (1991). There is something

---

5. Percentile summaries in BUGS are based on methods for continuous data, and should be used with caution in summarizing discrete distributions. In the present case, posterior percentiles 2.5, 50, and 97.5 of N are reported as 38, 40, and 45. In reality $\Pr(N \leq x | \text{Data}) = 0.210, 0.687,$ and 0.984 for $x = 38, 40,$ and 45.

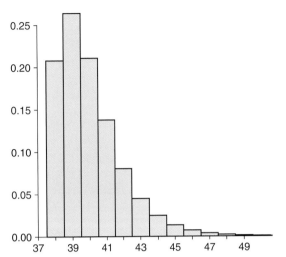

FIGURE 9.8   Posterior density for N, the number of deer mice in the example used by Huggins (1991).

unsettling about the approximate interval and its fractional endpoints: are we to round off these numbers? If so, what is the effect on confidence level?

The need to model the missing covariates highlights an important feature of the modeling approach adopted by Huggins (1989, 1991): the assumption that encounter histories are independent given the covariates that were measured. This assumption is rather strong and is untestable. In principle, we can extend the BUGS code of Panel 9.2 to include a random effect to account for unmeasured influences on detectability but, as with all models in the M_h family, we expect that this approach will require a lot of data to be useful.

9.4.2   Issues with Model M_h

As shown above, inference using model M_h may be influenced by the choice of prior parameters for the distribution governing the variation in p between individuals. This is not the only issue; inference under model M_h is also sensitive to the choice of family of distributions for modeling p. In fact, model M_h is not actually identifiable, in the sense that different choices of distribution for p lead to data distributions that are identical or are nearly identical, while indicating vastly different values of N (Link, 2003).

The nonidentifiability, or near nonidentifiability, of model M_h means that different choices for N and $[p|\theta]$ lead to an identical, or near identical likelihood, and we cannot discriminate among models solely on the basis of the data. Choice of the model $[p|\theta]$ must be guided by knowledge of the population being studied and the way in which the mark-recapture data were obtained. However, in many studies there will not be a lot of information on which to base model choices. In such cases, we recommend that researchers consider a variety of models and caution against the routine application of default methods. We offer further thoughts on this issue at the end of this chapter (Section 9.6).

Faced with closed-population mark-recapture data in which heterogeneity is suspected, there are a number of approaches that can be adopted. The approach discussed above used a continuous distribution to model p. Models based on dividing the population into discrete classes according to their catchability have also been proposed (Agresti, 1994; Norris and Pollock, 1996; Coull and Agresti, 1999; Pledger, 2000). These so-called latent-class or finite-mixture models can be fitted with relatively minor modification to the BUGS code. The finite-mixtures approach is quite general and can be extended to allow specific versions of the general model M_{tbh} in which time, behavior, and individual effects are all present (Pledger, 2000). Similar models for continuous mixtures have also been proposed by King and Brooks (2008) who also describe a reversible-jump algorithm for assessing the relative support for different models.

9.4.3 Loglinear Representation

A very general representation of mark-recapture models makes use of incomplete contingency table models to describe dependencies between the different samples (Agresti, 1994; Cormack, 1989; Fienberg, 1972). This approach also allows assessment of a variety of models with different assumptions about time, behavior, and individual effects on catchability. These models are also appropriate for multiple-list studies where there is no time ordering of the samples, as in the example discussed in Section 9.5.

Contingency tables are used for cross-classified data, and in most applications are used to assess the types of dependencies that exist between multivariate categorical data. Bayesian inference methods for general contingency table problems are well-developed (Albert, 2007; Gelman *et al.*, 2004), and inference typically focuses on the selection of appropriate models, with different models expressing different forms of dependence between variables. In a mark-recapture setting, inference focuses on N, which can be written as a function of table parameters. However, model-selection remains useful if a parsimonious model is desired.

Data from a k-sample mark-recapture study can be represented in the form of a 2^k contingency table with one structural zero corresponding to the cell for individuals that were available for capture during the study but never caught. The contingency table is constructed from a cross-classification of all individuals according to whether or not they were caught in each sample. Associated with each cell is the number of animals that have the corresponding capture history. For example, in a two-sample study, we obtain the incomplete 2×2 contingency as shown in Table 9.1. Note that x_{hl} in Table 9.1 is the number of animals with history $X_{i1} = h$ and $X_{i2} = l$.

TABLE 9.1 Example of an incomplete 2×2 contingency table for a two-sample mark-recapture study.

	Not caught in Sample 2	Caught in Sample 2	Total
Not caught in sample 1	—	x_{01}	$N - n_1$
Caught in sample 1	x_{10}	x_{11}	n_1
Total	$N - n_2$	n_2	N

One approach to modeling contingency tables is by loglinear models, with cell frequencies modeled as independent Poisson random variables.[6] For our two-way table, the observed cell counts under model M_t can be modeled as $P(\mu_{hl})$ random variables, where $\mu_{hl} = E[x_{hl}]$ is the expected cell count in row $h+1$ and column $l+1$ of the table. The natural logarithm of the expected cell count is modeled using a linear predictor. For the data in Table 9.1, the linear predictor can be written as:

$$\eta_{hl} = \ln(\mu_{hl}) = \begin{cases} \beta_0 & h=0, l=0 \\ \beta_0 + \beta_2 & h=0, l=1 \\ \beta_0 + \beta_1 & h=1, l=0 \\ \beta_0 + \beta_1 + \beta_2 & h=1, l=1 \end{cases}.$$

To see the relationship between the loglinear model parameters and the more usual capture probability formulation, we can compare them in terms of the cell probabilities of the contingency table (Table 9.1). For the cell corresponding to the count x_{11}, the expected value using the loglinear formulation is $e^{\beta_0+\beta_1+\beta_2}$ and for the usual capture probability formulation of the model it is Np_1p_2. For the cell corresponding to the count x_{01}, the expected value using the loglinear formulation is $e^{\beta_0+\beta_2}$ and for the usual capture probability formulation of the model it is $N(1-p_1)p_2$. It follows then that the expected number of animals caught in sample 2 is

$$Np_2 = E[x_{11} + x_{01} | \boldsymbol{\beta}] = e^{\beta_0+\beta_2}(1 + e^{\beta_1}).$$

Thus,

$$\frac{Np_1p_2}{Np_2} = p_1 = \frac{e^{\beta_0+\beta_1+\beta_2}}{e^{\beta_0+\beta_2}(1+e^{\beta_1})} = \frac{e^{\beta_1}}{1+e^{\beta_1}}$$

from which it follows that β_1 is the logit of the capture probability p_1. With a similar argument, we can show that β_2 is the logit of p_2.

With three observations x_{11}, x_{10}, x_{01}, and three parameters $\beta_0, \beta_1, \beta_2$, the model is saturated. Fitting model M_t has used up all the information in the data under the loglinear model. If we could observe the missing cell x_{00}, this would allow us to generalize the model to allow for a dependency between capture in sample 1 and capture in sample 2.

In the context of the loglinear model, any dependencies between the two samples are modeled using an interaction parameter. For our 2×2 example, we can write the linear predictor corresponding to the count x_{11} as:

$$\eta_{11} = \beta_0 + \beta_1 + \beta_2 + \beta_{12}.$$

Here, we can interpret the interaction parameter β_{12} as the amount that the logit of p_2 is changed by as a result of the animal having been caught in sample 1. In this context, it would be natural to think of β_{12} as an index of behavioral response to capture (i.e., trap-happiness or trap-shyness). However, $-\beta_{12}$ indexes the effect of capture in sample 2 on the probability of capture in sample 1; the two interpretations are equivalent. The best way to interpret the interaction is simply as a measure of dependence between samples. This might be caused

---

6. This approach is reasonable because collections of independent Poisson random variables, conditional on their totals, are multinomial random variables (see Appendix B.7).

by a behavioral response, but it could also be caused by individual heterogeneity in capture probabilities, a point we will discuss later.

With observed data the absence of the count x_{00} means that we must assume a value for β_{12}. Usually, we assume that the two samples are independent, given N, which is the same as assuming $\beta_{12} = 0$. This assumption is forced on us by the missing cell and is untestable. If we knew how many individuals had been missed, we would be able to test the hypothesis of independence, or more interestingly, we could estimate β_{12}.

As discussed above, the loglinear parameter β_1 corresponds to logit(p_1) and β_2 corresponds to logit(p_1). The third parameter β_0 also has a simple interpretation:

$$\beta_0 = \ln(E[x_{00}]).$$

Thus, a useful way to fit model M_t to two-sample data is to use a loglinear model to estimate the parameters β_0, β_1, and β_2. To estimate population size, we need a prediction for x_{00}, the number of animals in the missing cell. Given a posterior sample for β_0, or an explicit posterior distribution, we could carry out posterior prediction for x_{00} by modeling it as a Poisson random variable with mean e^{β_0}.

The loglinear model formulation generalizes to 2^k tables ($k > 2$) in which case terms for interactions between samples can be included up to the order of $k - 1$. For example, with three samples the second-order interactions $\beta_{12}, \beta_{13}, \beta_{23}$ are identifiable but not the three-way interaction β_{123}.

The loglinear modeling approach to mark recapture is very flexible and includes the standard time and behavior models as special cases. For example, if we include just the main effects $\beta_1, \ldots, \beta_k$ and set all interactions to zero, we obtain model M_t of Otis et al. (1978). To fit model M_b, we include $k - 1$ sample-specific indicators for whether the animal has previously appeared in a sample and then fit the model with no time effects on either the main effects or the effect of previous capture.

9.4.4 The Rasch Model and M_h

Loglinear models for incomplete contingency tables can also be used to model heterogeneity (Agresti, 1994). Intuitively, heterogeneity in sighting probabilities induces dependence between samples and so should lead to nonzero interaction terms. In a study where the same method was used to catch animals in all samples, we might expect that an animal that is easy to catch in one sample, would be easy to catch in all of them, assuming that we have no behavior effects. Thus, knowing that an animal was caught in the first sample tells us that we are more likely to see it in later samples than one not caught in the first sample.

This connection between interactions in contingency tables and model M_h can be illustrated using the Rasch (1961) model, originally developed in the context of item-response modeling in psychometrics (Agresti, 2002). In the Rasch model each animal has its own probabilities of detection, which are modeled as:

$$\text{logit}(p_{ij}) = \alpha_i + \gamma_j, \tag{9.5}$$

$$\alpha_i \sim N(0, \sigma^2)$$

where i ($i = 1, \ldots N$) indexes animal and j ($j = 1, \ldots, k$) indexes the sampling occasion (Agresti, 1994). The detection probabilities depend on the individual-specific parameter α_i, assumed to have been drawn from a normal distribution with mean 0 and variance σ^2, and an additive effect (on the logit scale) as determined by the parameter γ_j. Thus, the logits of the detection probability are individual-specific and vary among animals in a parallel manner.

There is an interesting connection between the Rasch model and loglinear models for contingency tables. Let ω denote a capture history, that is, a vector of capture indicators ω_j ($j = 1, \ldots, k$). Let x_ω be the number of animals with the capture history, ω. It can be shown that if we average across the animal population (i.e., take expectations over the $\{\alpha_i\}$), the expected numbers of animals with history ω is given by:

$$\mu_\omega = \beta_0 + \beta_1 \omega_1 + \cdots + \beta_k \omega_k + \lambda(f_\omega) \tag{9.6}$$

where $\lambda(f_\omega)$ is a parameter indexed by the number of times an animal with the history ω is caught (Agresti, 1994). Tjur (1982) showed that maximum likelihood estimates of the β parameters in Eq. (9.6) are conditional maximum-likelihood estimates in the Rasch model Eq. (9.6), where the conditioning is on the number of times each animal was caught (i.e., the individual contributions to the capture frequencies, f_ω). These numbers represent the sufficient statistics for the parameters α_i in Eq. (9.6). Agresti (1994) refers to this as a model for "quasi-symmetry." "Symmetry" corresponds to the case where all main effects are equal and also all sets of interactions are equal; "quasi-symmetry" corresponds to the case where all sets of interactions are equal but the main effects differ.

We can fit this loglinear model using standard statistical software, by specifying x_ω as a Poisson random variable with explanatory variables ω and a factor equal to $\sum_{j=1}^{k} \omega_j$. Unfortunately, this model is over-parameterized and as a consequence it is not possible to estimate the parameter β_0 and hence the population size N. Although no reliable inference for N is possible, it can still be useful to fit this model for goodness-of-fit testing purposes. Coull and Agresti (1999) provide interesting discussion of this model and its connection with finite mixture models as considered by Pledger (2000).

If all three-way and higher order interactions are set equal to 0, and the two-way interactions set equal (but not 0), then we obtain the model labeled as two-factor quasi-symmetric by Agresti (1994). The quasi-symmetry model is a special case of M_{th} in which capture probabilities vary between animals and over time. The parameter β_0 is identifiable in this model, and so we can estimate abundance. If we also constrain the main effects to be equal, then we obtain a version of model M_h corresponding to $\beta_1 = \ldots = \beta_k$ in the Rasch model Eq. (9.6).

9.5 EXAMPLE: KOALAS

The data in Table 9.2 are based on an Australian study, in which three different observers traversed a study forest and mapped the locations of koalas that they saw. This mapping was used to form a three-list mark-recapture study with sighting histories summarized using a three-way cross-classification of koala sightings.

TABLE 9.2   Counts of koalas by three observers represented by a $2 \times 2 \times 2$ cross-classification of koala sightings.

	C = 0		C = 1	
	B = 0	B = 1	B = 0	B = 1
A = 0	NA	7	6	7
A = 1	8	14	12	43

A, B and C, are indicators for sightings by the three observers.

To illustrate the flexibility of the loglinear approach, and also to provide another example illustrating the use of Bayesian multimodel inference, we fit 10 models, each of which describes a different set of dependencies among observers. The first eight models include distinct main effects and various combinations of two-way interactions. Another model is the version of model M_h obtained by setting the main effects to be equal and also setting the two-way interactions to be equal. The final model represents model M_0, and is obtained by setting the main effects to be equal and all interactions to zero. The BUGS data input file is given in Panel 9.4 and the BUGS code in Panel 9.5. A summary of model fitting output is given in Table 9.3.

The data statement includes structures X, Xmod, and PIM that together are used to set up the loglinear model and the different constraints considered for multimodel inference.[7] The structure X is used to define the linear predictor for each of the cells in the table, and corresponds to the design matrix Z in the matrix-based formulation of the model $\eta = Z\beta$. Rows of X correspond to cells in the contingency table and columns correspond to parameters, ordered $\beta_0, \beta_1, \beta_2, \beta_3, \beta_{12}, \beta_{13}, \beta_{23}$. For example, the first row corresponds to the cell for koalas seen by all three observers (A = 1, B = 1, C = 1). For this cell, all parameters appear, hence the row of 1's. The second row corresponds to the cell for koalas seen by observers A and B but not C. For this cell, the linear predictor is given by

$$\beta_0 + \beta_1 + \beta_2 + \beta_{12},$$

hence the row values 1 1 1 0 1 0 0.

Models are defined in the BUGS code in terms of the nodes that contribute to the likelihood function. In the BUGS code of Panel 9.5, this is accomplished through Xmod and PIM. The structure Xmod defines which parameters of a particular model are constrained to zero and which are not. Rows of Xmod correspond to models and columns to parameters, with the column orderings the same as for the X-matrix. For example, model M_t is defined by nonzero values for the parameters $\beta_0, \beta_1, \ldots, \beta_3$ and all interactions constrained to equal 0. Model M_t corresponds to the first row of Xmod, which equals 1 1 1 1 0 0 0.

To allow constraints of the form $\beta_h = \beta_l$, we use the structure PIM. The BUGS code is set up so that each of the seven nodes defined by beta is updated in every iteration. In the

7. Readers familiar with program MARK (White and Burnham, 1999) will recognize the strucure PIM as a *parameter index matrix*.

PANEL 9.4   BUGS data input file for fitting loglinear models to the koala data model including a stochastic model indicator for carrying out Bayesian multimodel inference.

```
list(K=10, Ncells=8, pi=c(0.016656, 0.052853, 0.045682, 0.087367,
0.091975, 0.221741, 0.19819, 0.284835, 0.0003505,0.0003505),

count=c(43,14,12,8,7,7,6,NA),

X=structure(.Data=c(
1,1,1,1,1,1,1,
1,1,1,0,1,0,0,
1,1,0,1,0,1,0,
1,1,0,0,0,0,0,
1,0,1,1,0,0,1,
1,0,1,0,0,0,0,
1,0,0,1,0,0,0,
1,0,0,0,0,0,0),.Dim=c(8,7)),

Xmod=structure(.Data=c(1,1,1,1,0,0,0,
                       1,1,1,1,1,0,0,
                       1,1,1,1,0,1,0,
                       1,1,1,1,0,0,1,
                       1,1,1,1,1,1,0,
                       1,1,1,1,1,0,1,
                       1,1,1,1,0,1,1,
                       1,1,1,1,1,1,1,
                       1,1,1,1,1,1,1,
                       1,1,1,1,0,0,0),.Dim=c(10,7)),


PIM=structure(.Data=c(1,2,3,4,5,6,7,
                      1,2,3,4,5,6,7,
                      1,2,3,4,5,6,7,
                      1,2,3,4,5,6,7,
                      1,2,3,4,5,6,7,
                      1,2,3,4,5,6,7,
                      1,2,3,4,5,6,7,
                      1,2,3,4,5,6,7,
                      1,2,2,2,5,5,5,
                      1,2,2,2,5,6,7),.Dim=c(10,7)))
#K = no. models, Ncells = no. cells in the table
```

terminology of Section 7.3.2, the "palette" of parameters is represented by the elements of the node beta. The bijection from the palette to the parameters in a particular model is defined in terms of the rows of PIM and Xmod.

The structure PIM assigns a particular element of beta to the node b, which is the same dimension as β. Each element of b is then multiplied by the appropriate element of Xmod, which indicates whether or not it is constrained to zero. If we look at the ninth row of PIM,

PANEL 9.5   BUGS code for fitting loglinear models to the koala data model including a stochastic model indicator for carrying out Bayesian multimodel inference.

```
model{
     Model~dcat(pi[])
     for(i in 1:K){
          MI[i]<-equals(Model,i)
     }
     for(i in 1:Ncells){
          count[i]~dpois(lam[i])
          log(lam[i])<-X[i,1]*b[1]+X[i,2]*b[2]+X[i,3]*b[3]
           +X[i,4]*b[4]+X[i,5]*b[5]+X[i,6]*b[6]+X[i,7]*b[7]
     }
     for(i in 1:Ncells-1){
          beta[i]~dnorm(0,0.1)
          b[i]<-Xmod[Model,i]*beta[PIM[Model,i]]
     }
     N<-sum(count[])
}
```

TABLE 9.3   Bayes factors (BF, scaled to model 1) and posterior model probabilities $Pr(M_i|Y)$ for 10 loglinear models fitted to the koala data.

Model	Interpretation	k	BF	$Pr(M_i|Y)$ Ock	Const	Comp
1. $\beta_0 + \beta_1 + \beta_2 + \beta_3$	M_t	4	1.000	0.002	0.012	0.039
2. $\beta_0 + \beta_1 + \beta_2 + \beta_3 + \beta_{12}$	{A,B}{C}	5	0.299	0.000	0.004	0.032
3. $\beta_0 + \beta_1 + \beta_2 + \beta_3 + \beta_{13}$	{A,C},{B}	5	0.337	0.000	0.004	0.036
4. $\beta_0 + \beta_1 + \beta_2 + \beta_3 + \beta_{23}$	{B,C},{A}	5	0.191	0.000	0.002	0.020
5. $\beta_0 + \beta_1 + \beta_2 + \beta_3 + \beta_{12} + \beta_{13}$	{A,B},{B,C}	6	0.164	0.000	0.002	0.047
6. $\beta_0 + \beta_1 + \beta_2 + \beta_3 + \beta_{12} + \beta_{23}$	{A,C}{B,C}	6	0.071	0.000	0.001	0.020
7. $\beta_0 + \beta_1 + \beta_2 + \beta_3 + \beta_{13} + \beta_{23}$	{A,B}{A,C}	6	0.079	0.000	0.001	0.023
8. $\beta_0 + \beta_1 + \beta_2 + \beta_3 + \beta_{12} + \beta_{13} + \beta_{23}$	{A,B,C}	7	0.052	0.000	0.001	0.041
9. $\beta_0 + \beta. + \beta. + \beta.. + \beta.. + \beta..$	M_h	3	36.341	0.242	0.453	0.522
10. $\beta_0 + \beta. + \beta. + \beta.$	M_0	2	41.626	0.754	0.519	0.220

Three prior model weights were used (Ock = "Ockham" with weights $\propto \exp(-k)$ favoring parsimony, Const = uniform prior model weights, and Comp = "Complexity" weights $\propto \exp(k)$ favoring complex models). The notation {A,B}{C} indicates that sightings are dependent between observers A and B but independent of C. The notation {A,B},{A,C} indicates that sightings by B and C are only related through the common association with A. The "dot" notation indicates a single parameter constrained across the subscript(s).

we see that beta[1] is assigned to the intercept parameter b[1]; beta[2] is assigned to the three main-effects parameters b[2], b[3] and b[4]; and beta[5] is assigned to the three interaction parameters b[5], b[6] and b[7]. Since the ninth row of Xmod is all 1's, we see that in the likelihood none of the parameters is constrained to 0, but the main effects are constrained to be equal, as are the interactions.

To fit the different models and to calculate the Bayes factors, we use the method outlined in Section 7.3 in which BUGS is first run to find an approximation to the Bayes factor. These approximations are then used to define model priors that lead to approximately equal posterior model weights. These prior model weights are the ones shown in Panel 9.4. The second set of prior and posterior model probabilities are then solved to get a better approximation to the Bayes factors.

For the koala data, most of the posterior support is for the last two models, regardless of which model prior was used, although the complexity prior spreads about 26% support roughly evenly among the other eight models (Table 9.3). For inference about the numbers of koalas, we can look at density summaries under each model (Fig. 9.9) and also average across models.

To obtain a posterior sample for N averaged over the different models, we made use of the posterior sample generated in the second run that was designed to have each model sampled a similar number of times. It is relatively straightforward to use R, or any similar computer package, to sample from each model with replacement, with the number of samples drawn in proportion to $\Pr(M_i|Y)$. From the model-averaged densities (Fig. 9.10), there is relatively little difference between the model-average posterior density estimates for N under the "Constant" and "Complexity" priors due to the relatively high weight placed on the heterogeneity model. In contrast, the "Ockham" prior places a lot of weight on the two-parameter model M_0, hence the shorter credible interval for N.

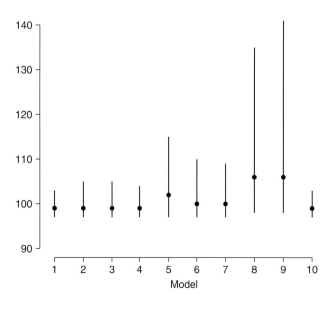

FIGURE 9.9 Posterior summaries for N under the 10 different models. The filled points indicate the median and the vertical lines span the 95% credible intervals.

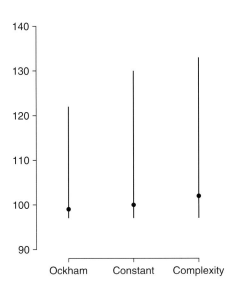

FIGURE 9.10   Posterior summaries for N averaging across the 10 different models under the three priors ("Ockham" with weights $\propto \exp(k)$ favoring parsimony, "Constant" = uniform prior model weights, and "Complexity" weights $\propto \exp(k)$ favoring complex models). The filled points indicate the median and the vertical lines span the 95% credible intervals.

9.6 AFTERWORD

For many years, the standard approach to analyzing closed population mark-recapture data has been to fit the suite of models outlined in Otis *et al.* (1978). Having fitted these models, model-selection methods are then used to choose among the various models. As we have demonstrated in this chapter, it is easy to write BUGS code for the CDL (Eq. 9.2), making use of a data-augmentation step to allow updating of the abundance parameter N. The model fitting problem then becomes one of specifying the relevant constraints for the model of interest.

Recent work on closed population mark recapture (e.g., Agresti, 1994; Huggins, 1991; Pledger, 2000; Royle *et al.*, 2007) has emphasized the development of better models to account for heterogeneity. The best hope for such modeling lies in the identification of measurable covariates affecting detection. The only alternative is to select models from the family M_h: one might choose to describe individual detection probabilities as draws from a beta distribution, a logit normal distribution, or any other distribution on [0, 1]. The problem is that the family M_h is not identifiable (Link, 2003), so that inference will be based on arbitrary and untestable assumptions.

Link (2006) illustrates with the following simple example. Suppose that we capture individuals on four sampling occasions, letting f_i denote the number of individuals captured i times. We might model individual heterogeneity by a two-point mixture (Pledger, 2000), supposing that $p = p_1$ with probability w, or $p = p_2$ with probability $(1 - w)$. If $p_1 = 1/4, p_2 = 3/4$, and $w = 3/4$ then the f_i occur in proportions 28:18:12:7. These are the exact same proportions as would occur under a $\beta(1/2, 3/2)$ model. However, the two models make vastly different predictions for N: the expected number of individuals detected is $0.508N$ under the beta model, and $0.762N$ under the two-point mixture model. Thus, the beta model indicates a population size, which is 50% larger than the two-point mixture model, and the data provide no basis for model selection.

We thus caution against the use of default models for M_h.

A better solution to the model identification problem is to obtain qualitatively different data that allows better discrimination among choices, or that shifts the focus to different parameters. In our view, closed-population mark-recapture studies that focus on N at a single point in time are of limited value. Of greater interest is the dynamics of population change. In Chapter 11, we discuss extensions of mark-recapture methods to open population. In these models, the focus of inference moves away from abundance onto survival probabilities, birth rates, and movement probabilities.

Latent Multinomial Models

OUTLINE

10.1 Model M_t 226   10.4 An Implementation of Model $M_{t,\alpha}$ 234
10.2 Model $M_{t,\alpha}$ 226   10.5 Extensions   236
10.3 Gibbs Sampling for Model $M_{t,\alpha}$ 232

It is often the case that the only observations available are corrupted or incomplete views of the processes which, ideally, would be the basis of scientific inference. The Bayesian paradigm is well suited to modeling such latent processes, and Markov chain Monte Carlo can be an invaluable tool for analysis.

In this chapter, we describe a Gibbs sampling approach for the analysis of frequency data arising as summaries of latent multinomial random variables. The approach we outline has broad applicability in mark-recapture analysis, being appropriate whenever observed capture histories are composite events. For instance, suppose that capture histories are recorded for T sampling occasions. A capture history ending with "not observed after occasion $T-1$" could mean that death occurred in the interval between occasions $T-1$ and T, or that the individual survived beyond occasion T without being detected at T. The observed frequencies of histories ending with "not observed after occasion $T-1$" are thus sums of latent frequencies.

The methods developed here are appropriate when "the data you wish you had" (Chapter 8) is a latent multinomial, of which the observed data are a summary. The methods we develop are suitable even when some of the latent categories contribute to multiple observed categories.

We consider a misidentification model, appropriate for capture–recapture data when individual identities are imperfectly determined using genetic material. The methods presented here extend the work of Lukacs and Burnham (2005) and Yoshizaki (2007). As background, and to establish notation, we once again consider Model M_t (Darroch, 1958; Otis et al., 1978, also Section 9.3.1) .

10.1 MODEL M_t

Suppose that we wish to estimate the number N of individuals in a closed population. The animals are individually identifiable, perhaps by natural markings, perhaps by the marks placed by investigators. We assume that each animal has probability p_t of being observed ("captured") on occasion t, for $t = 1, 2, \ldots, T$, and that these events are independent among individuals and through time. These assumptions constitute Model M_t.

The sequence of events associated with each animal is referred to as a capture history, and represented by a binary series of length T. For instance, an animal captured on the first and third of $T = 4$ sampling occasions and not captured on the second or fourth, is represented by the sequence $\boldsymbol{\omega} = (\omega_1, \omega_2, \ldots, \omega_T) = \{1, 0, 1, 0\}$, or simply 1010. It will be convenient to uniquely identify the capture histories on the basis of the values

$$j \equiv j(\boldsymbol{\omega}) = 1 + \sum_{t=1}^{T} \omega_t 2^{t-1}. \tag{10.1}$$

For ease of notation, we simply write j; thus history 1010 has $j = 6$.

There are 2^T distinct capture histories. Let f_j denote the number of individuals with capture history j, and let $\boldsymbol{f} = (f_1, f_2, \ldots, f_{2^T})'$. Under Model M_t, the vector $\boldsymbol{f}$ is a multinomial random variable with index N and cell probabilities

$$\pi_j = \prod_{t=1}^{T} p_t^{\omega_{j,t}} (1 - p_t)^{1 - \omega_{j,t}},$$

where $w_{j,t}$ is the tth entry in history j.

The vector $\boldsymbol{f}$ is not completely observed. We observe f_j for $j > 1$, but f_1 (the number of individuals never seen, i.e., with constant zero capture histories) is unknown. We know $n = N - f_1$, but N is unknown.

Statistical analysis is straightforward. Conditioning on n, the observed vector $\boldsymbol{f}^+ = (f_2, f_3, \ldots, f_{2^T})'$ is multinomial with cell probabilities

$$\pi_j^+ = \frac{\pi_j}{1 - \pi_1}.$$

The number of individuals observed, n, has a binomial distribution, $n \sim B(N, 1 - \pi_1)$. We thus decompose the multinomial distribution f as

$$[\boldsymbol{f}] \propto [\boldsymbol{f}^+ | n][n],$$

i.e., as the product of a multinomial and a binomial with unknown index. The observed frequencies $\boldsymbol{f}^+$ provides information about the parameters p_i, hence also π_1; using this information, the binomial distribution for n provides information about N. BUGS code is given in Panel 10.1.

10.2 MODEL $M_{t,\alpha}$

Implicit to Model M_t is the assumption that animals are correctly identified. This assumption can be violated because of observer error or due to loss or modification of marks. Here we consider a specific misidentification problem, and build its possibility into the model.

PANEL 10.1   BUGS code for model M$_t$.

```
Data
list(T=2,Cells=4,f=c(NA,22,363,174),
       omegas=structure(.Data=c(0,0,
                              1,0,
                              0,1,
                              1,1),.Dim=c(4,2)))
model{
    for (t in 1:T){ p[t] ~ dunif(0,1) }
    for (j in 1:Cells){
         for (t in 1:T){
                 c[j,t] <- pow(p[t],omegas[j,t])
                     *pow(1-p[t],1-omegas[j,t])
         }
         pi[j] <- prod(c[j,1:T])
    }
    for (j in 2:Cells){
         pi.obs[j] <- pi[j]/(1-pi[1])
    }
    n <- sum(f[2:Cells])
    f[2:Cells] ~ dmulti(pi.obs[2:Cells],n)
    pn <- 1-pi[1]
    n ~ dbin(pn,N)
    cN ~ dunif(n,10000)
    N <- round(cN)
}
```

Suppose that individuals are identified on the basis of genotypic information such as obtained from fur or scat samples. It is possible that genetic material obtained from an individual will be misidentified. A "ghost" record, of an animal that does not exist, enters the data set. Working under the assumptions of Model M$_t$, the (apparent) number of distinct animals encountered is inflated, and population size will be overestimated.

Model M$_{t,\alpha}$ generalizes Model M$_t$ by taking into account the possibility of ghost records. Since it is unlikely that the same misidentification will occur twice, it is assumed that ghosts are not resighted.

There are, then, three possible events for each animal on sampling occasion t. It might not be detected ($\tilde{\omega}_t = 0$), it might be detected and correctly identified ($\tilde{\omega}_t = 1$) or it might be detected but incorrectly identified ($\tilde{\omega}_t = 2$). Model M$_{t,\alpha}$ states that these events occur with probabilities $1 - p_t, \alpha p_t$ and $(1 - \alpha)p_t$. As under M$_t$, the samples are assumed to be independent, and the parameters to be constant among individuals. Parameter α, the correct identification probability, is treated as fixed through time.

Under Model M$_{t,\alpha}$, there are 3^T latent histories $\{\tilde{\omega}_1, \tilde{\omega}_2, \dots, \tilde{\omega}_T\}$. These are uniquely identified by indices

$$i = 1 + \sum_{t=1}^{T} \tilde{\omega}_t 3^{t-1}.$$

The challenge of Model $M_{t,\alpha}$ is that frequencies of these 3^T types of latent histories are confusingly summarized by $2^T - 1$ types of records. For instance, an individual with latent history 12021 is recorded as 3 distinct individuals, with histories 10001, 01000, and 00010. We thus distinguish *latent* histories (with generic index i) from these *recorded* histories, for which we use generic index j, and follow the numbering system at Eq. (10.1).

Data for Model $M_{t,\alpha}$ consist of the frequencies for recorded histories $j = 2, 3, \ldots, 2^T$. These data are a badly mangled and incomplete summary of the latent frequencies. For instance, recorded history 01000 not only includes individuals correctly identified on the second sampling occasion and otherwise never observed, but also all individuals with latent histories $\{\tilde{\omega}_1, 2, \tilde{\omega}_3, \tilde{\omega}_4, \tilde{\omega}_5\}$ and all individuals with latent histories $\{\tilde{\omega}_1, 1, \tilde{\omega}_3, \tilde{\omega}_4, \tilde{\omega}_5\}$ with $\tilde{\omega}_t \neq 1$ for $t \neq 2$ and at least one $\tilde{\omega}_t = 2$.

Analysis of Model $M_{t,\alpha}$ begins by describing the recorded frequency vector $f^+ = (f_2, f_3, \ldots, f_{2^T})$ as an affine transformation of latent history frequencies $X = (x_1, x_2, \ldots, x_{3^T})$. That is, we may write the recorded frequency vector as

$$f^+ = A'X, \tag{10.2}$$

where A is a $3^T \times (2^T - 1)$ matrix. This feature of the model becomes clear by looking at an example.

A complete list of latent and recorded histories, and of recorded histories resulting from the latent histories is given, for $T = 3$, in Table 10.1. For example, latent history 23 is 112; individuals with this history produce recorded histories 110 and 001, recorded histories 4 and 5, respectively.

Table 10.1 shows how the matrix A in Eq. (10.2) is constructed. Matrix A is $3^T \times (2^T - 1)$ with a 1 in the entry at row i, column $j - 1$ if latent history i gives rise to recorded history j; all of the other entries in A are zeros. To visualize this matrix for the case $T = 3$, simply replace all of the dots (.) in the center column of Table 10.1 with zeros, and all of the nonzero entries with ones.

Armed with knowledge of matrix A, we can use Eq. (10.2) to define a likelihood function. Since the probability structure for model $M_{t,\alpha}$ is expressed in terms of latent vectors X, and the observation is f^+, the likelihood is

$$L(p_1, p_2, \ldots, p_T, \alpha | f^+) = \sum_{X : A'X = f^+} \Pr(X | p_1, p_2, \ldots, p_T, \alpha),$$

the summation being over the set of all latent frequencies consistent with the recorded data. This looks great, but as soon as one gives some thought to computing the likelihood, some of the luster is lost.

For one thing, how does one enumerate the feasible set $\{X : A'X = f^+\}$ of latent frequency vectors consistent with the recorded frequencies?[1] And even if one were able to identify the set, one would soon realize that numerous complicated summations are needed for even a single calculation of the likelihood, and many such calculations are required for maximization. MLE seems far from straightforward, but a full Bayesian analysis can be implemented using MCMC, an analysis which is elegant in its simplicity.

---

1. The reader might consider constructing the feasible set of vectors x consistent with $f_2 = f_3 = \ldots = f_7 = 2$, for $T = 3$. It is clear that x_1 can be any nonnegative integer, and that for $2 \leq i \leq 27$, $x_i = 0, 1,$ or 2. There are over 1300 feasible combinations of $\{x_2, x_3, \ldots, x_{27}\}$.

TABLE 10.1   Latent histories (i) and recorded (j) histories under model $M_{t,\alpha}$, with $T = 3$.

i	Latent history	Contributed records, j from i	j	Recorded history
1	0 0 0	. . . .	1	0 0 0
2	1 0 0	2	2	1 0 0
3	2 0 0	2	3	0 1 0
4	0 1 0	. 3	4	1 1 0
5	1 1 0	. . 4	5	0 0 1
6	2 1 0	2 3	6	1 0 1
7	0 2 0	. 3	7	0 1 1
8	1 2 0	2 3	8	1 1 1
9	2 2 0	2 3		
10	0 0 1	. . . 5 . . .		
11	1 0 1	. . . . 6 . .		
12	2 0 1	2 . . 5 . . .		
13	0 1 1	. . . . . 7 .		
14	1 1 1	 8		
15	2 1 1	2 . . . . 7 .		
16	0 2 1	. 3 . 5 . . .		
17	1 2 1	. 3 . . 6 . .		
18	2 2 1	2 3 . 5 . . .		
19	0 0 2	. . . 5 . . .		
20	1 0 2	2 . . 5 . . .		
21	2 0 2	2 . . 5 . . .		
22	0 1 2	. 3 . 5 . . .		
23	1 1 2	. . 4 5 . . .		
24	2 1 2	2 3 . 5 . . .		
25	0 2 2	. 3 . 5 . . .		
26	1 2 2	2 3 . 5 . . .		
27	2 2 2	2 3 . 5 . . .		

Center column shows the recorded histories arising from specific latent histories. For example, latent history 16, 021, gives rise to recorded histories 010 and 001 for which $j = 3$ and 5.

The idea will be to use Gibbs sampling, treating the latent frequencies as unobserved quantities just like the the p_t and α. For given values of the latent frequencies, uniform (or other beta) priors on the p_t and α will induce corresponding beta full conditionals, which are easily sampled. The challenge for implementing MCMC is in sampling latent frequencies from the feasible set. Not only must the Markov chain be able to move around within the feasible set, but we must be able to guarantee that all points in the feasible set can be reached. This is a challenge, but not insurmountable.

First, however, we will need a brief digression into linear algebra.

Digression: Some Linear Algebra

Real numbers are nice. Think about the number 5, for instance. We know that if

$$5x = 5y,$$

then $x = y$. The only exception to the rule $ax = ay$ implies $x = y$ is when $a = 0$. With matrices, things aren't so nice. A nonzero matrix B can multiply vectors x and y with the result that

$$Bx = By$$

without it being the case that $x = y$; indeed, most matrices B "misbehave" like this. If the number of columns of B is greater than the number of rows of B, such behavior is inevitable. In particular, the matrix A' relating latent histories to recorded histories under Model $M_{t,\alpha}$ has this property: distinct latent history frequencies lead to the same recorded history frequencies.

Every matrix B has an associated set of vectors v, called its *null space*, denoted $null(B)$. The null space consists of all vectors v which B maps to the zero vector. That is, $v \in null(B)$ if and only if $Bv = 0$. Thus, while $Bx = By$ does not imply $x = y$, it does imply $B(x - y) = 0$, which is to say that $x - y$ is in the null space of B. In other words, $Bx = By$ if and only if

$$x = y + v, \quad \text{for some } v \in null(B).$$

Thus in Model $M_{t,\alpha}$, if we can find a single vector of latent frequencies X_0 satisfying $A'X_0 = f^+$, then we can describe the feasible set $\{X : A'X = f^+\}$ of latent frequencies as the set of vectors

$$\{X : X = X_0 + v, v \in null(A')\}. \tag{10.3}$$

One such vector is readily available: if $\alpha = 1$, latent frequencies of histories with 2's are zeros, and the remaining frequencies match up with recorded histories. For the case $T = 3$ displayed in Table 10.1, we set $x_2 = f_2$, $x_4 = f_3$, $x_5 = f_4$, $x_{10} = f_5$, $x_{11} = f_6$, $x_{13} = f_7$, and $x_{14} = f_8$. All other values x_i, $2 \leq i \leq 27$ would be set equal to zero, and x_1 can be any nonnegative integer.

The null space of a matrix B is spanned by r linearly independent basis vectors, $b_1, b_2, \ldots, b_r$. These vectors are the building blocks for $null(B)$: any vector $v \in null(B)$ can be written as

$$v = \sum_{k=1}^{r} a_k b_k,$$

where $a_1, a_2, \ldots, a_r$ are constants. Thus, given X_0 satisfying $A'X_0 = f^+$ and given the set of basis vectors b_k, we have a complete description of the feasible set at Eq. (10.3).

Computation of the set of basis vectors of a null space is a straightforward but tedious job, best left to software. The process is illustrated in the next section. As it turns out (for readers who might want to skip the details), $null(A')$ required for fitting Model $M_{t,\alpha}$ is spanned by

TABLE 10.2   Basis vectors for null space of matrix A' relating frequencies of latent and recorded histories under model $M_{t,\alpha}$, when $T = 3$.

1	$(1,0,0,0,0,0,0,0,0,0,0,0,0,0,0,0,0,0,0,0,0,0,0,0,0,0)'$
2	$(0,1,-1,0,0,0,0,0,0,0,0,0,0,0,0,0,0,0,0,0,0,0,0,0,0,0)'$
3	$(0,1,0,1,0,-1,0,0,0,0,0,0,0,0,0,0,0,0,0,0,0,0,0,0,0,0)'$
4	$(0,0,0,1,0,0,-1,0,0,0,0,0,0,0,0,0,0,0,0,0,0,0,0,0,0,0)'$
5	$(0,1,0,1,0,0,0,-1,0,0,0,0,0,0,0,0,0,0,0,0,0,0,0,0,0,0)'$
6	$(0,1,0,1,0,0,0,0,-1,0,0,0,0,0,0,0,0,0,0,0,0,0,0,0,0,0)'$
7	$(0,1,0,0,0,0,0,0,0,1,0,-1,0,0,0,0,0,0,0,0,0,0,0,0,0,0)'$
8	$(0,1,0,0,0,0,0,0,0,0,0,1,0,-1,0,0,0,0,0,0,0,0,0,0,0,0)'$
9	$(0,0,0,1,0,0,0,0,0,1,0,0,0,0,0,-1,0,0,0,0,0,0,0,0,0,0)'$
10	$(0,0,0,1,0,0,0,0,0,0,1,0,0,0,0,0,-1,0,0,0,0,0,0,0,0,0)'$
11	$(0,1,0,1,0,0,0,0,0,1,0,0,0,0,0,0,0,-1,0,0,0,0,0,0,0,0)'$
12	$(0,0,0,0,0,0,0,0,0,1,0,0,0,0,0,0,0,0,-1,0,0,0,0,0,0,0)'$
13	$(0,1,0,0,0,0,0,0,0,1,0,0,0,0,0,0,0,0,0,-1,0,0,0,0,0,0)'$
14	$(0,1,0,0,0,0,0,0,0,1,0,0,0,0,0,0,0,0,0,0,-1,0,0,0,0,0)'$
15	$(0,0,0,1,0,0,0,0,0,1,0,0,0,0,0,0,0,0,0,0,0,-1,0,0,0,0)'$
16	$(0,0,0,0,1,0,0,0,0,1,0,0,0,0,0,0,0,0,0,0,0,0,-1,0,0,0)'$
17	$(0,1,0,1,0,0,0,0,0,1,0,0,0,0,0,0,0,0,0,0,0,0,0,-1,0,0,0)'$
18	$(0,0,0,1,0,0,0,0,0,1,0,0,0,0,0,0,0,0,0,0,0,0,0,0,-1,0,0)'$
19	$(0,1,0,1,0,0,0,0,0,1,0,0,0,0,0,0,0,0,0,0,0,0,0,0,0,-1,0)'$
20	$(0,1,0,1,0,0,0,0,0,1,0,0,0,0,0,0,0,0,0,0,0,0,0,0,0,0,-1)'$

$r = 3^T - 2^T + 1$ vectors, each of which consists of nothing other than 0's, 1's, and (-1)'s. The 20 basis vectors corresponding to the case $T = 3$ are given in Table 10.2.

Sample Calculation of Null Space Basis Vectors

Here, we illustrate the computation of basis vectors for the null space of matrix A'. To keep things simple, we consider the case $T = 2$. The $3^T \times (2^T - 1)$ (i.e., 9×3) matrix A relating latent frequencies to recorded frequencies is constructed as previously described. We have

$$A' = \begin{bmatrix} 0 & 1 & 1 & 0 & 0 & 1 & 0 & 1 & 1 \\ 0 & 0 & 0 & 1 & 0 & 1 & 1 & 1 & 1 \\ 0 & 0 & 0 & 0 & 1 & 0 & 0 & 0 & 0 \end{bmatrix}.$$

This matrix is in reduced echelon form, meaning that the first nonzero entry in each row is a 1, and that the first such nonzero entry is farther to the right in each of the successive rows. This feature of matrix A' is a consequence of the numbering system we used for the latent and recorded histories. Generally, one must first attain reduced echelon form through elementary row operations.

Suppose that $x = (x_1, x_2, \ldots, x_9)'$ is in the null space of A'. Then $A'x = 0$ implies

$$x_2 + x_3 + x_6 + x_8 + x_9 = 0$$
$$x_4 + x_6 + x_7 + x_8 + x_9 = 0$$
$$x_5 = 0.$$

We treat these three equations as constraining x_2, x_4, and x_5, leaving the remaining six values x_i arbitrary. We conclude that vectors x in null(A') are of the general form

$$x = \begin{bmatrix} x_1 \\ x_2 \\ x_3 \\ x_4 \\ x_5 \\ x_6 \\ x_7 \\ x_8 \\ x_9 \end{bmatrix} = \begin{bmatrix} x_1 \\ -x_3 - x_6 - x_8 - x_9 \\ x_3 \\ -x_6 - x_7 - x_8 - x_9 \\ 0 \\ x_6 \\ x_7 \\ x_8 \\ x_9 \end{bmatrix}$$

$$= x_1 \begin{bmatrix} 1 \\ 0 \\ 0 \\ 0 \\ 0 \\ 0 \\ 0 \\ 0 \\ 0 \end{bmatrix} + x_3 \begin{bmatrix} 0 \\ -1 \\ 1 \\ 0 \\ 0 \\ 0 \\ 0 \\ 0 \\ 0 \end{bmatrix} + x_6 \begin{bmatrix} 0 \\ -1 \\ 0 \\ -1 \\ 0 \\ 1 \\ 0 \\ 0 \\ 0 \end{bmatrix} + x_7 \begin{bmatrix} 0 \\ 0 \\ 0 \\ -1 \\ 0 \\ 0 \\ 1 \\ 0 \\ 0 \end{bmatrix} + x_8 \begin{bmatrix} 0 \\ -1 \\ 0 \\ -1 \\ 0 \\ 0 \\ 0 \\ 1 \\ 0 \end{bmatrix} + x_9 \begin{bmatrix} 0 \\ -1 \\ 0 \\ -1 \\ 0 \\ 0 \\ 0 \\ 0 \\ 1 \end{bmatrix}.$$

The $6 (= 3^T - 2^T + 1)$ column vectors form a basis for the null space of A'; every vector x satisfying $A'x = 0$ is a unique linear combination of them.

10.3 GIBBS SAMPLING FOR MODEL $M_{t,\alpha}$

Under Model $M_{t,\alpha}$, the probability an individual will have latent history ω_i is

$$\pi_i = \prod_{t=1}^{T} p_t^{I(\omega_{i,t} > 0)} (1 - p_t)^{I(\omega_{i,t} = 0)} \alpha^{I(\omega_{i,t} = 1)} (1 - \alpha)^{I(\omega_{i,t} = 2)}, \qquad (10.4)$$

for $i = 1, 2, \ldots, 3^T$. Conditioning on there being N individuals in the population, the probability of a vector $X = (x_1, x_2, \ldots, x_{3^T})'$ of latent frequencies is

$$[X | N, p_1, \ldots, p_T, \alpha] = \left\{ \frac{N!}{\prod x_i!} \prod \pi_i^{x_i} \right\} I\left(\sum x_i = N \right). \qquad (10.5)$$

the sum and products being over indices $i = 1, 2, \ldots, 3^T$.

The probability of the recorded data f^+, conditional on the latent frequencies, is

$$[f^+|X,N,p_1,\ldots,p_T,\alpha] = \mathbf{I}(f^+ = A'X).\tag{10.6}$$

Completing the specification of a Bayesian model requires priors on p_t, α, and on N. We will assign independent priors, with the result that full conditional distributions will be proportional to the product

$$[f^+|X,N,p_1,\ldots,p_T,\alpha][X|N,p_1,\ldots,p_T,\alpha][N][p_1][p_t]\cdots[p_T][\alpha].\tag{10.7}$$

Priors and Full Conditionals for p_t and α

The beta family of distributions provides natural choices for priors for p_t and α; the uniform distribution reflects vague prior knowledge within this family. We let $Be(a_0^t, b_0^t)$ denote the prior on p_t and $Be(a_0^\alpha, b_0^\alpha)$ denote the prior on α.

From Eqs. (10.4) through (10.7), it follows that these priors lead to beta full-conditional distributions $p_t \sim Be(a^t, b^t)$ and $\alpha \sim Be(a^\alpha, b^\alpha)$, where

$$a^t = a_0^t + \sum_i x_i \mathbf{I}(\omega_{i,t} > 0),$$

$$b^t = b_0^t + \sum_i x_i \mathbf{I}(\omega_{i,t} = 0),$$

$$a^\alpha = a_0^\alpha + \sum_i \sum_t x_i \mathbf{I}(\omega_{i,t} = 1),$$

and

$$b^\alpha = b_0^\alpha + \sum_i \sum_t x_i \mathbf{I}(\omega_{i,t} = 2);$$

the summations are over the indices $i = 1, 2, \ldots, 3^T$, and $t = 1, 2, 3, \ldots, T$.

Prior and Full Conditional for X and N

It is possible to sample a new vector X for a fixed N, but it is not possible to sample a new value N for a fixed X; the value of N cannot change without changing X. The posterior distributions are inextricably linked. Thus, we treat the pair (X, N) as a single entity in a Gibbs sampling scheme. The joint full-conditional distribution is

$$[X,N|\cdot] \propto \left\{\frac{N!}{\prod x_i!}\prod \pi_i^{x_i}\right\}\mathbf{I}(A'X = f^+)\mathbf{I}\left(\sum x_i = N\right)g(N),\tag{10.8}$$

where $g(N)$ is a prior on N.

In the absence of prior knowledge regarding N, a discrete uniform prior on $\{1, 2, \ldots, M\}$, for some large value M may be reasonable. If one wishes a prior with infinite range, $[N] \propto 1/N$, while improper, nonetheless leads to a proper posterior distribution. As usual, we recommend that analysts evaluate the sensitivity of analytical results to choice of alternative vague priors.

The distribution (10.8) will be sampled using a Metropolis–Hastings approach. That is, we will generate candidate values $(X_{\text{cand}}, N_{\text{cand}})$ to be compared with current values $(X_{\text{curr}}, N_{\text{curr}})$. The Markov chain sampled will either move to the candidate value $(X_{\text{next}} = X_{\text{cand}},$

$N_{\text{next}} = N_{\text{cand}})$ or remain at the present value $(X_{\text{next}} = X_{\text{curr}}, N_{\text{next}} = N_{\text{curr}})$ based on the outcome of a Bernoulli trial, described subsequently.

The real challenge for the Metropolis–Hastings sampler is producing candidate values satisfying the requirement $A'X_{\text{cand}} = f^+$. Here, our foray into linear algebra stands us in good stead. Given that X_{curr} is in the feasible set, so also must be the candidate value:

$$X_{\text{cand}} = X_{\text{curr}} + c_k b_k,$$

where b_k is a basis vector for the null space of A', and c_k is a real number. We will sample c_k from discrete uniform distributions centered on (but not including) zero. We need only consider integer values for c_k, since frequency counts are integer valued. By cycling through the entire set of $3^T - 2^T + 1$ basis vectors, we ensure that any integer-valued feasible solution can be reached by our sampler. The candidate generating distribution is symmetric in its arguments, simplifying calculation of the Metropolis–Hastings ratio. Sampled values X_{cand} with negative frequencies are immediately rejected.

Gibbs Sampling

Gibbs sampling proceeds as follows:

Step 1: Initialize X from the feasible set of latent frequencies. Call this value X_{curr}.
Step 2: Calculate values a^α and b^α, and values a^t and b^t for $t = 1, 2, \ldots, T$ based on the value X_{curr}.
Step 3: Sample p_t's and α from their full conditional distributions.
Step 4: Set $k = 0$.
Step 5: Increment k by 1. Sample c_k from a discrete uniform distribution on the integers $\{-D_k, \ldots, D_k\}\backslash 0$, and generate a candidate latent frequency

$$X_{\text{cand}} = X_{\text{curr}} + c_k b_k$$

where b_k is the kth basis vector of null(A'). Note that the candidate value remains in the feasible set, because b_k is a basis vector of null(A). The values D_k are tuning parameters, chosen by the analyst.
Step 6: Calculate $r = \min\left\{\dfrac{[X_{\text{cand}}|\cdot]}{[X_{\text{curr}}|\cdot]}, 1\right\}$ using Eq. (10.8).[2]
Accept the candidate value (i.e., set $X_{\text{curr}} = X_{\text{cand}}$) with probability r.
Step 7: Repeat steps 5 and 6 for all of the $3^T - 2^T + 1$ basis vectors of null(A').
Step 8: Repeat steps 2 through 7 a large number of times.

10.4 AN IMPLEMENTATION OF MODEL $M_{t,\alpha}$

Here, we report an implementation of Model $M_{t,\alpha}$ for $T = 5$ using recorded frequencies $f^+ = \{17, 25, 3, 30, 6, 8, 3, 41, 7, 11, 4, 17, 6, 9, 4, 54, 11, 17, 7, 24, 9, 15, 5, 35, 13, 20, 8, 29, 11, 17, 6\}$.

---

2. The nature of the candidate generation renders this calculation easy. One need only keep track of incremental changes in the x_i, a^t, b^t, a^α, and b^α.

These frequencies arise from $N=400$ latent histories with frequencies approximating those expected with $p_1=0.3$, $p_2=0.4$, $p_3=0.5$, $p_4=0.6$, $p_5=0.7$, and $\alpha=0.90$.

Given these parameter values, we would expect $\sum p_i=2.5$ sightings per individual, with 10% misidentifications. The recorded data should then suggest 0.25 ghosts per individual, so that population size estimation might be anticipated to be inflated by roughly 25%, from 400 to 500. Indeed, fitting Model M_t with flat priors on p_t's and N, the posterior median for N was 511, and the 95% HPDI was [495, 525].

We generated a chain of length 110,000, and discarded the first 10,000 values as a burn-in. This calculation took just over a half hour on a 3.8 GHz Pentium processor. The long computation time is necessitated by the number ($3^5=243$) of latent histories consistent with the recorded ($2^T-1=31$) frequencies. The null space of A' has dimension $243-31+1=212$, so there are 212 basis vectors, directions in which to increment the latent history vector.

We used the first 5000 values of the chain to tune the candidate generating distributions. During this tuning period, we adjusted the values D_k in step (5) as follows. Beginning with $\delta_k=1$, we multiplied δ_k by 0.95 whenever a move in the direction of the kth null basis vector was rejected and divided by 0.95 when such a move was accepted. We set D_k in step (5) equal to the least integer greater than or equal to δ_k. The effect of this tuning is to allow larger or smaller increments in X as appropriate for the various basis vectors.

Posterior summaries are given in Table 10.3; the results are gratifying.

We note that the Markov chains produced have fairly long autocorrelations (Fig. 10.1). If we were analyzing a "real" data set, gathered at a cost of thousands of dollars and hundreds of hours of work, we would gladly multiply the computation time by 100 and leave the computer running over a weekend, thinning the chains as storage requirements dictated, and more precisely approximating the posterior distributions.

The long autocorrelation is likely due to a low movement rate for the Metropolis–Hastings sampler; 205 of the D_k were set equal to 1, five were set to equal to 2, one to 3, and one to 12 (this last being for the unique basis vector incrementing the constant zero latent history). In turn, the low movement rates are due to low frequencies for many of the latent histories: of the 243 latent histories, 145 had zero frequency, and 215 had frequencies less than or equal to three. Posterior percentiles 2.5, 50, and 97.5 are plotted for the remaining

TABLE 10.3   Posterior summaries for sample data of Section 10.4.

Parameter	Mean	SD	Percentiles		
			2.5th	50th	97.5th
α	0.910	0.016	0.879	0.910	0.940
p_1	0.302	0.026	0.254	0.301	0.354
p_2	0.407	0.029	0.350	0.406	0.465
p_3	0.499	0.032	0.437	0.499	0.562
p_4	0.596	0.034	0.530	0.596	0.663
p_5	0.704	0.036	0.632	0.704	0.773
N	399.4	16.03	370	399	432

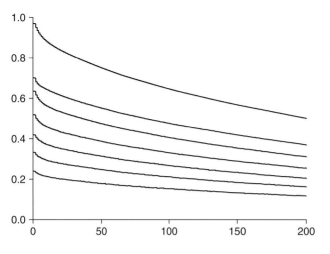

FIGURE 10.1 Lagged correlations of Markov chains for (highest to lowest) N, α, p_5, p_4, p_3, p_2, and p_1.

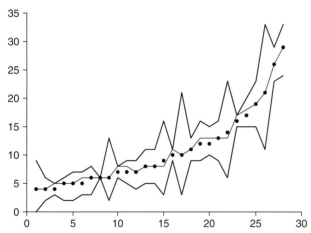

FIGURE 10.2 Posterior percentiles 2.5, 50, and 97.5 of latent frequencies for the 28 most common latent histories, along with true values. True values represented by unconnected dots, medians by thin line.

28 histories along with the actual latent frequencies (known, because this was a simulation) in Fig. 10.2.

10.5 EXTENSIONS

Most mark-recapture data can be described in terms of event histories, which are sequences of Bernoulli trials. In cases where complete event histories are observed, Gibbs sampling of model parameters is usually straightforward, since the probabilities of event histories are simply products of conditional probabilities.

However, most mark-recapture data consist of capture histories rather than event histories. Capture histories are composites of possible event histories; many of the basic events are not observed. In the Brownie models, and in the CJS model and its extensions, a typical composite event is "not seen again." These composite events lead to likelihood functions based on sums of probabilities of event histories, hence sums of products, which are not easily worked with.

The latent multinomial approach described in this chapter can be adapted for many such cases. We can conceive of latent multinomial random variables X describing possibly unobservable event histories, summarized to frequency vectors f (capture histories) by means of a known transformation $f = A'X$, where A is a matrix of zeros and ones. The Gibbs sampling scheme developed here for model $M_{t,\alpha}$ can then be adapted to many capture–recapture models.

An important difference between model $M_{t,\alpha}$ and other capture–recapture models is that the number of ones in a row of A sometimes exceeds one; that is, single latent histories spawn multiple observed frequencies. It can be shown that if $f = A'X$, and X is multinomial, then the conditional distribution of f given its total is also multinomial, *if and only if* none of the row totals of A exceeds 1. Thus, while most capture–recapture data are summarized by multinomial vectors f, the data of model $M_{t,\alpha}$ are not. Acknowledgment of the latent multinomial structure is therefore a necessity in analyzing model $M_{t,\alpha}$. For other capture–recapture models, recognition of this structure is not essential, but provides conceptual simplicity and an omnibus framework for analysis.

11

Open Population Models

O U T L I N E

11.1 Continuous-Time Survival
Models   240
   11.1.1 *Right Censoring and the CDL*   241
   11.1.2 *Interval Censoring, Staggered Entry, and Known Fates*   243
11.2 Open Population Mark-Recapture – Band-Recovery Models   244
11.3 Open Population Mark-Recapture – The CJS Model   247
   11.3.1 *A Gibbs Sampling Algorithm for the CJS Model*   250

11.4 Full Open Population Modeling Formalities   252
11.5 CMSA Model and Extensions   255
   11.5.1 *Complete Data Likelihood*   255
   11.5.2 *Example: Nonmelanic Gonodontis*   256
11.6 Multiple Groups   260
11.7 Robust Design   261
   11.7.1 *Example: Meadow Voles*   262
   11.7.2 *Temporary Emigration*   264
11.8 Multistate Models and Other Extensions of the CJS Model   268
11.9 Afterword   269

In Chapters 9 and 10, we considered a variety of models appropriate for making inference about abundance in closed animal populations. In this chapter, we consider open population models in which animals may enter or leave the population during the study. We begin with survival models and then progress to mark-recapture models for both survival and abundance.

Closed population models describe a fixed population of size N; N is a fixed parameter to be estimated. Open population models, on the other hand, describe dynamic populations; there is no single N, but a vector $= (N_1, N_2, \ldots, N_k)'$ of population sizes, with N_j describing the population at the jth sampling occasion. The values N_j can be thought of as a sequence of random variables, rather than parameters, and our interest is in the process generating these random variables and the parameters that govern this process.

In the spirit of "modeling the data you wish you had" it is instructive to consider what the ideal data would be for a study of population dynamics. The best we could ask for would be a complete description of the times of birth, death, immigration, and emigration, for all individuals in the study population between times t_1 and t_k that define the start and end times of the study. Failing a complete description for each member of the population, we would want a complete description for a random sample of individuals from that population.

With ecological data, we rarely know the exact times of birth or death. Some animals will have been born before the start of the study at time t_1; their data is said to be *left censored* at t_1. Other animals will survive beyond the completion of the study at time t_k; their data is said to be *right censored* at t_k. Because times of birth and death are known only partially a key part of modeling involves completing the data (i.e., formulating the CDL). For now we will ignore animal movement and discuss open population modeling assuming that the population is closed geographically.

11.1 CONTINUOUS-TIME SURVIVAL MODELS

A common statistical problem is to fit a lifetime distribution to a sample of subjects all alive at a starting time t_0.[1] Models used for such data are known as "survival models." Ideally, we observe lifetimes $T_1, T_2, \ldots, T_n$ for a sample of n individuals. These are then modeled assuming a suitable lifetime distribution with probability distribution function (*pdf*) denoted $f(t)$. Survival distributions are sometimes defined in terms of the *hazard function* $h(t)$ instead of the *pdf*, with the relationship between $h(t)$ and $f(t)$ given by

$$h(t) = \frac{f(t)}{1 - F(t)},$$

where $F(t)$ is the cumulative distribution function $F(t) = \int_0^t f(s)ds$. The hazard rate $h(t)$ can be interpreted as the instantaneous risk of dying or failure at time t, given survival up until time t.

A simple example is the exponential survival model in which $f(t) = \lambda e^{-\lambda t}$ and $h(t) \equiv \lambda$ is a constant for all values of t.[2] This model rarely provides an adequate fit to lifetime data because of the restrictive assumption of a constant hazard rate. The "force of mortality" (another name for the hazard function) can be expected to vary with the individual's age, and in association with seasons and temporal effects. Furthermore, hazard rates vary among individuals, in association with gender, size, and many other factors.[3] The constant hazard exponential model has been modified in a variety of ways. Weibull models allow monotone increasing

---

1. Continuous-time models are typically indexed to starting time t_0 rather than t_1. If the focus of the study is individual survival, rather than population change, the starting time is often the time of an individual's birth, rather than the beginning of the study.

2. See Appendix B.8 for more details on hazard functions and exponential distributions.

3. Grimly fascinating evidence of variation in human hazard rates can be found in extensive tables of mortality provided by the US National Center for Health Statistics (www.cdc.gov/nchs/datawh/statab/unpubd/mortabs.htm). There, one finds evidence for variation by age, gender, and race, and can uncover such arcana as that far more deaths occur on Wednesdays and Saturdays than other days of the week.

or decreasing hazard functions. Frailty models assign individual specific random effects to hazard rates. Other models describe variation in hazard functions through use of covariates.

The most widely-used covariate model is Cox's proportional hazard model (Cox, 1972), in which it is assumed that the hazard function for individual i can be written as:

$$h_i(t) = \lambda_0(t)e^{z_i'\beta}$$

for a set of covariates z_i and a parameter vector β. The term $\lambda_0(t)$ is a continuous function of time and defines the baseline hazard rate which we can interpret as the hazard function for an individual (possibly hypothetical) whose covariate values are all zero. In Cox's (1972) formulation the baseline hazard is allowed to take arbitrary values and in this sense the model is regarded as nonparametric. The standard frequentist approach is to fit the model using a partial likelihood (Cox, 1975), i.e., by retaining factors of a full likelihood that depend on the parameters of interest (in this case β) and discarding factors in which the parameter of interest is inextricably entangled with nuisance parameters (in this case parameters in the function $\lambda_0(t)$). A fully Bayesian approach to fitting this model is given by Chen *et al.* (2006). An example can be found in the BUGS help file for a set of data on leukemia remission times, from McCullagh and Nelder (1989).

Survival models must accommodate censoring. We have mentioned left- and right-censoring; another possibility is interval censoring. Observations are taken at discrete times. If we index the sample times by i, an individual observed alive at time t_i but dead at time t_{i+1} is known to have died in the interval $[t_i, t_{i+1})$ but the exact time of death is unknown. In many applications the intervals are small relative to the study period and the discrete nature of the data is ignored, other than it introduces a complication due to the presence of tied times of death. For the moment we will ignore left- and interval-censoring and concentrate on right censoring. Right-censored observations are modeled as $Y = \min\{T, C\}$, where T is the true survival time and C is the time of censoring.

11.1.1 Right Censoring and the CDL

Censored data are an example of *nonignorable* missing data (Gelman *et al.*, 2004), because whether T_i was observed or not is related to its magnitude: individuals with longer lifetimes are more likely to be subject to right censoring than individuals with short lifetimes. We cannot simply ignore the missing data without biasing our inference.

Instead, we describe the model in terms of a complete data likelihood (CDL). The complete data consist of $T = \{T^{\text{obs}}, T^{\text{mis}}\}$ (vectors of fully observed and censored lifetimes, respectively), C (the vector of censoring times), and D, a vector of indicators D_i for whether corresponding observations T_i are censored. Letting θ denote a vector of parameters governing the distributions of the T_i, the CDL is

$$\mathcal{L}(T^{\text{mis}}, \theta | C, T^{\text{obs}}, D) \propto [T, C, D | \theta] \tag{11.1}$$

$$= [D | T, C, \theta][T | C, \theta][C | \theta] \tag{11.2}$$

$$\propto [D | T, C][T | \theta]. \tag{11.3}$$

Equations (11.1) through (11.3) require some comment. In passing from Eq. (11.1) to Eq. (11.2), we have simply made use of the laws of conditional probability. In passing from Eq. (11.2) to Eq. (11.3), we have assumed that survival times and censoring times are independent, so that $[T\,|\,C,\boldsymbol{\theta}]\equiv[T\,|\,\boldsymbol{\theta}]$. This is a standard assumption, and generally necessary for inference, which we discuss subsequently. We have also assumed that the distribution of C does not depend on the parameters governing the distribution of T, so that $[C\,|\,\boldsymbol{\theta}]\equiv[C]$. Because we have omitted C^{mis} from the CDL, we may simply absorb $[C]$ into the proportionality constant: it has no bearing on the variables T^{mis} or $\boldsymbol{\theta}$. Finally, in passing from Eq. (11.2) to Eq. (11.3), we made use of the fact that D is completely determined by T and C, so that we need not condition on $\boldsymbol{\theta}$.

Consider the first term on the right-hand side of Eq. (11.3). Each D_i is an indicator of the event that $T_i\leq C_i$. Conditional on T_i and C_i, we can think of D_i as a Bernoulli trial with success parameter equal to 0 or 1, based on whether $T_i\leq C_i$; thus $[D_i|T_i,C_i]=B(1,\mathbf{I}(T_i\leq C_i))$. The CDL is simply the product of the likelihood for these Bernoulli trials and the likelihood for the T_i, observed or missing.

The simplicity of the CDL approach is evident if we consider what is required in writing BUGS code. We have illustrated the approach in Panel 11.1, with a sample of size 30 from an exponential distribution, with $C_i\equiv20$. Only the slightest modifications would be necessary for more general applications; we would need to replace T[i]~dexp(theta) with whatever alternative distribution were desired, and to replace delta[i]<-step(T[i]-20) with delta[i]<-step(T[i]-C[i]), with vector C[] of censoring times supplied as data.[4]

In contrast, the usual method of fitting this model is through use of the observed data likelihood (ODL) which is given by

$$\text{ODL}\propto[T^{\text{obs}},d\,|\,\boldsymbol{\theta},C]$$
$$=\int_{T^{\text{mis}}}[T^{\text{mis}},T^{\text{obs}},d\,|\,\boldsymbol{\theta},C]dT^{\text{mis}}$$

PANEL 11.1   BUGS code for fitting an exponential survival model to 30 survival times censored at $t=20$. The variable y contains the survival times with censored values indicated by "NA," meaning that for these cases y > 20. The variable d contains the censoring indicators.

```
model{
    for(i in 1:n){
        T[i] ~ dexp(theta)
        delta[i] <- step(T[i]-20)
        d[i] ~ dbern(delta[i])
    }
    theta ~ dunif(0,10)
}

Data
list(T=c(1.84,0.61,1.65,3.36,7.37,NA,16.73,0.27,4.98,8.97,
    7.73,12.51,17,7.78,NA,1.12,4.51,5.97,1.64,10.46,14.39,
    4.42,7.04,18.13,NA,NA,7.19,14.92,0.66,9.62),d=c(0,0,0,
    0,0,1,0,0,0,0,0,0,1,0,0,0,0,0,0,0,0,1,1,0,0,0,0))
```

4. One would set C[i]=K, with K a very large number, for uncensored observations.

and for a random sample of survival times:

$$[\boldsymbol{y}^{\text{obs}}, \boldsymbol{d} \mid \boldsymbol{\theta}, C_i] = \prod_{i=1}^{n} f(y_i)^{1-d_i}(1 - F(C_i))^{d_i}$$

where $f(y)$ is the *pdf* and $F(y)$ the *cdf*.

For the exponential distribution, as in the example of Panel 11.1, computation of the *cdf* is not difficult, but in many cases the *cdf* cannot be calculated in closed form.

In formulating the model given earlier, we treated the censoring times C_i as independent of the survival times. In many wildlife studies, censoring results from causes that are beyond the control of the researcher. For instance, studies that use radio-telemetry to obtain survival data may have censored data due to the failure of radio transmitters. In such cases it is often reasonable to assume that the censoring events are "uninformative," that the event of censoring carries no prognosis for future survival, and that the censored observations carry no information other than that the animal was alive up until time C_i.

If censoring is connected with the fate of the animal, for example if radio-failure results from a predator destroying the radio while dining on the subject of our study, then we must account for this additional information in our modeling.

11.1.2 Interval Censoring, Staggered Entry, and Known Fates

Survival models have been popular for analyzing animal survival data obtained by radio-telemetry. Two common features of this type of data are that times of death are interval censored and that animals often enter the study at different times. This entry of animals into the study at different times is referred to as "staggered entry."

Interval censoring results from having discrete sampling periods. If the times between sampling periods are short relative to the study period, for example if death is recorded to the nearest day in a study extending over many months, continuous time models can be fitted as a reasonable approximation. Otherwise, modeling should be based on multinomial distributions, as we now describe.

We will begin by ruling out staggered entries, as well as left- and right-censoring. Suppose that we have a sample of n independent lifetimes T_i, all beginning at time t_0. We are able to determine whether each individual is alive at each of k sampling occasions, $t_1, t_2, \ldots, t_k$, but the exact times of death are unknown. Thus every observation is interval-censored.

Let X_j denote the number of individuals that died in time interval $[t_j - 1, t_j)$, for $j = 1, 2, \ldots, k, k+1$; here we set $t_{k+1} = \infty$, to that X_{k+1} is the number of individuals surviving beyond the last sampling period. Our data can be summarized by a multinomial vector $\boldsymbol{X} = (X_1, X_2, \ldots, X_{k+1})'$, and cell probabilities π_j, $j = 1, 2, \ldots, k+1$, summarized by a vector $\boldsymbol{\pi}$. Our model is $\boldsymbol{X} \sim M_{k+1}(n, \boldsymbol{\pi})$.

This formulation is very general, and nonparametric in the sense that we can make inference about the parameters π_j without having to assume any model for the lifetime distribution of T_i. If we wish, we can constrain the lifetime distribution by modeling π_i as

$$\pi_i = \int_{t_{i-1}}^{t_i} f_\theta(s) \, ds$$

assuming that we are willing to specify a family of distributions $f_\theta(\cdot)$. This sort of constraint, while not essential for inference about π_j, would be expected to lead to improved estimation.

Now let us suppose that instead of following all n individuals from time t_0, we sample n_j staggered entries at sampling occasion t_j, $j = 0, 1, \ldots, k-1$. The foregoing model is easily adapted. Let $X_{j,h}$ denote the number of individuals sampled at time t_j which die in the interval $[t_h, t_{h+1})$, for $h = j, j+1, \ldots, k$. Because the staggered entries at time t_j are known to have survived at least to time t_j, the probability of death in interval $[t_h, t_{h+1})$ is $\pi_{j,h} = \pi_h/(1-\pi_j)$. Letting $\boldsymbol{\pi}_j = (\pi_{j,j}, \pi_{j,j+1} \ldots, \pi_{j,k})'$, we describe the staggered entry data by k multinomials $\boldsymbol{X}_j = (X_{j,j}, X_{j,j+1} \ldots, X_{j,k})' \sim M_{k+1-j}(n_j, \boldsymbol{\pi}_j)$, for $j = 0, 1, \ldots, k-1$.

This combination of interval-censored data and staggered entry is referred to in the wildlife literature as the "known fate" model (White and Burnham, 1999). The key features of this design are that new animals can be introduced to the study population for monitoring at any sampling occasion (staggered entry) and that the interval of death (or "failure" event) is known exactly. In wildlife studies this information is usually obtained by some sort of remote telemetry, say through use of radio-tagging. A conceptually simple variation on this is that, by chance, we fail to determine the fate of some animals at a sampling occasion. This variation leads to the Cormack–Jolly–Seber (CJS) model (Cormack, 1964; Jolly, 1965; Seber, 1965) and the band recovery models of Seber (1970) and Brownie *et al.* (1985).

11.2 OPEN POPULATION MARK-RECAPTURE – BAND-RECOVERY MODELS

We turn our attention to mark-recapture models, and their Bayesian analysis, considering a model for band-recovery data. Not only are band-recovery models of interest as a popular means of making inference about animal survival, they are also useful for demonstrating the role that data augmentation can play in simplifying MCMC.

As an example we revisit the study considered in Section 7.3.1, a simple 3-year banding study in which 400 birds were released at the beginning of each year. Data for this type of study are often summarized in a dead recovery array (Table 11.1), a data summary that is appropriate when there are no individual covariates. The bold entries in the table indicate the values for statistics r_{ij} which are the number of birds in release cohort

TABLE 11.1   Dead recovery array for a bird-banding study with 400 birds released each year for 3 years and recoveries recorded for 3 years.

Year released i	No. released	Year found dead j			Never seen again
		1	2	3	
1	400	50	22	14	314
2	400		28	21	351
3	400			25	375

TABLE 11.2   Complete data array for a bird-banding study with three release cohorts followed for three complete years.

i	Dies and recovered			Dies and not recovered			Survives study
	1	2	3	1	2	3	
1	r_{11}	r_{12}	r_{13}	$\bar{r}_{11}$	$\bar{r}_{12}$	$\bar{r}_{13}$	w_1
2		r_{22}	r_{23}		$\bar{r}_{22}$	$\bar{r}_{23}$	w_2
3			r_{33}			$\bar{r}_{33}$	w_3

i that were found dead between the start of year j and the start of year $j+1$. The size of each cohort is denoted by R_i, which in this case is 400 for each i. Thus Table 11.1 tells us that of the 400 birds released at the start of year 1, 50 birds were recovered before the start of year 2, 22 were recovered in the second year, and 14 were recovered in year 3.

For each bird that contributes to an r_{ij} the interval of death is known exactly. A complication is that the interval of death, if indeed the bird died, is unknown for each of the birds that were never seen again. The numbers of birds in this category are indicated by the entries in italics in Table 11.1 with one such entry for each release cohort.

To model the data we wish we had, we need to complete the information about the fate of each bird as in Table 11.2. Here we use the notation $\bar{r}_{ij}$ to denote the number of birds from release cohort i that died in the interval $(j, j+1]$ but were not recovered. These numbers, and w_j, the numbers that survived the study, are not observed. All we know is that $w_i + \sum_j \bar{r}_{ij}$ is equal to the numbers never seen again in each release cohort, and these numbers are known. If we could complete this table then inference about survival probabilities and recovery rates would be straightforward.

A Reparameterization

For ease of calculation we reparameterize in terms of the reporting rate λ_j instead of the recovery rate f_j, with the relationship

$$f_j = (1 - S_j)\lambda_j \tag{11.4}$$

for $0 < \lambda_j < 1$. The reporting rate λ_j is defined to be the probability that an animal is reported dead given that it died in the interval $[j, j+1)$. An advantage of this parameterization is that the λ_j parameters are functionally distinct from the survival probabilities. Equation (11.4) makes it clear that the recovery rate and the survival rate probabilities are not functionally distinct: If we fix λ_j, high values for S_j are associated with low values for f_j and vice versa.

As in the analysis of the observed data we treat each release cohort as a multinomial experiment. The cell probabilities associated with the multinomial data in Table 11.2 are given in Table 11.3 with $S'_j = 1 - S_j$ and $\lambda'_j = 1 - \lambda_j$.

TABLE 11.3   Cell probabilities for a 3-year banding study.

	Dies and recovered			Dies and not recovered			Survives study
i	1	2	3	1	2	3	
1	$S_1'\lambda_1$	$S_1 S_2' \lambda_2$	$S_1 S_2 S_3' \lambda_3$	$S_1' \lambda_1'$	$S_1 S_2' \lambda_2'$	$S_1 S_2 S_3' \lambda_3'$	$S_1 S_2 S_3$
2		$S_2' \lambda_2$	$S_2 S_3' \lambda_3$		$S_2' \lambda_2'$	$S_2 S_3' \lambda_3'$	$S_2 S_3$
3			$S_3' \lambda_3$			$S_3' \lambda_3'$	S_3

Full Conditional Distributions for Latent Variables

From the statistics in Table 11.2 we can identify all those marked animals that died in the interval $[j, j+1)$ regardless of whether they were recovered. If we denote this statistic by d_j then

$$d_j = \sum_{i=1}^{j} r_{ij} + \bar{r}_{ij}.$$

Of these d_j animals, the number that were recovered is r_{ij} and, conditional on d_j, r_{ij} is a binomial random variable with index d_j and probability λ_j. This result follows from the multinomial distribution of $\{r_{ij}, \bar{r}_{ij}\}$ within each cohort and the multinomial factorization theorem (Appendix B.7).

Now consider all those animals alive at the start of year j. This number, which we denote by A_j, is given by

$$A_j = \sum_{i=1}^{j} \left(w_i + \sum_{h=j}^{k} r_{ih} + \bar{r}_{ih} \right),$$

where k is the number of years in the study. We can also define A_j recursively as:

$$A_j = \begin{cases} R_1 & j=1 \\ A_{j-1} - d_{j-1} + R_j & j=2,\dots,k \end{cases}.$$

As an example, A_2 is calculated by summing up the statistics from Table 11.2 that are shown in the following table, with the bold entries indicating the contributions to d_2, the number of birds that died in year 2. The statistics in this table correspond to all the birds that were know for sure were marked and alive immediately following the release of the year 2 cohort.

r_{12}	r_{13}	$\bar{r}_{12}$	$\bar{r}_{13}$	w_1
r_{22}	r_{23}	$\bar{r}_{22}$	$\bar{r}_{23}$	w_2

With a little effort we can show that given A_j, d_j is a binomial random variable with parameter $1 - S_j$ and index A_j. If we knew the values for $\bar{r}_{ij}$ and w_j it would then be straightforward to carry out posterior sampling for inference about S_j and λ_j. For example, if we adopt a $Be(\alpha_S, \beta_S)$ prior for S_j then the posterior for S_j is a $Be(\alpha_S + A_j - d_j, \beta_S + d_j)$ distribution. Similarly, if we adopt a $Be(\alpha_\lambda, \beta_\lambda)$ prior for λ_j, then the posterior distribution for λ_j is a

$Be(\alpha_\lambda + \sum_{i=1}^{j} r_{ij}, \beta_\lambda + \sum_{i=1}^{j} \bar{r}_{ij})$ distribution. These densities would exactly describe the respective posterior distributions if we observed the complete data.

Even though $\bar{r}_{ij}$ and w_j are unobserved we can take advantage of these results in developing a Gibbs sampler for the bird-banding model provided we are able to sample the latent variables from their full-conditional distributions. We show how this can be done, using as an example the birds in our first release cohort. Given r_{11}, r_{12}, r_{13}, and R_1, and also given the parameters, the distribution of the vector of unknowns $(\bar{r}_{11}, \bar{r}_{12}, \bar{r}_{13}, w_1)'$ is multinomial with index equal to their total, which is the number from cohort 1 that were never seen after release. The cell probabilities for this multinomial distribution are the corresponding cell probabilities in Table 11.3 scaled to sum to 1. That is, $S_1'\lambda_1'/\psi_1$ for $\bar{r}_{11}$, $S_1 S_2'\lambda_2'/\psi_1$ for $\bar{r}_{12}$, $S_1 S_2 S_3'\lambda_3'/\psi_1$ for $\bar{r}_{13}$, and $S_1 S_2 S_3/\psi_1$ for w_1, where $\psi_1 = S_1'\lambda_1' + S_1 S_2'\lambda_2' + S_1 S_2 S_3'\lambda_3' + S_1 S_2 S_3$. With similar arguments we can identify the required multinomial distributions for the other cohorts.

We construct a Gibbs sampler as follows:

Step 1: Initialize the complete array (Table 11.2) by allocating the numbers never seen again in each cohort to $\bar{r}_{ij}$ and w_j (any starting allocation will do).

Step 2: Draw samples from the full-conditional distributions for S_j by sampling from $Be(\alpha_S + A_j - d_j, \beta_S + d_j)$ distributions.

Step 3: Draw samples from the full-conditional distributions for λ_j by sampling from $Be(\alpha_\lambda + \sum_{i=1}^{j} r_{ij}, \beta_\lambda + \sum_{i=1}^{j} \bar{r}_{ij})$ distributions.

Step 4: Draw samples for all $\bar{r}_{ij}$ and w_j by sampling from the appropriate multinomial distributions conditioning on the values for S_j and λ_j sampled at the previous steps.

Step 5: Repeat steps 2 to 4 a large number of times.

Constrained Models

A nice feature of this Gibbs sampler is that it is easy to incorporate constraints such as $S_1 = S_2 = S_3 \equiv S.$ (constant survival rate) or $\lambda_1 = \lambda_2 = \lambda_3 \equiv \lambda.$ (constant reporting rate). For example, if we wish to fit a constant reporting rate model we simply make use of the fact that under this constraint, the full conditional distribution for the total number of animals recovered dead during the study, which is given by $r.. = \sum_i \sum_j r_{ij}$ is binomial with probability $\lambda.$ and index $\sum_j d_j$, the total number of animals that died during the study.

Instead of drawing from separate beta distribution for λ_j, under the constant reporting rate model we draw a value for $\lambda.$ from a $Be(\alpha_\lambda + r.., \beta_\lambda + \bar{r}..)$ distribution and set $\lambda_j = \lambda.$ for each j. All other aspects of the Gibbs sampler remain the same. Note that here we have used the notation $\bar{r}.. = \sum_i \sum_j \bar{r}_{ij}$. Similarly, if we wish to fit a constant survival model the only modification needed is to replace the separate draws for S_j with a single draw for $S.$ taken from a $Be(\alpha_S + \sum_j d_j, \beta_S + \sum_j A_j - d_j)$ distribution.

11.3 OPEN POPULATION MARK-RECAPTURE – THE CJS MODEL

The CJS model (discussed in Section 5.5) is closely related to the band-recovery model of Section 11.2. In fact, although the types of data to which the two models are applied are quite distinct it is easy to show that their likelihood functions are algebraically equivalent.[5]

The CJS model is suitable when a population is studied by tagging and releasing animals at a series of discrete sampling occasions and then recording subsequent recaptures at later sampling occasions. An important feature of the CJS model, and one that distinguishes it from the Jolly-Seber (JS) model discussed later in Section 11.4, is that we condition on the releases and make no attempt to model the first capture of each animal; we only model recaptures. The focus in the CJS model is on survival of animals.

Because not all animals are caught in each sample there is a partial confounding of survival and recapture processes in the CJS model. We can see this by noting that when we fail to catch an animal in a sample we cannot tell whether the animal was alive at the time of that sample and we did not catch it, or it was dead. Although a subsequent recapture will tell us that the animal must have been alive, we obtain no recaptures from dead animals. The fact that we never saw an animal again could mean that it is dead but it could also mean that it has successfully avoided recapture. The challenge is to disentangle the survival and recapture processes.

The basic structure of the data is the same as for closed mark-recapture models in that we summarize the results of the experiment in an array X^{obs} of capture indicators, with rows corresponding to individual animals and columns corresponding to sampling occasions. An important difference between the closed models and the CJS model is that we condition on the first release of each animal and, where applicable, on any losses on capture. Therefore, the X^{mis} component discussed in Section 9.2 is not included in the model.

Losses on capture are a complication that can arise in mark-recapture studies either because animals die in traps or because their tags are removed, and they are released unmarked. Provided that:

(i) loss on capture in sample j is unrelated to earlier events, other than survival, and
(ii) loss on capture parameters are distinct from other parameters in the model

the term for losses on capture factors out of the CJS model and need not be considered in making inference about survival or capture probabilities.

An Observed Data Likelihood

If we let y denote a vector indicating time of first capture for each individual, with one element y_i for each distinct animal caught in the study, then the ODL based on all the information in X^{obs}, conditional on first capture, can be written as:

$$\mathcal{L}(\phi,p,v|X^{obs},y) \propto [X^{obs}|\phi,p,v,y]$$
$$= [X^{obs}|R,r,m] \times [R|n,v] \times [r|R,\phi,p][m|T,\phi,p] \qquad (11.5)$$

with parameters and statistics defined in Table 11.4.

5. Data obtained under the CJS model can be summarized in what is known as an $m_{ij}-array$, which is of the same form as the band-recovery matrix (Table 11.1). The CJS likelihood is obtained through a judicious reparameterization of the banding model.

TABLE 11.4   Parameters and statistics used in the CJS model.

Parameters

ϕ_j	The probability that an animal alive and in the population at the time of sample j, is alive and in the population at the time of sample $j+1$, $(j=1,\ldots,k-1)$.
p_j	The probability an animal that is available for recapture at the time of sample j is caught in sample j, $(j=2,\ldots,k)$.
v_j	The probability that an animal caught in sample j is released (i.e., not lost on capture), $(j=1,\ldots,k)$.

Statistics

n_j	The number of animals caught in sample j, $(j=1,\ldots,k)$.
R_j	The number of animals caught in sample j and released with a mark, $(j=1,\ldots,k)$.
r_j	The number of animals marked and released in sample j that were ever caught again, $(j=1,\ldots,k-1)$.
m_j	The number of marked animals caught in sample j, $(j=2,\ldots,k)$.
T_j	The number of marked animals in the population immediately before sample j that were caught in sample j or later, $(j=2,\ldots,k)$.

Note that the survival probability is now denoted by ϕ_j instead of the S_j used earlier for the bird-banding model. The parameter ϕ_j is in fact an "apparent" survival. If animals can permanently move away from the study site then this movement is indistinguishable from death.[6]

Of the four terms in Eq. (11.5), $[R\,|\,n,v]$ deals with losses on captures and is the least interesting: this term models the number released in sample i conditional on the number caught and depending only on the nuisance parameter v_i.

The third and the fourth terms correspond to the ODL for the CJS model expressed in terms of the minimal sufficient statistics. Somewhat remarkably, the CJS model can be expressed as the product of binomial distributions:

$$[r\,|\,R,\phi,p] = \prod_{j=1}^{k-1} B(r_j; R_j, \lambda_j)$$

and

$$[m\,|\,T,\phi,p] = \prod_{j=2}^{k} B(m_j; T_j, \tau_j)$$

with λ_j and τ_j defined in Section 5.4. A description of a Gibbs sampler for this formulation of the CJS model is also given in Section 5.4.

An interesting observation is that the first term $[X^{\mathrm{obs}}\,|\,R,r,m]$ in the ODL (11.5) is free of the parameters v, ϕ, and p. In fact, this term can be expressed as the product of multiple

---

6. The CJS likelihood is valid under alternative permanent emigration or random temporary emigration assumptions; these ideas are explored further by Burnham (1993).

hypergeometric distributions, so that $[X^{\mathrm{obs}}\,|\,R,r,m]$ can be used to develop goodness of fit tests based on contingency tables (for examples, see Pollock *et al.*, 1990).

As discussed in Section 5.4, the complicated form of the ODL means that known full conditional distributions for ϕ and p are unavailable if this formulation of the model is used. However, a useful alternative formulation is to complete the data with a latent vector d, which has one element for each distinct animal caught in the study. If animal i dies between the time of samples h and $h+1$ ($h<k$) then $d_i=h$. If animal i is still alive at the time of sample k then $d_i=k$.

If we use the vector v, with the same dimension as y, to indicate any times of loss on capture, then we can write a complete data model, conditional on times of first release and losses on capture, as

$$[X^{\mathrm{obs}},d\,|\,y,v,\phi,p]=[X^{\mathrm{obs}}\,|\,d,y,v,\phi,p]\times[d\,|\,y,v,\phi,p]$$
$$=[X^{\mathrm{obs}}\,|\,d,y,v,p]\times[d\,|\,y,v,\phi].$$

This formulation leads to a Gibbs sampler which is based on known full-conditional distributions for the unknown and censored elements of d provided that conjugate priors are used for ϕ and p (Dupuis, 1995; Schofield, 2007).

11.3.1 A Gibbs Sampling Algorithm for the CJS Model

Using a $\mathrm{Be}(\alpha_\phi,\beta_\phi)$ prior for ϕ_j the full-conditional distribution is also a beta distribution

$$[\phi_j\,|\,\cdot]=\mathrm{Be}(M_j^+-D_j+\alpha_\phi,D_j+\beta_\phi)$$

where M_j^+ is the number of marked animals in the population immediately following sample j and D_j is the number of these that died between samples j and $j+1$. Note that any animals lost on capture are not regarded as having died in the interval between samples j and $j+1$; these are not counted in computing the statistics M_j^+ and D_j.

Using a $\mathrm{Be}(\alpha_p,\beta_p)$ prior for p_j, the full-conditional distribution is

$$[p_j\,|\,\cdot]=\mathrm{Be}(m_j+\alpha_p,M_j-m_j+\beta_p)$$

where M_j is the number of marked animals in the population immediately before the sample at time j and m_j is the number of these that were caught.

The full-conditional distributions for unknown and censored elements of $\{d_j\}$ are derived from the model $[d\,|\,y,\phi]$. For individual i, $[d_i|y_i,\phi]$ is a categorical distribution with sample space $\{y_i,y_i+1,\ldots,k\}$ and parameter vector $\{\xi_{y_i,y_i},\xi_{y_i,y_i+1},\ldots,\xi_{y_i,k}\}$ that depends on the survival probabilities

$$\xi_{ij}=\begin{cases}1-\phi_i & i=j\\ \phi_{k-1} & i=k-1;\,j=k\\ \xi_{i+1,j}\phi_i & i=1,\ldots,k-1;\quad j=2,\ldots,k\end{cases}$$

for $i=1,\ldots,k-1$. As an example, values of ξ_{ij} are given in Table 11.5 for a study with $k=4$ sampling occasions. Updating the latent vector d then proceeds according to one of the following:

TABLE 11.5   Values of ξ_{ij} for a CJS model with $k = 4$.

	$j = 1$	2	3	4
$h = 1$	$1 - \phi_1$	$\phi_1(1 - \phi_2)$	$\phi_1\phi_2(1 - \phi_3)$	$\phi_1\phi_2\phi_3$
2	–	$1 - \phi_2$	$\phi_2(1 - \phi_3)$	$\phi_2\phi_3$
3	–	–	$1 - \phi_3$	ϕ_3

Here h indexes the sample in which the animal was first caught and j indexes the sample that marks the beginning of the interval in which the animal died.

(i) If the animal was caught in sample k, d_i is set equal to k because we know death occurred after sample k.

(ii) For an animal last caught in sample l_i, d_i is censored at l_i. Thus, we draw a value for d_i from a categorical distribution with sample space $\{l_i, l_i + 1, \ldots, k\}$ and parameter vector $(\xi_{l_i,l_i}, \xi_{l_i,l_i+1}, \ldots, \xi_{l_i,k})'$.

(iii) If the animal was lost on capture in sample l_i then we can, if we wish, draw a value for d_i as under option (ii). We can regard this as the predicted time of death for animal i had it remained in the study. Note that updating d_i for these animals is not essential; inference about all other unknowns is valid regardless.

An efficient Gibbs sampler based on the above mentioned full-conditional distributions can be readily programmed using a language such as R. However, the CJS model including latent times of death can be also be easily implemented in BUGS (Schofield *et al.*, 2008) as shown in Panel 11.2.

Features of this BUGS code are:

- `first[i]` is data and indexes the first sample in which animal i was caught and released.

- `AvailUntil[i]` is data and indicates the last sample in which the animal was known to have been available for capture. This is assigned the value h if animal i was removed in sample h, or is set to k if the animal was never removed. Capture probabilities for occasions after this time are set to zero. In the CDL analysis a death time is imputed for these animals and is interpreted as the predicted time of death had the animal remained in the study.

- The node `a[i,j]`, which is a latent variable, has the value 1 if animal i is available for capture in sample j and 0 otherwise. Note that `a[i, first[i]]` is data and is assigned the value 1.

- The nodes `asuse[]`, `sv[]`, `apuse[]`, and `pcap[]` are used to ensure that animals that have not yet been marked or that have died do not contribute to the likelihood.

- The node `avail[]` keeps track of whether the animal is available for recapture (i.e., marked and not removed from the population).

The CDL specified in Panel 11.2 is written in terms of Bernoulli distributions and probabilities that depend on whether or not the animal has survived and thus is available for capture.

PANEL 11.2   BUGS code for the CJS model fitted using data augmentation for the unknown times of death.

```
model{
  for (i in 1:udot) {
      for (j in first[i]+1:k){
          asuse[i,j] <- a[i,j-1] + 1
          apuse[i,j] <- a[i,j] + 1
          a[i,j] ~ dbern(sv[asuse[i,j],i,j-1])
          avail[i,j]<-a[i,j]*step(AvailUntil[i]-j)
      }
      for (j in first[i]+1:AvailUntil[i]){
          X[i,j] ~ dbern(pcap[apuse[i,j],i,j])
      }
      for(j in first[i]:k-1){
          sv[1,i,j] <- 0
          sv[2,i,j] <- S[j]
          pcap[1,i,j+1] <- 0
          pcap[2,i,j+1] <- p[j+1]
      }
  }

  for(j in 1:k-1){
      S[j] ~ dbeta(1,1)
      p[j+1] ~ dbeta(1,1)
  }
}
```

Although this code does not explicitly include the latent vector d of animals' death times, the vector can be easily derived from the matrix a[].

The main benefits of writing out the CJS model in terms of the CDL, including latent times of death, are computational convenience and faster mixing in MCMC. The latent times of death can also be used to form a prediction for the number of marked animals remaining in the population but this will rarely be of interest.

The CJS model does not include population sizes or birth rates; it is limited in this regard by conditioning on the numbers of animals released at sampling occasions. To make inference about abundance and birth, we must use a likelihood for the full capture–recapture data set, and in particular, we must include a model for the first captures.

11.4 FULL OPEN POPULATION MODELING FORMALITIES

In a closed population experiment, the population is described by a single population size, which does not change during the experimental period. Descriptions of open populations are move complicated, reflecting the dynamic nature of populations. Despite these differences, models for closed and open populations can be described in the same general way, the main difference being the many different summaries that we can use to characterize changes in the population during the experiment.

The quantity of interest in a closed population experiment is the population size. An analogue in open population experiments is the number of distinct animals ever alive during the study. It is tempting to refer to the totality of these animals as "the population" under consideration, but a difficulty presents itself. Some animals are born and die between sampling periods, hence are invisible to the experiment. Including such animals in models would require untestable assumptions. It seems advisable to exclude them from consideration; judicious choice of sampling periods may render their numbers negligible. Thus we will use the phrase "the population" to describe all individuals alive and available for capture on at least one of the sampling occasions.[7]

Using Λ to denote the number of animals ever available for capture in the population, the CDL can be represented by

$$\text{CDL} \propto [X^{\text{obs}}, X^{\text{miss}}, I \mid \theta^{\text{obs}}, \theta^{\text{miss}}, \Lambda].$$

This is essentially the same general representation that we used for closed populations except now the inclusion indicator is a $\Lambda \times t$ matrix with $I_{ij} = 1$ if animal i was available for capture and caught in sample j and zero otherwise. Also, because the population is open, the parameter vector θ now includes components used to model entry and exit of animals from the population as well as capture probabilities.

An animal is available for capture at j if it entered the population before sample j and left the population after sample j. Thus, we can summarize the information in the inclusion indicator I using the $\Lambda \times 1$ vectors b and d, where elements b_i and d_i indicate the birth and death intervals for individual i.

As discussed at the beginning of this chapter, all birth and death times are censored in mark-recapture studies. Birth times b for individuals present at the start of the study are left-censored; all we know is that their time of birth was before t_1. Times of death d of those individuals alive at the end of the study are right-censored; all we know is that death occurred (or will occur) after t_k. Even the birth and death times that occur during the study are interval-censored.

Open population models focus on birth and death times b_i, and d_i, and relationships among them. Using the CDL allows investigation of derived quantities of interest, such as N_j, the number of animals available for capture in sample j, and B_j, the number born between samples j and $j+1$ which survive at least until $j+1$.

Figure 11.1 depicts the lifespans of 18 animals in relation to a four-period study. Sixteen were alive during the study period, but only 13 are taken account of in the matrix X; three that were born and died between consecutive sampling occasions are excluded from the model. Each of the $\Lambda = 13$ animals can be described by when they were born, either (i) before t_1, (ii) between t_1 and t_2, (iii) between t_2 and t_3, or (iv) between t_3 and t_4. Denoting the number born before t_1 by B_0, we have $B_0 = 1$, $B_1 = 5$, $B_2 = 4$, and $B_3 = 3$ in Fig. 11.1. We can also count $N_1 = 1$, $N_2 = 6$, $N_3 = 7$, and $N_4 = 8$.

7. Like "supermodel," the term "superpopulation" used by Schwarz and Arneson (1996), is ambiguous; to us it suggests something hypothetical. We prefer the term "population" as it describes a real collection of individuals.

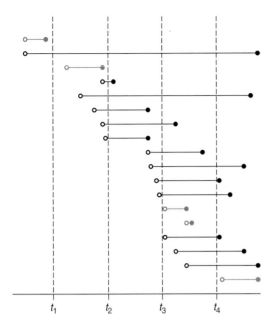

FIGURE 11.1   Lifetimes for 18 hypothetical individuals, 16 of which were alive during a mark-recapture experiment carried out from time t_1 to t_4. The open circles indicate the time of birth and the closed circles the time of death. The light-gray lines correspond to individuals that are excluded from inference either because they did not live during the study period or because they were born and died between consecutive samples.

Jolly–Seber Model

The first full open population model was the Jolly–Seber (JS) model (Jolly, 1965; Seber, 1965) in which the model $[X^{obs} \mid \phi, p, \nu, y]$ of Section 11.3 is augmented by a model for the first captures $[y \mid p, U]$, where the vector $U = (U_1, \ldots, U_k)'$ with U_j defined as the number of unmarked animals in the population at the time of sample j. Remarkably, Jolly and Seber proposed the same model independently and published side-by-side in *Biometrika* in 1965.

In the JS model, the information contained in y on when animals are first caught is summarized by the vector $u = (u_1, \ldots, u_k)'$, with element u_j the number of unmarked animals caught in sample j. The model for first captures is then expressed as:

$$[y \mid p, U] \propto \prod_{j=1}^{k} [u_j \mid p_j, U_j]$$

where $[u_j \mid p_j, U_j]$ is a binomial distribution with index U_j and parameter p_j.

Notice that the value of U_i cannot be deduced from Fig. 11.1 as each U_i includes not only the B_{i-1} animals newly born at the time of sample i, and hence unmarked, but animals born before t_{i-1} that are as yet uncaught. Because of this confounding of birth and capture processes the parameters U are not really of any particular biological interest. However, they do have value in that given estimators for U_j and other parameters in the model, it is possible to find estimators for N_j and B_j. A drawback of parameterizing the likelihood in terms of U is that it makes hierarchical extensions of the model difficult; it is hard to conceive of a reasonable process prior for this parameter because U's reflect a complicated interaction of the population with the experiment. Process priors are of interest because we would like to be able to model the more biologically meaningful quantities N_j and B_j.

11.5 CMSA MODEL AND EXTENSIONS

An important variation on the JS model is one developed initially by Crosbie and Manly (1985) and then extended by Schwarz and Arneson (1996). We refer to this model as the CMSA parameterization of the JS model, or the CMSA model for short. An advantage of this formulation is that it is amenable to hierarchical extension, as discussed by Link and Barker (2005).

The CMSA model is parameterized in terms of Λ and the vector $\boldsymbol{\beta} = (\beta_0, \ldots, \beta_{k-1})'$, where β_j is the probability that an animal ever available for capture enters the population between samples j and $j+1$. Under the assumption that birth events are independent, and that the vector $\boldsymbol{\beta}$ is the same for each animal, the vector B of birth frequencies can be modeled as a multinomial random variable $B | \Lambda, \boldsymbol{\beta} \sim M_k(\Lambda, \boldsymbol{\beta})$.

This is the basis of the CMSA as an ODL extension of the CJS.

A detailed description of the ODL for the CMSA model is given by Link and Barker (2005) along with discussion of the relationship between this model and the model used by Pradel (1996). Here, we base inference on a CDL that is expressed in terms of the latent times of birth b and latent times of death d. With this approach, inference about summaries such as N_i or B_i is straightforward, because these appear as simple functions of b and d. Working with the CDL can also lead to more efficient Gibbs sampling, and is relatively easy to implement in BUGS as we show in code appearing subsequently.

11.5.1 Complete Data Likelihood

The CDL for the CMSA model is proportional to

$$[X, b, d \mid \Lambda, \boldsymbol{\beta}, p, \boldsymbol{\phi}] = [X \mid p, \Lambda, b, d][d \mid \boldsymbol{\phi}, b, \Lambda][b \mid \boldsymbol{\beta}, \Lambda],$$

where X is the complete $(\Lambda \times k)$ X-matrix

$$\begin{pmatrix} X^{\text{obs}} \\ 0 \end{pmatrix}.$$

For simplicity, we have ignored losses on capture.

The term $[X \mid p, \Lambda, b, d]$ is similar to the term $[X | p, N]$ for a closed capture model, except that each animal has capture probability 0 prior to birth and after death. Similarly, the term $[d \mid \boldsymbol{\phi}, b, \Lambda]$ is the same as the equivalent term in the CJS model, except we condition on when the animal was born, not when it first entered the study.

To model births in the CDL we use

$$[b \mid \boldsymbol{\beta}, \Lambda] = \prod_{i=1}^{\Lambda} [b_i \mid \boldsymbol{\beta}]$$

where $[b_i \mid \boldsymbol{\beta}]$ is a categorical distribution with sample space $\{1,\dots,k\}$ and parameter vector $\boldsymbol{\beta}$. Here, $b_i = h$ indicates that the animal was born between samples $h-1$ and h $(h=2,\dots,k)$, with the special case of $b_i = 1$ indicating birth before sample 1. We model the times of death d_i in the same way that we did for the CJS model except that now we condition on the time of birth, not the time at which the animal first entered the study, and we model the times of death for all Λ animals. That is, we model d_i $(i=1,\dots,\Lambda)$ by specifying $[d_i|\boldsymbol{\phi},b_i]$ as a categorical distribution with sample space $\{b_i,b_i+1,\dots,k\}$ and parameter vector $(\xi_{b_ib_i},\xi_{b_i,b_i+1},\dots,\xi_{b_i,k})'$. The elements ξ_{ij} are defined in Section 11.3.1.

As in the CJS model the times of death are partially observed: we know that death must have occurred after the last time we caught the animal, if we caught it. Similarly the times of birth are also partially observed: we know that the time of birth occurred before the first occasion on which we caught the animal.

It is relatively straightforward to extend the BUGS code in Panel 11.2 in order to fit the CMSA model. We do this by

1. Adding a data-augmentation step to account for the dimension change associated with updating Λ.
2. Adding a censored multinomial term to account for the time of birth.
3. Adapting the model for d by taking the possible times of death from when the animal was born rather than from when it first entered the study.

We also add a $(\Lambda \times 1)$ vector (`birthCT`) to the data containing the censored times of birth, with entries set to k for the animals that were never captured. We include a partially observed $(\Lambda \times k)$ data matrix (`notyetdead`) to indicate whether a member of the superpopulation has yet died. This is assigned the value 1 for all animals at the start of the study and has element ij set to 1 for all samples between the first and last sighting of animal i and set to `NA` for all samples after the last occasion on which it was caught.

Using this code (Panel 11.3) it is also easy to construct predictions for N_i and B_i as these are simple functions of the matrices `avail[]` and `a[]` (BUGS code in Panel 11.4). However, simply estimating values for the parameters $\boldsymbol{\phi},\boldsymbol{\beta}$ and making predictions about N or B is unlikely to satisfy most researchers. Of more interest is modeling these quantities either as a device for obtaining higher-level summaries or to answer biologically interesting questions.

11.5.2 Example: Nonmelanic *Gonodontis*

Bishop *et al.* (1978) reported results from a study in which 689 nonmelanic male *Gonodontis bidentata* were captured, marked and released daily at Cressington Park near Liverpool in England. *Gonodontis* come in two forms, melanic and nonmelanic, with an increasing dominance of the melanic form due to industrial pollution being one of the classic instructional tools for teaching about natural selection.

In Section 5.5.2 we fitted a CJS model to the *Gonodontis* data. Here we fit the CMSA model to the *Gonodontis* data using the BUGS code in Panel 11.3 with posterior summaries in Fig. 11.2 for the daily abundances, births, and survival probabilities.

A feature of Bayesian inference is the ease with which we can express inference using different parameterizations as discussed in Section 5.2. Instead of describing population

PANEL 11.3   BUGS code for the CMSA model fitted using data augmentation for the unknown times of birth and death and for the superpopulation size Λ.

```
model{
    for(i in 1:M){
        w[i]~dbern(psi)
        b[i]~dcat(beta[])I(,birthCT[i])
    }
    Lambda<-sum(w[])

for (i in 1:M) {
    a[i,1]<-w[i]*(1-notyetborn[i,1])
    avail[i,1]<-a[i,1]*step(AvailUntil[i]-1)
    for (j in 1:k){
        notyetborn[i,j]<-step(b[i]-j-1)
    }
    for (j in 1:AvailUntil[i]){
            pcap[i,j]<-a[i,j]*p[j]
            X[i,j] ~ dbern(pcap[i,j])
    }
    for (j in 2:k){
        a[i,j]<-w[i]*(1-notyetborn[i,j])*notyetdead[i,j]
        avail[i,j]<-a[i,j]*step(AvailUntil[i]-j)
        asuse[i,j] <- notyetdead[i,j-1] + 1
        notyetdead[i,j] ~ dbern(sv[asuse[i,j],i,j-1])
    }
    for(j in 1:k-1){
        sv[1,i,j] <- 0
        sv[2,i,j] <- notyetborn[i,j]+(1-notyetborn[i,j])*S[j]
    }
}

for(j in 1:k){
    p[j]~dbeta(1,1)
}
for(j in 1:k-1){
    S[j]~dbeta(1,1)
}
beta[1:k]~ddirch(alpha[])
psi~dbeta(1,1)
}
```

demographics in terms of N and B we can construct posterior densities for quantities such as per capita birth rate f (note that this is not the same as the recovery rate parameter used earlier in the bird-banding model) or population growth rate λ using the relationships given in Link and Barker (2005). The reparameterization is accomplished by noting that

$$f_j = \frac{\beta_j}{d_j} \quad j=1,\ldots,k-1,$$

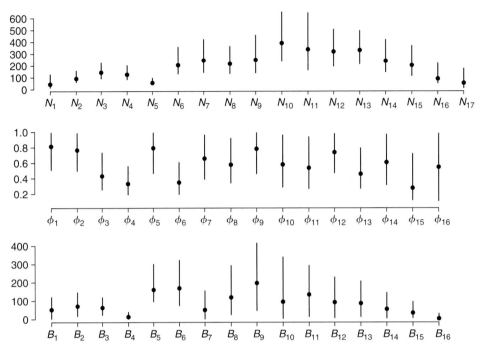

FIGURE 11.2   Posterior summaries for the parameters N (abundance), ϕ (survival), and B (births) for the Crosbie–Manly–Schwarz–Arnason model fitted to the *Gonodontis* data of Bishop *et al.* (1978).

PANEL 11.4   BUGS code for predicting values for N (abundance) and B (births).

```
for(j in 1:k){
        N[j]<-sum(avail[1:M,j])
}

for(j in 1:k-1){
    B[j]<-sum(bij[1:M,j])
}

for(i in 1:M){
    for(j in 1:k-1){
        bij[i,j]<-equals(a[i,j],0)*equals(a[i,j+1],1)
    }
}
```

where

$$d_j = \begin{cases} \beta_0 & j=1 \\ d_{j-1}S_{j-1} + \beta_{j-1} & j=2,\ldots,k-1, \end{cases}$$

and

$$\lambda_j = S_j + f_j \quad j=1,\ldots,k-1.$$

To generate samples from the posterior densities for f and λ, we simply translate the Markov chain for S and β into a Markov chain for f and λ using the above functional relationships. Posterior summaries for f and λ for the *Gonodontis* analysis are shown in Fig. 11.3.

On the basis of the posterior densities for the daily population growth rates, the male nonmelanic *Gonodontis* population at Cressington Park appears to have been mostly increasing through this study with some reduction in the population evident toward the end. Although a similar conclusion is evident from looking at the numbers of moths born (B_j) and the abundances (N_j), an advantage of the CMSA parameterization is that we can readily adapt the model to include random effects on parameters such as λ_j as a device for summarizing the behavior of the population in terms of mean growth rate and also to introduce some parsimony.

A further advantage, one we have stressed several times in this book, is that Bayesian hierarchical modeling can greatly extend the types of questions addressed by researchers. One example is the ability to model relationships between sets of parameters. Reasoning that survival probabilities and per capita birth rates may well be related, possibly due to a common environmental influence, Link and Barker (2005) placed a multivariate normal random effects model on the parameters ϕ and f using the stochastic constraint

$$\left(\text{logit}(\phi_j), \ln(f_j)\right)' \sim \text{MVN}(\mu_j, \Sigma)$$

where the elements of μ_j are the expected value of $\text{logit}(\phi_j) = \mu_{1j}$ and the expected value of $\ln(f_j) = \mu_{2j}$. The parameter Σ is the variance–covariance matrix that represents the covariation in the joint distribution of the parameters.

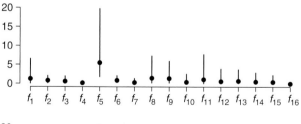

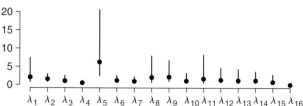

FIGURE 11.3 Posterior summaries for the parameters f (per capita recruitment rate) and λ (population growth rate) for the Crosbie–Manly–Schwarz–Arnason model fitted to the *Gonodontis* data of Bishop *et al.* (1978).

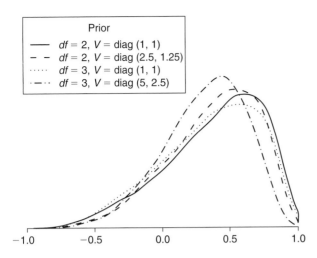

FIGURE 11.4 Posterior density for the parameter ρ which measures the strength of linear association between logit(ϕ) and log(f) in the model (logit(ϕ), log(f))$' \sim$ MVN(μ, Σ) for the *Gonodontis* data under four different priors for Σ. From Link and Barker (2005).

This "process covariance" describes biologically relevant association between the demographic parameters and is not to be confused with biologically irrelevant sampling covariance that between-parameter *estimates*, which results from the process of estimation.

To fit this model we can adapt the BUGS code in Panel 11.3 by reparameterizing the model in terms of f instead of β and then adding a multivariate normal prior for the joint distribution of ϕ and f. Link and Barker (2005) used vague normal prior distributions for the elements of μ, and for [Σ] they used an inverse-Wishart $IW_2(V, df)$ distribution.

Because of the hierarchical relationship this code is slow to mix. Being impatient, Link and Barker (2005) instead made use of a partial likelihood derived by conditioning on $u.$, the number of animals that were ever caught during the study, a device also used by Pradel (1996). By conditioning on $u.$, which is data, it appears that important information has been discarded from the model. However, Link and Barker (2005) showed that $u.$ is effectively ancillary, meaning that it includes almost no useable information for inference about the identifiable parameters in the model. After including the hierarchical model relating ϕ and f, Link and Barker (2005) found some evidence of a positive association between the two sets of parameters (Fig. 11.4).

A possible explanation for the apparent association between ϕ and f is density dependence, with both sets of parameters depending on the abundance of moths. Because the CDL can be used to predict the abundances N_j it is also possible to include the abundances as predictors for the survival probability and per capita birth rate, an approach described by Schofield *et al.* (2008) for this same set of data.

11.6 MULTIPLE GROUPS

It is common for study populations to be grouped into fixed strata, for example male vs. female, with no movement between strata, as for sex, or deterministic movement between strata, as for age. We can regard group membership as a categorical covariate. In the case of the CJS model where we condition on first release, this categorical covariate will usually

PANEL 11.5   Modification to the BUGS code for fitting the CMSA model (Panel 11.3) to allow for multiple groups.

```
for(i in 1:M){
    w[i]~dbern(psi)
    b[i]~dcat(beta[])I(,birthCT[i])
    grp[i]~dcat(pi_grps[b[i],])
}

for(i in 1:k){
    pi_grps[i,1:ngroups]~ddirch(alpha_grp[i,])
}
```

be completely known as it can be measured for each animal that appears in the experiment. In the case of the full open population models, where entry of animals into the study population is modeled, group classification is unknown for the animals that were never caught and so must be modeled.

Fortunately, modeling the unknown group for animals that were never caught involves a simple extension of the data augmentation technique used to model the unknown number of animals in the population. All we require is the addition of a group variable to the data statement, with values set as NA for the $M - u.$ animals that were never caught. Group is then modeled as a categorical random variable with the number of categories equal to the number of groups. Parameters are indexed by group as appropriate.

To illustrate we use an example from Nichols (2005) of a study on male and female meadow voles at the Patuxent Wildlife Research Center. The model we fitted was a two-group CMSA model using the BUGS code in Panel 11.3, but with the modification to the data augmentation step for abundance shown in Panel 11.5 along with the specification of the prior for the interval-specific allocation of new individuals to groups. Results from fitting this model to the meadow vole data are illustrated in Fig. 11.5 for sex-specific abundance, births, and survival probabilities. Modifications for groups are made to related models, such as the robust design model discussed in the next section, in exactly the same way.

11.7 ROBUST DESIGN

The CMSA model provides a convenient starting point for a model for Pollock's robust design (Pollock, 1982). In the robust design, the population is regarded as closed between some capture occasions and open between others. During the closed intervals, closed population models such as M_t or M_h may be used for inference about abundance, with encounters across the open periods providing information on survival probabilities.

A simple modification of the CMSA model is to restrict the survival probabilities ϕ_j to 1 and the entry probabilities β_j to 0 for each of the closed periods. This leads to the equivalent of model M_t being applied during each closed period, with simple time dependence on survival probabilities and birth rates for the open periods. For fitting in BUGS it is more efficient to modify the code in Panel 11.3 by adding to the data list a variable K, the number of primary periods, and a vector Prim, a k-dimensioned vector with element j the index of the primary period that

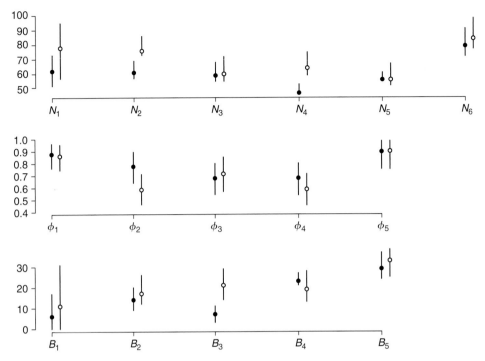

FIGURE 11.5 Posterior density summaries for abundance (N), births (B), and survival rates from fitting a sex-specific CMSA model to the male and female meadow vole data of Nichols (2005). The closed circles are for males and the open circles for females.

the sampling occasion belongs to. Recall that here, k is the total number of sampling occasions. The amended code is given in Panel 11.6.

11.7.1 Example: Meadow Voles

Williams *et al.* (2002) describe a study of adult male meadow voles (*Microtus pennsylvanicus*, in which there were six primary periods ($K=6$) at half-yearly intervals, with each primary period comprising five secondary occasions. Thus $k=30$. A further feature of this study is that a small number of voles were lost on capture. Because a racoon disrupted the traps during the last 2 days of primary period two, Williams *et al.* (2002) analyzed the data using just the first 3 days in primary period two. Using conventional closed-capture modeling they found it necessary to model heterogeneity in detection in each of the primary occasions.

To carry out a similar analysis, we fitted a model in which

$$\text{logit}(p_{ij}) = \gamma_j + \varepsilon_i$$

where ε_i is an individual-specific random effect modeled as a normal random variable with mean zero and variance σ^2. In this model, which we label as $p(\text{primary}+\text{individual})$, the capture probabilities are allowed to vary between primary occasions and between individ-

PANEL 11.6   BUGS code for the robust design model fitted using data augmentation for the unknown times of birth and death and for the superpopulation size Λ.

```
model{
    for(i in 1:M){
        w[i]~dbern(psi)
        b[i]~dcat(beta[])I(,birthCT[i])
    }
    Lambda<-sum(z[])
    for (i in 1:M) {
    for(j in 1:AvailUntil[i]){
        aCap[i,j]<-a[i,Prim[j]]
        pcap[i,j]<-aCap[i,j]*p[j]
        X[i,j] ~ dbern(pcap[i,j])
    }

        a[i,1]<-w[i]*(1-notyetborn[i,1])
        avail[i,1]<-a[i,1]*step(Prim[AvailUntil[i]]-1)
    for (j in 1:K){
        notyetborn[i,j]<-step(b[i]-j-1)
    }
    for (j in 2:K){
        a[i,j]<-w[i]*(1-notyetborn[i,j])*notyetdead[i,j]
        avail[i,j]<-a[i,j]*step(Prim[AvailUntil[i]]-j)
        asuse[i,j] <- notyetdead[i,j-1] + 1
        notyetdead[i,j] ~ dbern(sv[asuse[i,j],i,j-1])
    }
    for(j in 1:K-1){
        sv[1,i,j] <- 0
        sv[2,i,j] <- notyetborn[i,j]+(1-notyetborn[i,j])*S[j]
        }
    }
    # prior distributions for S and p
    for(j in 1:k){
        p[j]~dbeta(1,1)
    }

    for(j in 1:K-1){
        S[j]~dbeta(1,1)
    }

    beta[1:K]~ddirch(alpha[])
    psi~dbeta(1,1)
}
```

uals but the effects are additive. Within a primary occasion, the individual-specific capture probability is fixed. BUGS code for this prior is shown in Panel 11.7.

Posterior summaries from fitting this model are provided in Fig. 11.6 for *N*, *S*, and *B*, and in Fig. 11.7 for the per capita birth rate *f* and population growth rate λ.

PANEL 11.7   BUGS code for the prior on the capture probabilities p_{ij} for the model p(primary+individual).

```
for(i in 1:M){
    eps[i]~dnorm(0,tau)
    for(j in 1:k){
        logit(p[i,j])<-gam[Prim[j]]+eps[i]
    }
}
for(j in 1:K){
    gam[j]~dnorm(0,0.01)
}
sdP~dunif(0,10)
tau<-1/pow(sdP,2)
```

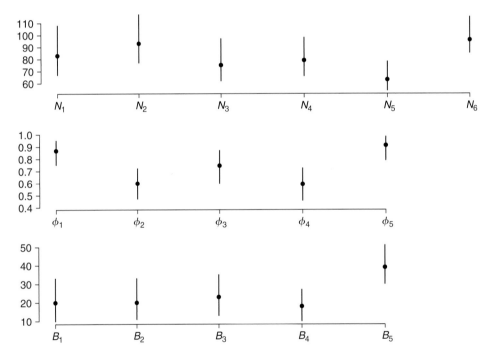

FIGURE 11.6   Posterior summaries for the parameters N (abundance), S (survival), and B (births) for the robust design and meadow vole data of Williams *et al.* (2002) with the model p(primary+individual) for capture probabilities in which there are additive effects of primary occasion and individual on logit(p_iij).

11.7.2 Temporary Emigration

A common feature of mark-recapture studies based on fixed trapping sites is that the area traversed by animals may not coincide exactly with the area sampled (Williams *et al.*, 2002). The models described earlier can be adapted to allow for temporary unavailability of animals and

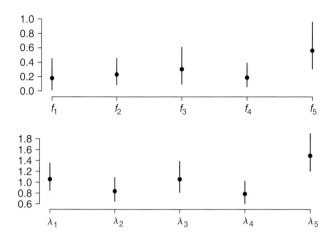

FIGURE 11.7   Posterior summaries for the parameters f (per capita recruitment) and λ (population growth rate) for the robust design model p(primary+individual) in which there are additive effects of primary occasion and individual on logit(p_iij) fitted to the meadow vole data of Williams *et al.* (2002).

to disentangle the confounding of mortality and movement. Kendall and Nichols (1995) and Kendall *et al.* (1997) show how to include temporary emigration in the robust design, but their likelihood conditions on first captures. In this approach the availability of animals is modeled as a first-order Markov chain with two states: "available for capture" and "unavailable for capture." All animals start out in the available state the first time they are released and are also known to be in this state when they are caught. Transitions between the states between samples j and $j+1$ are governed by the transition matrix Ψ_j, which in the notation of Kendall and Nichols (1995) is given by

$$\Psi_j = \begin{pmatrix} 1-\gamma_j'' & \gamma_j'' \\ 1-\gamma_j' & \gamma_j' \end{pmatrix}$$

where γ_j'' is the probability that an animal available for capture in sample $j-1$ is unavailable for capture in sample j and γ_j' is the probability that an animal unavailable for capture in sample $j-1$ is still unavailable for capture in sample $j+1$.

To include a model for the first capture in the robust design we need to introduce parameters governing the allocation of new animals to the states "available for capture" and "unavailable for capture." If δ_j is used to denote the probability that an animal born between samples $j-1$ and j is available for capture at j, then we can write the availability model in terms of a partially observed Markov chain. Letting Z_{ij} denote the state of the chain for individual i alive at the time of sample j, with state 1 denoting "available for capture" and state 2 "unavailable for capture," and letting b_i index the interval of birth and d_i the interval of death, then the Markov chain $\{Z_{i,b_i}, \ldots, Z_{i,d_i}\}$ has probabilities,

$$\Pr(Z_{i,b_i} = s | b_i) = \begin{cases} \delta_{b_i} & s=1 \\ 1-\delta_{b_i} & s=2 \end{cases}$$

and

$$\Pr(Z_{i,j}=s\,|\,Z_{i,j-1}=r,b_i,d_i)= \begin{cases} 1-\gamma_j'' & r=1,s=1 \\ \gamma_j'' & r=1,s=2 \\ 1-\gamma_j' & r=2,s=1 \\ \gamma_j' & r=2,s=2 \end{cases}$$

for $j=b_i+1,\ldots,d_i$.

To illustrate, probabilities for various Markov chains describing the availability of individuals for capture in a study with four primary periods ($K=4$) are given in Table 11.6.

Example: Meadow Vole Robust Design with Heterogeneity and Temporary Emigration

To fit the temporary emigration robust design model in BUGS, we need to add an individual covariate Z_{ij} indicating the status of animal i (available or unavailable) in sample j, for each of the samples following when it was born and before it died. We fitted this model to the meadow vole data, constraining the parameters δ, γ'', and γ' to be constant over time. Fitting this model by maximum-likelihood usually requires some sort of time-constraints as discussed in Kendall *et al.* (1997). We also modeled heterogeneity in capture probabilities fitting the model p(primary+heterogeneity) for capture probabilities that was discussed in Section 11.7.1.

Because animals cannot be seen in the "unavailable" state, not all of the parameters are identifiable, including Λ, the abundances N_j, and the numbers born B_j. However, we can estimate the portion of the population that is at risk of capture, which we denote by N'. Also, the birth rate parameters are identifiable and so provided the assumption that the birth and survival rates are the same for the two components of the population ("available" and "unavailable") we can make valid inference about survival and birth rate.

Posterior summaries for the temporary emigration analysis with heterogeneity are given in Fig. 11.8 for N_j', the numbers of animals available for capture in sample j, and S_j the survival probability from primary occasion j to primary occasion $j+1$. Observant readers may notice that the values for N_j' are smaller than the equivalent values for N_j in Fig. 11.6. This is because

TABLE 11.6   Some Markov chains describing animal availability for capture, and their associated probabilities, conditional on time of birth and death, for a study with four primary periods. State 1 = "available," state 2 = "unavailable," and a value of "NA" is used to represent states that are unassigned, either because the animal is not yet born or because it has died.

Chain	b_i	d_i	$\Pr(Z_{i,j}|b_i,d_i)$
{1,1,2,1}	1	4	$\delta_1(1-\gamma_1'')\gamma_2''(1-\gamma_3')$
{2,1,2,2}	1	4	$(1-\delta_1)(1-\gamma_1')\gamma_2''\gamma_3'$
{NA,1,2,1}	2	4	$1\times\delta_2\gamma_2''(1-\gamma_3')$
{NA,2,2,2}	2	4	$1\times(1-\delta_2)\gamma_2'\gamma_3'$
{NA,1,1,NA}	2	3	$1\times\delta_2(1-\gamma_2'')\times 1$
{NA,2,2,NA}	2	3	$1\times(1-\delta_2)\gamma_2'\times 1$

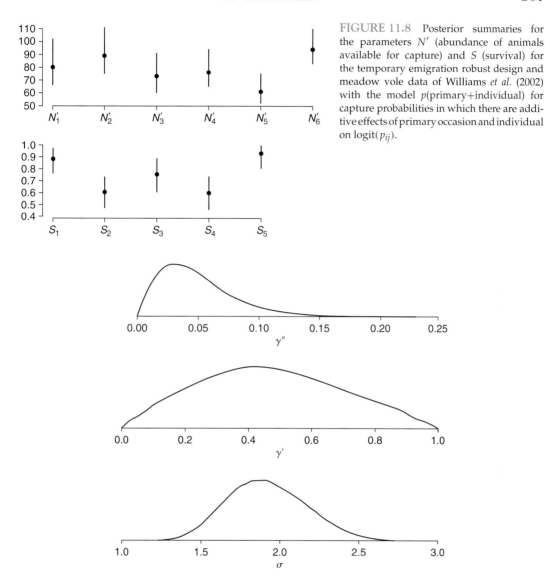

FIGURE 11.8 Posterior summaries for the parameters N' (abundance of animals available for capture) and S (survival) for the temporary emigration robust design and meadow vole data of Williams *et al.* (2002) with the model p(primary+individual) for capture probabilities in which there are additive effects of primary occasion and individual on logit(p_{ij}).

FIGURE 11.9 Posterior summaries for the parameters γ'' (probability of moving from the available state), γ' (probability of remaining in the unavailable state), and σ_p (standard deviation of logit(p_{ij})) for the temporary emigration robust design and meadow vole data of Williams *et al.* (2002) with the model p(primary+individual) for capture probabilities in which there are additive effects of primary occasion and individual on logit(p_{ij}).

the abundances in Fig. 11.8 are only for the animals available for capture, whereas the abundances in Fig. 11.6 are for all animals alive in the population at the time of sample j, regardless of whether they are available for capture.

Summaries for γ'', γ', and σ_p are in Fig. 11.9. The posterior density for γ'' has most of its mass well below 0.10 indicating a relatively small amount of temporary emigration from the "available for capture" to "unavailable for capture" state. The posterior density for γ' indicates that the data do not provide a lot of information on this parameter. This is unsurprising given the small amount of movement out of the population at risk of capture. Even when there is a lot of movement, inference about γ' still tends to be imprecise because we never have animals in any sample j where we know for certain that they were alive but unavailable for capture at the time of sample $j-1$.

11.8 MULTISTATE MODELS AND OTHER EXTENSIONS OF THE CJS MODEL

The last example had animals in one of two states, either "available" or "unavailable" for capture. Studies of animal populations often require modeling the changing state of individuals, whether these relate to geographical location, breeding or health condition, size class, etc. These states may be of interest in themselves, or as explaining demographic parameters or detection rates.

The most widely used multistate model has been the Arnason–Schwarz model (Arnason, 1972; Brownie *et al.*, 1993; Schwarz *et al.*, 1993) which is a multistate extension of the CJS model. Two important features of the Arnason–Schwarz model are that (i) it conditions on the first release of the animal, and (ii) states are observed at each capture. However, because the state of the animal is unobserved when we do not catch it, "state" can be considered as a partially-observed categorical covariate. The model of Kendall *et al.* (1997) that extended the robust design to include the unavailable state, can also be considered as a multistate model with two partially observed states: "available for capture" and "unavailable for capture." In this model animals in the "unavailable for capture" state can never be observed.

A fully Bayesian treatment of the Arnason–Schwarz model was given by Dupuis (1995) who developed a Gibbs sampling algorithm that uses data-augmentation to predict values for the state occupied by animals when they are not caught. A similar approach was used by King and Brooks (2002), who also described a reversible jump MCMC algorithm to allow for model discrimination. The procedure used by Dupuis (1995) and King and Brooks (2002) updates Z_{ij}^{mis}, the unknown state of individual i at time j, by sampling from the full conditional distribution $[Z_{ij}^{\text{mis}} \mid \cdot]$. This procedure was implemented by Schofield *et al.* (2008) who also provide BUGS code.

More recently, Dupuis and Schwarz (2007) extended the multistate model to model the births using the Crosbie–Manly–Schwarz–Arnason formulation of the JS model. To fit a multistate model in BUGS it is easy to modify our code for either the CMSA model (Panel 11.3), or the robust-design extension of it (Panel 11.6), to include an individual covariate Z_{ij} for the state of animal i in sample[8] j, as discussed in Section 11.7.1 for modeling temporary emigration.

---

8. Primary sample j if it is a robust design model.

Dupuis and Schwarz (2007) also provide important theoretical results on the ergodicity of the MCMC procedure, meaning its propensity to converge to the required distribution. In particular, Dupuis (1995) showed that when the Z_{ij}^{mis} are updated one at a time, the resulting chain should converge to the target density provided that the transition probabilities for the movement of animals between states do not include zeros. As we are often interested in cases where movement is restricted it may be possible that the resulting chain is non-ergodic. Dupuis and Schwarz (2007) outline a Gibbs sampling algorithm that ensures that this potential problem does not arise. Their algorithm makes use of block updating in which groups of missing covariates Z_{ij}^{mis} are updated using multinomial sampling that is carried out conditional on Λ and sufficient statistics that are multistate analogs of the JS sufficient statistics R, r, and m.

An interesting extension of the multistate model is to the case where there is uncertainty about the observed states but where this uncertainty is informed by covariates. An example of state uncertainty is given by Nichols *et al.* (2004) for a study of roseate terns (*Sterna dougalii*) where the sex of the animal can be predicted using behavioral observations but often not definitively. Cases where the predictor variables for the missing state are also categorical correspond to the multievent model of Pradel (2005); however, the model provided by Schofield *et al.* (2008) allows these covariates to be continuous. The states themselves can also be considered as categorical covariates for survival and capture probabilities. Bonner and Schwarz (2006) extended the CJS model to allow for partially observed continuous covariates, and this model and code are also provided by Schofield *et al.* (2008).

11.9 AFTERWORD

The censoring models, known-fate models, and open population mark-recapture models with imperfect detection all share a common feature: the data we get to see are not quite the data we wish we had. The analyses presented here have been based on CDL's. The benefit of CDL's is two-fold. First, they lead to more intuitive and efficient computer code for fitting models. Implementation of MCMC via Gibbs sampling is generally very easy, especially in contrast to likelihood-based methods relying on ODL's, involving complicated integrals and sums.

Second, the use of CDL's concentrates inference on aspects of our models that are of biological interest. Much of the "data we wish we had" describes demographic phenomena of primary interest, and about which we wish to make inference. The use of CDL's in a Bayesian setting makes predictions of these quantities a routine feature of analysis.

Likelihood functions that have appeared in the literature are almost exclusively ODL's and the complexities of the confounding between birth, death, and capture processes can make these models difficult to understand. Also, many of the standard parameterizations of mark-recapture models, particularly those that model births and abundance such as the JS model, are not amenable to hierarchical extensions. It is these hierarchical extensions that give us the ability to model latent variables that form the data that we wish we had been able to collect. This additional structure is easily accommodated in Bayesian analysis.

We have shown that it is relatively straightforward to extend the CDL for basic models such as the CJS model to include births, group dependencies, multiple-strata and covariates that change over time, and to include robust design aspects. Although not discussed in this chapter, it is also straightforward to extend the CDL to include auxiliary data such as may be provided by resightings or dead recoveries between sampling occasions, as in the models of Burnham (1993), Barker (1997), and Barker *et al.* (2004). These ideas are explored further by Schofield *et al.* (2008).

Individual Fitness

OUTLINE

12.1 **Population Fitness**	272	12.6 **Population Summaries of Fitness** 281
12.2 **Individual Fitness**	273	*12.6.1 Estimating the Population*
12.3 **Realized Individual Fitness**	274	*Distribution of Individual*
12.4 **Individual Fitness in Group**		*Fitness* 282
Context	275	*12.6.2 Estimating the Population*
12.5 **Analysis of Individual Fitness:**		*Fitness* 285
An Example	276	12.7 **Afterword** 286

A population's fitness is its capacity to persist. It is a summary of potentialities, rather than a summary of outcomes. We can imagine a perfectly "fit" population (however defined) becoming extinct through extraordinary events, such as an asteroid strike. We understand that even in stable environmental circumstances, a population's persistence is subject to stochasticity. But fitness is a feature of the population that exists before the outcomes. Thus in discussing fitness, we require a mathematical model of population change. Fitness is a latent feature of the population and is defined as a function of model parameters.

For a geographically closed population, persistence depends on individual survival and on recruitment of new individuals. We naturally conceive of survival and recruitment rates as ratios of numbers of individuals: if X individuals from a population of size N survive from one time period to the next, the period survival rate is X/N; if Y new individuals have been born and survived into the population, the recruitment rate is Y/N. But these rates do not define fitness, since they have to do with realizations of events rather than a priori capacities. Given that no asteroids have struck, these rates might be used to estimate parameters of mathematical models, with fitness being a function of the parameters.

The distinction between realization and capacity, between estimates and parameters, is one that matters, if only in keeping our thinking straight. For example, we might begin by treating X as a binomial random variable with index N, and success parameter P, and Y as a Poisson random variable with mean $F \times N$. But, then we realize that this model is an oversimplification of reality, that (for instance) there is likely to be age-specific variation among individuals in the population, so we extend our model description using age-class-specific parameters, F_i and P_i.

Population fitness is described in terms of F_i and P_i, as described in Section 12.1. One might question whether realized survival and recruitment rates depend on factors other than age class. Temporal events, even those less dramatic than asteroid strikes, may influence realized rates; we might require a mathematical model for temporal variation, and to modify our definition of fitness accordingly. Furthermore, it is usually to be expected that there will be individual heterogeneity in factors affecting survival and recruitment.

In this chapter, we consider individual fitness, defined by analogy to population fitness, in terms of individual-specific survival and reproductive parameters, determined at birth. In this context, the distinction between realization and capacity is especially important since individuals only live once. Individual fitness is a derived parameter, one that is best evaluated in context of a hierarchical model, and naturally evaluated under the Bayesian māramatanga.

12.1 POPULATION FITNESS

For a geographically closed, age-structured population with maximum lifetime T, with $n_i(t)$ individuals in age class i at time t, the change in population structure through time is represented by the matrix equation

$$
\begin{bmatrix}
F_1 & F_2 & F_3 & \cdots & F_{T-1} & F_T \\
P_1 & 0 & 0 & \cdots & 0 & 0 \\
0 & P_2 & 0 & \cdots & 0 & 0 \\
0 & 0 & P_3 & \cdots & 0 & 0 \\
0 & 0 & \cdots & \ddots & 0 & 0 \\
0 & 0 & 0 & \cdots & P_{T-1} & 0
\end{bmatrix}
\begin{bmatrix}
n_1(t) \\ n_2(t) \\ n_3(t) \\ \cdots \\ n_T(t)
\end{bmatrix}
=
\begin{bmatrix}
n_1(t+1) \\ n_2(t+1) \\ n_3(t+1) \\ \cdots \\ n_T(t+1)
\end{bmatrix},
\tag{12.1}
$$

compactly expressed as $M\,n(t) = n(t+1)$.

In considering the matrix M, it will be useful to define cumulative survival probabilities $S_1 = 1$, and $S_i = P_{i-1}S_{i-1}$ for $i = 2,3,\ldots,T$. Matrix M has a unique positive eigenvalue provided that at least one of the terms S_iF_i is positive. This value is the unique positive solution in x of

$$
\sum_{i=1}^{T}(S_iF_i)x^{T-i} = x^T.
\tag{12.2}
$$

We will write $\{SF\}$ for the element by element product of vectors $S = (S_1,S_2,\ldots,S_T)'$ and $F = (F_1,F_2,\ldots,F_T)'$. Using this notation, we denote the solution of Eq. (12.2) as $\lambda(\{SF\})$.

An eigenvalue λ and an eigenvector x of a matrix A have the property that $A\,x = \lambda\,x$. Thus, if the population vector $n(t)$ is an eigenvector of M, so will be the population vector $n(t+1)$ at the next time period. The two vectors will be proportional, elements of the latter having been

PANEL 12.1   R code for calculation of fitness.

```
Lambda=function(S,F){
        SF=S*F
        T=length(SF)
        h=function(x){
            sum(SF*(x^seq(T-1,0,-1)))-x^T
        }
        if (max(SF)==0) 0
        else uniroot(h,c(1E-10,30))$root
}

S=c(1.0000,0.8000,0.6400,0.5120,0.4096)
F=c(0,1,2,3,2)
Lambda(S,F)
```

scaled by a factor $\lambda(\{SF\})$. The population is said to be in a stable age distribution, and $\lambda(\{SF\})$ is the growth rate.

Regardless of whether the population is in stable age distribution, the quantity $\lambda(\{SF\})$ is a convenient and widely accepted scalar-valued summary of vectors S and F. Values greater than 1 suggest a growing population, values less than 1 suggest a declining population.[1] Thus, $\lambda(\{SF\})$ is considered a reasonable measure of a population's fitness, reflecting the combined effects of survival and reproduction on population change.[2] Calculation of $\lambda(\{SF\})$ is straightforward using standard math packages; R code is given in Panel 12.1. For a comprehensive discussion of matrix models of population change, we refer the reader to the authoritative text of Caswell (2001).

12.2 INDIVIDUAL FITNESS

In genetic terms, fitness is the capacity of a genotype to be propagated into future generations. Given a population of genetically identical individuals described by rates S and F, the value of $\lambda(\{SF\})$ could be taken as a measure of the genotype's fitness. If $\lambda(\{SF\}) > 1$, the genotype would tend to be successfully propagated into future generations; if $\lambda(\{SF\}) < 1$, the genotype would tend to disappear.

It is more natural, however, to conceive of genotypic fitness in terms of individuals; genetically identical populations are the exception rather than the rule. It makes sense to think of individual-specific fecundity rates and survival rates, latent features associated entirely with an individual's genetic make-up.

Thus, we imagine that individual j is born with a set of parameters governing survival and a set of parameters governing breeding success. The event of survival from age class i

---

1. It should be noted that $\lambda(\{SF\}) > 1$ does not guarantee an increasing population; the growth rate interpretation depends on the population being in (or at least near) stable age distribution. Even then, the parameter describes an expected rate of change, rather than an actual change.
2. The eigenvalue definition of $\lambda(\{SF\})$ assumes at least one of the values $S_iF_i > 0$. If all values $S_iF_i = 0$, individuals all die before breeding; in this case $\lambda(\{SF\})$ is defined to be zero.

to $i+1$ is a Bernoulli trial with success parameter $P_i^{(j)}$. The number of young produced in age class i is a random variable $f_i^{(j)}$ with expected value $F_i^{(j)}$. As before, we define cumulative survival probabilities $S_i^{(j)} = S_{i-1}^{(j)} P_{i-1}^{(j)}$, with $S_1^{(j)} = 1$, and summarize the collections of parameters by vectors $\mathbf{S}^{(j)}$ and $\mathbf{F}^{(j)}$.

The important feature of all these parameters is that they are individual specific and completely determined at birth; $F_i^{(j)}$ and $P_i^{(j)}$ exist even if the animal does not survive to age class i. The parameters are consequences of the genotype. It is natural, then, to define individual fitness in analogy to population fitness as $\lambda^{(j)} = \lambda(\{\mathbf{S}^{(j)} \mathbf{F}^{(j)}\})$. Individual fitness is thus a latent quantity, a summary of latent features associated with the genotype.

To make the notation a bit easier, we drop the superscript $^{(j)}$ in the remainder of this chapter, except where it is necessary for clarity; unless otherwise indicated, it is to be understood that the quantities discussed are individual-specific.

It is natural to ask whether individual fitness is a measurable quantity, or whether it only exists as a theoretical construct. After all, individual fitness is determined by parameters governing a single life history, without any hope of replication.[3] Nevertheless, the answer is "yes": individual fitness can be estimated, given the information in a single life history. The reason is that individual fitness is a function of the products $S_i F_i$, and these values are estimable. Letting I_i be an indicator variable for whether the individual is alive in age class i, the quantity $I_i f_i$ is an unbiased estimator of $S_i F_i$.[4] One might then simply substitute the estimates of $S_i F_i$ in Eq. (12.2), solve for the eigenvalue, and use the result as the estimate of individual fitness.

McGraw and Caswell (1996) took essentially this approach in defining individual fitness as the eigenvalue of individual-specific Leslie matrices similar to $\mathbf{M}$ in Eq. (12.1), but with 1 and f_i replacing S_i and F_i in years survived, and 0 replacing S_i in the year of death. Suppose that an individual survives into age class X, then dies. Writing $\mathbf{1}_X$ for a vector of X ones followed by $T - X$ zeros, and writing $\mathbf{f}$ for the vector consisting of $F_1, F_2, \ldots, F_x$ followed by $T - X$ zeros, McGraw and Caswell's measure of individual fitness can be written using the notation established above, as $\lambda(\{\mathbf{1}_X \mathbf{f}\})$.

McGraw and Caswell did not emphasize the distinction between latent individual fitness, and an estimate of the quantity. For reasons soon to become apparent, we believe the distinction is important, and describe $\lambda(\{\mathbf{1}_X \mathbf{f}\})$ as the "realized individual fitness" to distinguish it from the latent quantity $\lambda(\{\mathbf{S}\mathbf{F}\})$. The important difference between these quantities is that an individual's latent fitness is fixed at birth, but the individual's realized fitness is a random outcome yet to occur. Realized fitness is a random variable with distribution determined by latent parameters; latent fitness is a deterministic function of the latent parameters.

12.3 REALIZED INDIVIDUAL FITNESS

Realized fitness $\lambda(\{\mathbf{1}_X \mathbf{f}\})$ is a quantity of interest in its own right, and is clearly related to latent individual fitness $\lambda(\{\mathbf{S}\mathbf{F}\})$, but realized fitness is not a very good estimator of latent

3. One exception is when the individual can be cloned, or is naturally part of an identifiable clone. Individual life histories within the clone could then be treated as replicates, and one could do a pretty good job of estimating 'individual' fitness, given enough replicates. But this circumstance is rare, and the 'solution' essentially consists in defining away the problem!
4. This is a bit sneaky, of course. The random variable f_i is not observed after the individual dies, but at that point $I_i = 0$, so we can still calculate $I_i f_i = 0$. It should be noted that $I_i f_i$ is an estimator of $S_i F_i$, not of F_i.

TABLE 12.1 Distribution of realized fitness.

X	1_X	$\lambda(\{1_X f\})$	Pr(X)
1	(1,0,0,0,0)'	0	0.2000
2	(1,1,0,0,0)'	1.0000	0.1600
3	(1,1,1,0,0)'	1.5214	0.1280
4	(1,1,1,1,0)'	1.7614	0.1024
5	(1,1,1,1,1)'	1.8241	0.4096

individual fitness. It is easy to construct examples of latent parameter vectors leading to equal latent fitnesses, but producing quite different realized fitnesses. That is, two genotypes having the same latent fitness could result in quite different life histories, with realized fitnesses that were not at all similar, even on the average. In such a case, realized individual fitness would do a poor job of representing the latent fitness associated with the genotypes.

Here is an example. Suppose that $T = 5$, and that latent annual survival rates are 80%, so that $S = (1.0000, 0.8000, 0.6400, 0.5120, 0.4096)'$. For simplicity, suppose that f is not random, but deterministic, with $f = F = (0, 1, 2, 3, 2)'$. The latent fitness is 1.567, calculated using the R code of Panel 12.1. Realized fitness takes one of five distinct values, as summarized in Table 12.1; its mean value is 1.282 and its standard deviation is 0.702.

This bias might not be so bad, especially if it were close to constant for a fixed level of latent fitness. But it is possible to construct another pair of vectors S and F leading to the same latent fitness, but having a quite different distribution of realized fitness. To do so, we halve the first year's survival rate, and double subsequent fecundities. Thus, $S = \{1, 0.4000, 0.3200, 0.2560, 0.2048\}$ and $F = \{0, 2, 4, 6, 4\}$.

The latent fitness is unchanged because $\{SF\}$ is the same in both cases. However in the second case, the probabilities in the final column of Table 12.1 are changed to 0.6000, 0.0800, 0.0640, 0.0512, and 0.2048, so that the mean and standard deviation of realized fitness become 0.833 and 1.032, respectively. Repeating the process twice more, the average realized fitness drops first to 0.54, then to 0.35. Thus realized fitness is an unreliable index to latent fitness.

12.4 INDIVIDUAL FITNESS IN GROUP CONTEXT

The problem with realized individual fitness is that it is an estimate based on a sample of size 1, the single realization of an individual's life history. Fortunately, we need not operate this way, but rather may (and in our view, ought to) consider individuals in a group context, seeing individual-specific parameters as stochastically related.

Suppose that we can model $P_i^{(j)}$ and $F_i^{(j)}$ in context of a group of individuals. We could then posit a hierarchical model and perform a Bayesian analysis, obtaining posterior distributions for $P_i^{(j)}$ and $F_i^{(j)}$ informed not only by the data for individual j, but by data for all of the individuals in the group. The analysis would be a compromise between extremes: on one hand, of treating all individuals as distinct and unrelated (as when using realized fitness), on the other, of treating all individuals as identical (when computing population fitness).

Individual fitness could then be treated as a derived parameter, and samples of its posterior distribution calculated from posterior samples of $P_i^{(j)}$ and $F_i^{(j)}$. These quantities are far better estimated in the group context than otherwise, with the result that we also do a much better job of examining individual fitness in the group context than by simply using realized fitnesses, which ignore any such structure. We illustrate this subsequently with analysis of a simple data set.

Cam *et al.* (2002) presented an analysis of survival and breeding for kittiwakes (*Rissa trydactyla*) with individual effects. Because individuals produced either zero or one offspring, the reproductive event $f_i^{(j)}$ was modeled as a Bernoulli trial, with success parameter $F_i^{(j)}$. Cam *et al.*'s model specified that $\mathrm{logit}(P_i^{(j)}) = \alpha_p^{(j)} + x_{i,j}' \beta_p$, and that $\mathrm{logit}(F_i^{(j)}) = \alpha_f^{(j)} + y_{i,j}' \beta_f$, where $x_{i,j}$ and $y_{i,j}$ are individual-specific covariates, with effects modeled at the group level; individual effects $\alpha_p^{(j)}$ and $\alpha_f^{(j)}$ were modeled as sampled from a bivariate normal distribution. Parameter vectors, β_p and β_f, described models for year effects and effects of senescent decline. The posterior distribution for the correlation parameter of individual effects had mean of 0.67, with 95% credible interval (0.28 0.98), strong evidence that birds with higher survival probabilities also had higher conditional probabilities of breeding, given survival. There was substantial variation among individuals in survival and breeding rates; in a subsequent reanalysis of the data, Link *et al.* (2002) examined the effects of this variation on individual fitness.

Our reason for including this example is to emphasize the point that given the posterior distribution of all unknown quantities, $[\theta | \text{Data}]$, the problem of estimating fitness for individual j is transparently easy. Given a value of θ, then we have all the information needed to calculate vectors $S^{(j)}$ and $F^{(j)}$ for individual j, hence to calculate individual fitness $\lambda(\{S^{(j)} F^{(j)}\})$. Indeed, it is appropriate to write

$$\lambda^{(j)} \equiv \lambda^{(j)}(\theta) = \lambda(\{S^{(j)} F^{(j)}\});$$

that is, that individual j fitness is a known function of the unknown parameters comprising θ.

Thus, given a sample from $[\theta | \text{Data}]$, say a Markov chain of values $\theta_1, \theta_2, \ldots, \theta_B$, one may calculate $\lambda_1^{(j)} = \lambda^{(j)}(\theta_1), \lambda_2^{(j)} = \lambda^{(j)}(\theta_2), \ldots, \lambda_B^{(j)} = \lambda^{(j)}(\theta_B)$; these values are a sample from the posterior distribution $[\lambda^{(j)} | \text{Data}]$, thus providing the basis for Bayesian inference about individual fitness. Because the posterior distributions of $\lambda^{(j)}$ tend to be skewed, we recommend the posterior median as a point estimator.[5] The posterior distribution provides a natural description of the uncertainty associated with an individual's fitness.

12.5   ANALYSIS OF INDIVIDUAL FITNESS: AN EXAMPLE

Table 12.2 gives 25 simulated life histories over a 15-year period. These data were generated in accordance with the kittiwake model of Cam *et al.* (2002) described in Section 12.4 but without year effects or effects describing senescent decline. Individuals do not breed in their first year, then produce zero or one offspring in subsequent years until age 15, after which breeding ceases. Missing values " · " in the table indicate that the individual had died. For

---

5. That is, if a point estimator is really needed. As usual, a single point is a rather limited summary of a posterior distribution.

TABLE 12.2 Simulated life history data. Rows are individuals, columns are years. Zeros and ones indicate number of young produced if alive; " · " indicates the individual did not survive to that year.

Case	Year														
	1	2	3	4	5	6	7	8	9	10	11	12	13	14	15
1	0	.	.	.	.	.	.	.	.	.	.	.	.	.	.
2	0	.	.	.	.	.	.	.	.	.	.	.	.	.	.
3	0	0	.	.	.	.	.	.	.	.	.	.	.	.	.
4	0	0	.	.	.	.	.	.	.	.	.	.	.	.	.
5	0	0	0	.	.	.	.	.	.	.	.	.	.	.	.
6	0	0	0	1	.	.	.	.	.	.	.	.	.	.	.
7	0	0	0	0	0	.	.	.	.	.	.	.	.	.	.
8	0	0	0	0	0	0	.	.	.	.	.	.	.	.	.
9	0	0	0	0	0	0	0	0	.	.	.	.	.	.	.
10	0	0	0	0	1	0	1	0	0	0	.	.	.	.	.
11	0	0	0	0	0	0	1	0	0	0	0	.	.	.	.
12	0	1	0	1	0	0	1	0	0	0	0	1	.	.	.
13	0	0	0	0	0	0	0	0	0	0	0	0	0	.	.
14	0	0	0	0	0	0	0	0	0	0	0	0	0	.	.
15	0	0	0	0	0	0	0	0	0	0	0	0	0	.	.
16	0	1	0	0	0	1	0	0	0	0	0	0	0	.	.
17	0	0	0	0	0	0	0	0	0	0	0	0	0	0	0
18	0	0	0	0	0	0	0	0	0	0	0	0	0	0	0
19	0	0	0	0	0	0	0	0	0	0	0	0	0	0	1
20	0	0	0	0	1	0	0	0	0	0	0	0	0	0	0
21	0	0	0	0	0	0	0	1	0	0	0	0	0	0	0
22	0	0	0	0	0	0	0	1	1	0	0	0	0	0	0
23	0	1	0	0	0	0	0	0	1	0	0	0	0	0	0
24	0	0	0	0	0	1	1	0	0	0	0	0	1	0	0
25	0	0	1	0	0	0	1	1	0	1	1	0	1	0	1

example, individual no. 6 survived into age class 4 without producing offspring, produced one that year, and died without reaching age class 5.

We generated the data with individual-specific parameters $\text{logit}(P_i^{(j)}) \equiv \alpha_p^{(j)}$ and $\text{logit}(F_i^{(j)}) \equiv \alpha_f^{(j)}$, the pair of individual effects having been sampled from a bivariate normal distribution with means $\mu_p = 3$ and $\mu_f = -2$, standard deviations $\sigma_p = \sigma_f = 0.50$, and correlation $\rho = 0.50$.

For analysis, we used vague normal priors (mean $= 0$, standard deviation $= 1000$) for μ_p and μ_f. We assigned $U(0,5)$ priors to σ_p and σ_f, and a uniform prior $U(-1,1)$ for ρ.

Given real data (in particular, a larger sample), we would typically include temporal effects, some relating to observed covariates (e.g., weather features), others in the form of time-stationary random effects when analyzing data. These temporal effects might be regarded as irrelevant to consideration of individual fitness (this being a summary of latent features prior to the realization of the life history) and set to zero in calculation of individual fitness. Alternatively, one might consider individual fitness given the particular realization of temporal effects, or even averaged over random replicates of the temporal effects, although this latter would entail considerably complex computations. One might also model effects of aging on survival and breeding, although to distinguish these from irrelevant temporal effects, data from multiple cohorts would be required.

The example considered here is included for its simplicity, to illustrate inference with a derived parameter. We leave to the reader the question of how one would carry out such analyses using frequentist methods. In considering the question, one should keep track of the number of approximations required and the associated uncertainties.

The model we are considering posits the existence of two random variables associated with each of the 25 individuals in Table 12.2. Because survival rates are not age-specific, individual j's lifespan s_j is a geometric random variable with parameter $P^{(j)}$; similarly, the total number of offspring produced by individual j is a binomial random variable with index $n^{(j)} = min\{s_j, 14\}$.[6]

BUGS code for the analysis is given in Panel 12.2. Note that we made use of the handy fact that if X_p and X_f are independent, standard, normal random variables, then

$$\alpha_p = \mu_p + \sigma_p X_p$$

and

$$\alpha_f = \mu_f + \sigma_p \left(\rho X_p + \sqrt{1 - \rho^2} X_f \right)$$

have a bivariate normal distribution with means μ_p and μ_f, standard deviations σ_p and σ_f, and correlation ρ.

Note also that binomial likelihoods have been used for geometric random variables. Individuals $j = 1, 2, \ldots, 16$ survived s_j times, then died, hence have the same likelihood as a binomial random variable with index $s_j + 1$ and observed value s_j. Individuals $j = 17, 18, \ldots, 25$ survived s_j times and did not die; their geometric likelihood is the same as a binomial random variable with index s_j and observed value s_j. Individuals $j = 1, 2$ died before the possibility of breeding, hence provide no information about breeding rates.

We used BUGS to produce Markov chains of length 20,000 after a burn-in of length 1000. We then calculated sets of 20,000 values $\lambda(\{S^{(j)} F^{(j)}\})$, one for each individual.[7] The posterior distributions of latent fitness for three individuals (Cases #1, #18, and #25) are plotted in Fig. 12.1.

Referring back to Table 12.2, we note that individual #1 died as a juvenile without breeding, that individual #18 survived through to the maximum breeding age but never produced young, and that individual #25 survived to the maximum breeding age and produced the most young.

---

6. 14, because no breeding occurs in age class zero or beyond age class 15.
7. Posterior samples of $S^{(j)}$ and $F^{(j)}$ were exported using BUGS' **coda** facility and values of λ were computed using other software. See, for example, Panel 12.1.

PANEL 12.2 BUGS code for individual fitness data.

```
s=c(0,0,1,1,2,3,4,5,7,9,10,11,12,12,12,12,14,14,14,14,14,14,14,
  14,14)
f=c(NA,NA,0,0,0,1,0,0,0,2,1,4,0,0,0,2,0,0,1,1,1,2,2,3,7)

####### Data ######
for (j in 1:16){
    ns[j] <-  s[j]+1
    s[j] ~ dbin(P[j],ns[j])
}
for (j in 17:25){
    s[j] ~ dbin(P[j],s[j])
}
for (j in 3:25){
    f[j] ~ dbin(F[j],s[j])
}

####### Parameters ######
for (j in 1:25){
   x[j] ~ dnorm(0,1)
   y[j] ~ dnorm(0,1)
   logit(P[j]) <- mu.p + sigma.p*x[j]
   logit(F[j]) <- mu.f + sigma.f*(rho*x[j]+sqrt(1-rho*rho)*y[j])
}

####### Priors ######
mu.p ~ dnorm(0,1.0E-6)
mu.f ~ dnorm(0,1.0E-6)
sigma.p ~ dunif(0,5)
sigma.f ~ dunif(0,5)
rho.f ~ dunif(-1,1)
```

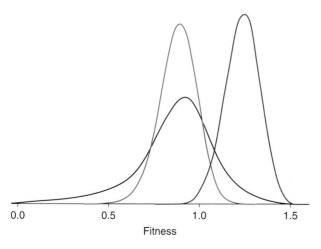

FIGURE 12.1 Posterior distribution of individual fitness. Cases 1 (black), 18 (red), and 25 (blue).

TABLE 12.3   Simulated life history summary. Parameters used in simulation (P, F, and resulting latent fitness $\lambda_{SF} = \lambda(\{SF\})$), data summaries ($s$, $Y = \sum f_i$, and realized fitness $\lambda_{1f} = \lambda(\{1xf\})$ calculated with $X = s+1$), and summaries of posterior distributions (2.5th, 50th, and 97.5th percentiles).

	Parameters			Data			Posterior		
Case	P	F	λ_{SF}	s	Y	λ_{1f}	2.5	50	97.5
1	0.947	0.129	1.028	0	0	0	0.323	0.894	1.264
2	0.954	0.085	0.980	0	0	0	0.323	0.894	1.264
3	0.948	0.073	0.956	1	0	0	0.454	0.893	1.204
4	0.949	0.131	1.032	1	0	0	0.454	0.893	1.204
5	0.900	0.080	0.924	2	0	0	0.516	0.888	1.173
6	0.979	0.182	1.112	3	1	1	0.686	1.002	1.305
7	0.880	0.043	0.844	4	0	0	0.577	0.885	1.128
8	0.918	0.096	0.962	5	0	0	0.590	0.884	1.119
9	0.905	0.098	0.952	7	0	0	0.615	0.880	1.092
10	0.925	0.093	0.965	9	2	1.124	0.828	1.025	1.235
11	0.979	0.157	1.088	10	1	1	0.747	0.952	1.149
12	0.955	0.105	1.007	11	4	1.350	0.930	1.124	1.327
13	0.919	0.048	0.886	12	0	0	0.639	0.872	1.062
14	0.908	0.044	0.870	12	0	0	0.639	0.872	1.062
15	0.924	0.064	0.921	12	0	0	0.639	0.872	1.062
16	0.957	0.143	1.053	12	2	1.211	0.825	1.006	1.197
17	0.947	0.070	0.951	14	0	0	0.669	0.889	1.074
18	0.981	0.238	1.163	14	0	0	0.669	0.889	1.074
19	0.988	0.076	0.997	14	1	1	0.776	0.956	1.136
20	0.926	0.072	0.935	14	1	1	0.776	0.956	1.136
21	0.943	0.136	1.032	14	1	1	0.776	0.956	1.136
22	0.957	0.120	1.028	14	2	1.085	0.845	1.013	1.191
23	0.975	0.149	1.076	14	2	1.162	0.845	1.012	1.191
24	0.954	0.211	1.114	14	3	1.146	0.903	1.064	1.248
25	0.954	0.162	1.069	14	7	1.282	1.041	1.237	1.412

The overlap in these marginal posterior distributions can be somewhat misleading. There is very strong evidence that $\lambda^{(25)} > \lambda^{(18)}$, and reasonably strong evidence that $\lambda^{(25)} > \lambda^{(1)}$. To evaluate the evidence, we consider some additional derived parameters, namely $\lambda^{(25)} - \lambda^{(18)}$ and $\lambda^{(25)} - \lambda^{(1)}$. We find that $\Pr(\lambda^{(25)} - \lambda^{(18)} \geq 0 | \text{Data}) = \Pr(\lambda^{(25)} \geq \lambda^{(18)} | \text{Data}) = 0.993$, and similarly, that $\Pr(\lambda^{(25)} \geq \lambda^{(1)} | \text{Data}) = 0.954$.

Another derived parameter of interest is the rank of an individual's fitness relative to the rest of the individuals. Thus, to examine the magnitude of $\lambda^{(25)}$ relative to other $\lambda^{(j)}$, we simply compute the rank of $\lambda^{(25)}$ among the $\lambda^{(j)}$ (for each of the 20,000 values in our Markov chain

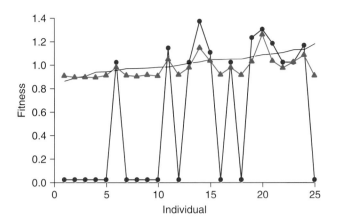

FIGURE 12.2 Estimates of individual fitness. True latent fitness (sorted, black), Bayes estimates (posterior medians, red), and realized fitnesses (blue).

Monte Carlo sample). We find that

$$\Pr\big(\lambda^{(25)} = max\{\lambda^{(j)} : j = 1,2,\dots,25\} \mid \text{Data}\big) = 0.616,$$

and that $\Pr(Rank(\lambda^{(25)}) \geq 22|\text{Data}) = 0.948$.

We compare Bayesian estimates with realized fitness in Table 12.3 and Fig. 12.2. Both correlate reasonably well with the true values (which were known, since the data were simulated): the correlation between true latent fitness and realized fitness is 0.51, between true latent fitness and its posterior median is 0.48. However the root mean squared error for the Bayes estimates is 0.098, versus 0.701 for the realized fitness. The Bayes estimate is closer to the true value in 21 of the 25 cases.

12.6 POPULATION SUMMARIES OF FITNESS

We have considered individual fitness in context of a population model for individual heterogeneity, focusing on individual values. In this section, we broaden our scope, looking at the collection of individual fitnesses, and at the overall population fitness implied by the model of individual heterogeneity.

Known Parameters

Given a model for individual heterogeneity in demographic parameters, and known population hyperparameters $\boldsymbol{\beta}$, there is an implied distribution of individual fitnesses. For example, we might ask what proportion of individuals have fitness less than 1.0; we might ask what are the mean, standard deviation, and various percentiles of individual fitness. These questions all have to do with the population distribution of individual fitness, namely $\Lambda_{\boldsymbol{\beta}}(t) = \Pr(\lambda \leq t|\boldsymbol{\beta})$.

The distribution cannot be calculated analytically, since the eigenvalue function $\lambda(\cdot)$ is only defined implicitly, as the solution to Eq. (12.2). However, we can obtain arbitrarily precise evaluations of the distribution through simulation.

Using the data generating model and parameters for the example in Section 12.5, we generated a population of 100,000 pairs $\{S^{(j)}, F^{(j)}\}$, and corresponding values $\lambda^{(j)}$. One might think of each pair as corresponding to a hypothetical individual in a large population, a population having identical patterns of variation to those underlying the sample of 25 we have considered. The difference is that for these 100,000 individuals, we *know* the latent fitness. From these, we conclude that the population of individual fitnesses has mean $=1.021$, standard deviation $=0.075$, and 5th and 95th percentiles $=0.900$ and 1.148; the large sample size allows us to estimate these quantities to three decimal place precision.[8]

Note that $\Lambda_\beta(t)$ is entirely determined by the hyperparameter β; thus, we may think of the population distribution of individual fitness, and its features (mean, percentiles, and so forth) as derived parameters. The importance of this observation is that it indicates how Bayesian analysis of these quantities is conducted: given a sample from the posterior distribution β, we need only calculate the corresponding values of the derived parameter to obtain a sample from its posterior distribution. The simulation needed to examine $\Lambda_\beta(t)$ is best regarded as a calculation, albeit approximate. Viewed thus, the Bayesian approach to inference about $\Lambda_\beta(t)$ falls naturally into the category of derived parameters considered in Section 5.1.1. We illustrate the process using the example of Section 12.5, below.

First, however, we make reference to another derived parameter of interest, the overall population fitness. Population fitness is $\lambda(\{SF\})$, calculated using S and F, the population mean values of $S^{(j)}$ and $F^{(j)}$. It is typically larger than the average individual fitness, even substantially so, in consequence of the nonlinear nature of the function $\lambda(\cdot)$. That is,

$$\lambda(\{SF\}) = \lambda(\{E(S^{(j)})E(F^{(j)})\}) > E(\lambda\{S^{(j)}F^{(j)}\}). \tag{12.3}$$

Using the 100,000 pairs $\{S^{(j)}, F^{(j)}\}$ described earlier, we calculated the mean values of $S^{(j)}$ and $F^{(j)}$, then found the population fitness to be $\lambda(\{SF\}) = 1.533$. The large discrepancy between individual fitness values and the overall population fitness suggests a need for caution in interpreting individual fitness values: their main value is comparative.

12.6.1 Estimating the Population Distribution of Individual Fitness

How does one go about estimating the population distribution of individual fitness?

At first, one might think the solution is to estimate individual fitnesses (using realized fitnesses, or posterior medians as in Table 12.3), then to summarize features of the collection of estimates, such as percentiles and means, perhaps drawing a histogram to estimate a density function. The flaw in this approach is that the results estimate the *sample* distribution of fitness *estimates*, not the *population* distribution of fitness *parameters*.

A better approach is to calculate the population distribution of individual fitness using point estimates of the hyperparameters β, say $\hat{\beta}$; that is, we estimate $\Lambda_\beta(t)$ by $\Lambda_{\hat{\beta}}(t)$. For the example of Section 12.5, the posterior medians of the hyperparameters were $\hat{\mu}_p = 2.72$, $\hat{\mu}_f = -2.52$, $\hat{\sigma}_p = 0.64$, $\hat{\sigma}_f = 1.22$, and $\hat{\rho} = 0.26$.

So once again we sampled 100,000 pairs $\{\text{logit}(P^{(j)}), \text{logit}(F^{(j)})\}$ from bivariate normal distributions, this time with mean vector $(\hat{\mu}_p, \hat{\mu}_f)$, standard deviations $\hat{\sigma}_p$ and $\hat{\sigma}_f$, and correlation $\hat{\rho}$, and computed associated fitness values $\lambda^{(j)}$. A smoothed histogram of the 100,000 $\lambda^{(j)}$ values is given in Fig. 12.3. We have included for comparison a normal distribution with mean $=0.958$

8. For example, the standard error of the mean is $0.075/\sqrt{100,000} = 0.0002$.

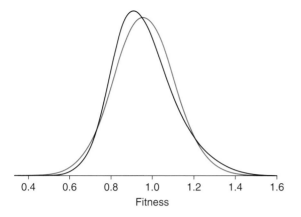

FIGURE 12.3   Estimated population distribution of individual fitness (black), based on point estimates of population distribution parameters. Red curve is approximating normal distribution.

and standard deviation $=0.147$; these were the values calculated for the 100,000 hypothetical individuals. Comparison of the two curves in Fig. 12.3 shows that the model and the estimated parameter values we have used indicate a population distribution of individual fitness that is approximated by a normal distribution, although perhaps slightly skewed to the right. The 5th and 95th percentiles of $\Lambda_{\hat{\beta}}(t)$ were approximated as 0.745 and 1.230.

The foregoing calculations of mean, standard deviation, and percentiles, and the density in Fig. 12.3 correspond to the estimated distribution of population fitness, $\Lambda_{\hat{\beta}}(t)$. There are two sources of uncertainty in these as estimates of features of the true distribution of population fitness $\Lambda_{\beta}(t)$. Take, for instance, the mean of 0.958. There is some imprecision in this estimate, based merely on the fact of simulation approximation. This imprecision is negligible since the standard error of the mean is $\sigma/\sqrt{n}=0.147/\sqrt{100{,}000}=0.0005$. The more important source of uncertainty has to do with $\hat{\beta}$ as an estimator of β.

Here, the ease (conceptual ease, anyway) of Bayesian inference comes into play. Although we might never know the actual value of the hyperparameter β, we do have a Markov chain of sampled values, a sample of the posterior distribution $[\beta|\text{Data}]$ described in Section 12.5. Instead of hanging all our hopes and placing all our confidence on a single estimate $\hat{\beta}$, we can carry out the same calculations 20,000 times, each time replacing μ_p, μ_f, σ_p, σ_f, and ρ with one set of values sampled from the posterior distribution. We will have obtained a sample of size 20,000 from the posterior distribution of $\Lambda_{\beta}(t)$; we will have posterior samples of size 20,000 of the corresponding mean, standard deviations, and percentiles of individual fitness and the means to evaluate the variation in the estimates.

The obvious problem with this is the size of the computational burden. Generating 100,000 individuals, and calculating $\lambda^{(j)}$ for each, took almost 4.8 minutes on a 3.2 GHz Pentium system; repeating the process 20,000 times would take roughly 1600 hours, about nine and a half weeks.

A solution is to sacrifice some of the precision in the simulations. To do so, we first got a rough estimate of the amount of variation "within" simulations (i.e., for a fixed set of population parameters, one drawn from the Markov chain) relative to the variation "among" simulations (i.e., the variation in mean individual fitnesses). We did this by sampling 10 sets of parameters from the Markov chain and sampling 10,000 individuals per set of parameters.

TABLE 12.4   Population fitness distribution features and estimates.

	Mean	SD	5th	95th
Truth: $\Lambda_\beta(t)$	1.022	0.075	0.900	1.148
Sampled $\lambda^{(j)}$	0.998	0.081	0.852	1.092
Realized fitness	0.534	0.574	0	1.329
Bayes estimate fitness	0.952	0.092	0.871	1.203
Estimated: $\Lambda_{\hat\beta}(t)$	0.959	0.147	0.710	1.290
Bayes estimate of $\Lambda_\beta(t)$	0.955	0.155	0.728	1.215
95% CI	(0.83,1.05)	(0.06,0.27)	(0.46,0.87)	(1.09,1.40)

Sample statistics indicated a typical standard deviation of 0.17 "within" simulations, and a standard deviation of 0.06 "among" means. The indicated ratio of variance within to among is thus about $(0.17/0.06)^2 \approx 8$, suggesting that a good allocation of computational effort puts the number of β and the number of simulations per β in a ratio of 1:8.

We settled on choosing 399 sets of parameters (sampled randomly from the Markov chain of 20,000) and 2999 individuals per set of parameters. We chose these numbers because $399 \times 2999 \approx 1/1600 \times 20,000 \times 100,000$, so that the computation would take about an hour.[9] Each of the 399 mean individual fitnesses was estimated with a standard error less than 0.0057; the average value was 0.950, and 95% fell in the range (0.832, 1.049). The standard deviation of these mean values was 0.054. We conclude that the posterior mean of the mean individual fitnesses is 0.950, plus or minus approximately $0.054/\sqrt{399} = 0.003$.[10] The interval (0.832, 1.049) approximates a 95% credible interval for the mean population fitness. It includes the true value, previously calculated as 1.021.

The standard deviation and 5th and 95th percentiles of individual fitness were similarly evaluated. The results, and a summary of the various analyses described in this section are given in Table 12.4.

The first two rows of Table 12.4 present information that usually would not be available, the first row being based on the hyperparameter values used in generating the data, the second being summaries of the latent features of the sampled individuals. The next two rows summarize the collections of 25 estimates available from the data set.

The second through fourth rows of Table 12.4 summarize samples of size 25. Since the ith order statistic in a sample of size n estimates percentile $100 \times i/(n+1)$, the smallest value of each estimates the $100 \times 1/26 = 3.84$th percentile, and the second smallest value estimates the 7.68th percentile. Since 5% is 30% of the way from 3.84 to 7.68%, we estimated the 5th percentile as 0.7 times the smallest, plus 0.3 times the second smallest of the 25 values for these

---

9. But why not 400 sets of 3000? The reason is that the ith order statistic in a sample of size n estimates the $100\, i/(n+1)$th percentile. Thus, the 5th percentile is estimated by the value with rank 150 among 2999, and so forth.

10. Strictly speaking, the formula $\mathrm{SE}(\bar X) = \sigma/\sqrt{n}$ applies only for independent samples, and the values we consider are samples from a Markov chain, hence not independent. Given the length of the chain, its low autocorrelation at long lags, and the fact that we have only taken 399 of 20,000 values, the approximation seems reasonable.

two summaries. Similarly, the 95th percentile was estimated by 0.7 times the largest plus 0.3 times the second largest value.

The collection of realized fitnesses performs dismally in estimating features of $\Lambda_\beta(t)$. As indicated at the end of Section 12.5, realized fitnesses correlated reasonably well with the latent values, but their mean and variance are poorly related to the population values.

Typically, we would not encourage the use of collections of Bayes estimates for estimation of population features of corresponding collections of parameters. Individually, Bayes estimates are not unbiased; collectively they typically underestimate the variation in the set of parameters. Nevertheless, the sample mean and standard deviation, and 5th and 95th percentiles do a reasonable job of estimating the population parameters, in this case.

The estimates based on posterior medians of the hyperparameters, $\Lambda_{\hat{\beta}}(t)$ agrees reasonably well with the Bayes estimate. However, none of the estimates except the Bayes estimate provides a measure of estimation uncertainty. The final row of Table 12.4 gives 95% credible intervals for the features of $\Lambda_\beta(t)$; three of the four include the true values of the quantities estimated.

12.6.2 Estimating the Population Fitness

We conclude our analysis of the data in Section 12.5, and our discussion of derived parameters relating to individual fitness, by considering population fitness as defined in context of individual heterogeneity at Eq. (12.3).

The procedure is the same as for all derived parameters. Since $\lambda(\{SF\})$ is determined by $S = E(S^{(j)})$ and $F = E(F^{(j)})$, and since these quantities are determined by the hyperparameters β, $\lambda(\{SF\})$ is a function of β, say $\psi(\beta)$. Given the true value of β, the true value of $\lambda(\{SF\})$ is obtained by a calculation. Applying the same calculation to values β sampled from the posterior distribution $[\beta|\text{Data}]$, we obtain a sample from the posterior distribution $[\lambda(\{SF\})|\text{Data}]$.

As with estimating features of the population distribution of individual fitness, the calculation $\psi(\beta)$ is difficult, requiring approximation through simulation.[11] Our only concern in performing these approximate calculations is with regard to their precision, and this we can control by our choice of simulation size.

In calculating the population fitness under the data-generating model of Section 12.5, we generated 100,000 pairs $\{S^{(j)}, F^{(j)}\}$, calculated the mean values of $S^{(j)}$ and $F^{(j)}$, then found the population fitness to be $\lambda(\{SF\}) = 1.533$. This simulation took about 1/5th of a second. We repeated it 20 times, and noted a standard deviation of less than 0.0001. Reducing the number of simulations to 10,000 reduced the computation time by a factor of 10, while only boosting the standard deviation to 0.0004, an acceptable level of precision.

We thus calculated population fitness values corresponding to all 20,000 values in our Markov chain of values β, using 10,000 simulations per value. The calculation took 10 minutes, rather than the hour and a half that would have yielded unnecessarily precise results.

---

11. Philosophically, this should not be all that troubling. Most commonly used functions cannot be exactly calculated. Whether we are aware of it or not, trigonometric functions, exponential functions, many statistical functions, even simple square root functions return approximate rather than exact answers.

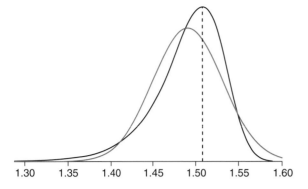

FIGURE 12.4   Posterior distribution of population individual fitness (black), based on point estimates of population distribution parameters. Red curve is approximating normal distribution (included to indicate skew in posterior); posterior mode at 1.508 indicated by dashed line.

The posterior median value was 1.498, with 95% credible interval (1.393, 1.551). A smoothed histogram of the 20,000 values is given in Fig. 12.4.

12.7 AFTERWORD

Estimation of individual fitness is most appropriately conducted in the context of a hierarchical model, if for no other reason than that individual lifetimes can only be observed once. While it is remarkable that SF can be estimated based on a sample of size one (by $\mathbf{1}_X f$), it is not at all surprising that realized fitness $\lambda(\mathbf{1}_X f)$ is an unreliable estimator of $\lambda(SF)$. Hierarchical structure is a necessity for reliable investigation of individual fitness.

Our evaluation of the data in Table 12.2 has included investigations of individual fitness values, differences between individual fitness values, and their ranks; we have investigated features of the population distribution of individual fitness and have estimated population fitness. All of these quantities are derived parameters; the Bayesian framework provides a conceptually simple, mathematically clean and relatively straightforward basis for inference.

13

Autoregressive Smoothing

O U T L I N E

13.1 Dove Data and Preliminary
Analyses 288
13.2 Modeling Differences in Parameter
Values 291

13.3 Results for Dove Analysis 292
13.4 Higher Order Differences 293
13.5 Afterword 295

Hierarchical modeling allows us to simultaneously evaluate patterns in data and patterns in parameters, describing parameters as signal plus noise. The process allows for a formal evaluation of latent processes of scientific interest; furthermore, the richer model structure increases precision in individual parameter estimates.

How do we go about choosing a model for pattern in parameters? If the parameter values are indexed by time, we might consider a linear regression of parameter on year. Or, we might examine whether a simple change in levels of the parameter had occurred, with early years represented by one value and later years by another. We might fit a "broken-stick" (piecewise linear) regression model, if we had reason to believe that a trend in early years had been replaced with a new trend. We might consider a Fourier model, if we anticipated cyclical fluctuations among the parameters. Ideally, the pattern modeled will have a scientific basis. As part of the prior specification, it represents knowledge the analyst brings to bear on the inferential process, and its effect may be profound.

The choice among models for pattern in parameters may be very difficult, and it is frequently the case that the data provide scant basis for evaluating the choice. There are two principle reasons for this. First, even fairly large data sets may be described by relatively few parameters; thus, even if we knew the exact values of the parameters, it would be difficult to select a model for variation among them. To take a case in point, the data set to be considered in this section consists of banding data for 14,087 doves, with 410 recoveries over a 9-year period. Measured by the sample sizes, or by the effort involved in collecting the data, this is a substantial data set. Yet when we come to consider the parameters of interest (survival rates, indexed by year) we have a sample of size 8. In terms of the information available for examining annual variation in survival rates, the data set is actually quite small, and our ability to legitimately evaluate pattern is limited.

A second difficulty in choosing models for variation in parameters is that we never actually get to see their true values. If we could observe, without error, the true values of the 8 survival rates, we might be able to evaluate whether a linear regression of rate on year were appropriate. But instead, we are attempting to evaluate pattern among 8 quantities, all of which are latent, unobserved.

There are times when we anticipate the existence of a pattern or signal in parameters, but not be sure which form to specify, having no a priori reason for the choice. We might anticipate that parameter values in one year are more similar to rates in adjacent years than in more distant years, but not be comfortable constraining our analysis by the choice of a specific pattern. We might desire an analysis that produces a smoothly varying though nonlinear pattern of change. This desire could be motivated by needs for efficient estimation, allowing information about individual parameter to be enhanced by considering them in the context of related parameters, or it may simply be that a smooth pattern of change in parameters is better for descriptive purposes.

A number of different techniques exist for modeling smooth patterns of change among parameters. Here we describe one of the simplest approaches, based on an autoregressive model. The model describes changes in parameter values (rather than parameter values themselves) as independent random variables. This model may possibly be of interest in its own right, on a priori grounds, as for instance if applied to data on population change. Here however, we present the model for its operational properties: it produces nicely smoothed estimates of collections of parameters, and may provide a reasonable basis for postulating specific patterns of variation among parameters.

13.1 DOVE DATA AND PRELIMINARY ANALYSES

Table 13.1 presents band recovery data for mourning doves (*Zenaida macroura*) banded and released in Oklahoma, USA in 1968 through 1973, along with the numbers of individuals recovered in 1968 through 1976 (Lewis and Morrison, 1978).

TABLE 13.1   Numbers of mourning doves banded, M_i, in Oklahoma (years $i + 1967$) and reported recoveries (1968–1976).

i	M_i	1968	1969	1970	1971	1972	1973	1974	1975	1976
1	2219	43	12	8	6	3	1	0	0	0
2	2143	.	23	19	11	4	3	3	2	0
3	3110	.	.	40	29	4	7	3	1	1
4	1360	.	.	.	13	12	3	4	0	2
5	3015	.	.	.	.	46	27	9	2	1
6	2240	.	.	.	.	.	52	11	4	1

Analysis of these data begins with a standard bird-banding model (see Section 11.2), in which recovery year ($j = 1, 2, \ldots, 9$) is treated as an independent multinomial random variable for each release year cohort ($i = 1, 2, \ldots, 6$). Let M_i denote the number of individuals in cohort i, and let X_{ij} denote the number of these individuals recovered in year j. Then the likelihood function is

$$\prod_{i=1}^{6} \prod_{j=i}^{9} \pi_{ij}^{X_{ij}} \times \prod_{i=1}^{6} \left(1 - \sum_{k=i}^{9} \pi_{ij}\right)^{M_i - \sum_{k=i}^{j} X_{ij}},$$

with π_{ij} being the probability that a bird released in year i is recovered and reported in year j. Cell probabilities π_{ij} are modeled in terms of recovery probabilities $f_j = \text{Pr}(\text{Bird recovered in year } j \mid \text{Bird alive in year } j)$ and annual survival probabilities $S_j = \text{Pr}(\text{Bird alive in year } j+1 \mid \text{Bird alive in year } j)$, with $\pi_{ii} = f_i$ and

$$\pi_{ij} = f_j \times \prod_{k=i}^{j-1} S_k, \tag{13.1}$$

for $j = i+1, i+2, \ldots, 9$.

Our goal in this analysis is to examine temporal pattern in the survival rates. As a first step, we might simply inspect the maximum likelihood estimates of the estimable parameters (Table 13.2). We note at the outset that $S_1, S_2, \ldots, S_5$ are rather poorly estimated, with asymptotic standard errors (ASE) roughly equivalent to what one would obtain in analysis of roughly 25 Bernoulli trials.[1]

Note that our lack of knowledge about parameters f_7, f_8, and f_9 prevents us from estimating the survival rates S_6, S_7, and S_8. One expedient is to set $f_7 = f_8 = f_9 = \bar{f}$, the mean value of f_i, $i = 1, 2, \ldots, 6$. Then, the MLE's (and ASE's) of S_i are 0.438 (0.089), 0.300 (0.114), and

TABLE 13.2   Maximum likelihood estimates (MLE) and ASE of estimable parameters under model 13.1.

Parameter	MLE	ASE	Parameter	MLE	ASE
f_1	0.0194	0.0029	S_1	0.4457	0.0974
f_2	0.0112	0.0020	S_2	0.7009	0.1264
f_3	0.0126	0.0018	S_3	0.5881	0.1262
f_4	0.0132	0.0025	S_4	0.4196	0.0939
f_5	0.0141	0.0019	S_5	0.4643	0.0843
f_6	0.0206	0.0027	$(f_7/f_6)\, S_6$	0.3226	0.0677
			$(f_8/f_7)\, S_7$	0.3000	0.1140
			$(f_9/f_8)\, S_8$	0.5556	0.3099

1. Based on a sample of size n, the squared standard error of an estimated Bernoulli rate $\hat{p}$ is $SE^2 = p(1-p)/n$. Thus it is reasonable to describe the precision of these estimated survival rates in terms of equivalent sample sizes, $n \approx \hat{S}(1 - \hat{S})/ASE^2$; the five values are 26, 13, 15, 28, and 35.

0.556 (0.310), suggesting a declining pattern in the survival rates. However, the estimated survival rates are sensitive to the choice of recovery rates: it could be that $f_7 = 0.00943, f_8 = 0.00357$, and $f_9 = 0.00218$, so that we would estimate $S_6 = 0.70, S_7 = 0.80$, and $S_9 = 0.90$. But of course these values of f_7, f_8, and f_9 are much lower than for the previous years, and perhaps not reasonable. The point is, that some prior knowledge is necessary for S_6, S_7, and S_8 to be estimated.

Rather than assume that $f_7 = f_8 = f_9 = \bar{f}$, we make a weaker assumption, that the f_i are a sample from a common distribution. We will treat the set of logit(f_i) as a sample from a normal distribution with mean μ_f and variance σ_f^2. Our analyses will use a vague normal prior for the mean μ_f, and a vague uniform prior for the precision $\tau_f = 1/\sigma_f^2$.

We conducted a preliminary analysis in which we imposed no structure or pattern on the survival rates, instead using independent $U(0,1)$ priors. BUGS code for this preliminary analysis is given in Panel 13.1.

Two comments on the code: first, that for $j > i$ the cell probabilities π_{ij} could be defined recursively and compactly, on the basis of the relation

$$\pi_{i,j} = \left(\frac{S_{j-1}f_j}{f_{j-1}}\right)\pi_{i,j-1}.$$

Indeed this formulation might appear more efficient than the coding in Panel 13.1, requiring fewer multiplications. However, the resulting code makes it more difficult for BUGS to choose an efficient MCMC algorithm, and more than doubles the run time; the efficiency loss would be even greater if there were more years of data, and a more highly parameterized model. Our experience in using BUGS is that recursive descriptions of parameter values should be avoided. Second, note that the indexing of values π_{ij} is for $j = i, i+1, \ldots, 9$. There is no need to

PANEL 13.1   BUGS code for preliminary analysis of dove data.

```
for (i in 1:6){
    x[i,i:10] ~ dmulti(pi[i,i:10],n[i])
    pi[i,10] <- 1-sum(pi[i,i:9])
}
for (i in 1:9){
    pi[i,i] <- f[i]
    for (j in i+1:9){
        pi[i,j] <- prod(S[i:j-1])*f[j]
    }
}
for (i in 1:8){
    S[i] ~ dunif(0,1)
}
for (i in 1:9){
    logitf[i] ~ dnorm(mu.f,tau.f)
    logit(f[i]) <- logitf[i]
}
mu.f ~ dnorm(0.0,1.0E-6)
tau.f ~ dunif(0,1000)
sd.f <- 1/sqrt(tau.f)
```

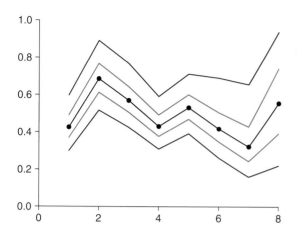

FIGURE 13.1   Posterior medians (black), quartiles (red), 5th and 95th percentiles (blue) of annual survival rates for dove data.

begin array indices at 1, nor to define array values that are not used in the model description, such as $\pi_{3,1}$ or $\pi_{4,3}$.

Posterior medians for the survival rates are plotted against year, along with 50% and 90% credible intervals, in Fig. 13.1. Would we say that the rates were declining? There is an apparent abrupt increase in the last year, but S_8 is poorly estimated. The apparent increase between years 4 and 5 could also be an artifact of estimation. Examining the Markov chain output, we find that the posterior probability of $S_4 > S_5$ is 24%; similarly, the posterior probability of $S_7 > S_8$ is 25%. On the other hand, it appears fairly certain that S_1 is less than S_2: the posterior probability of $S_1 < S_2$ is 94%. It may be that the rates increased at the outset, then declined over the years of the study.

It would be useful to have a model allowing for gradual changes in parameter values, without imposing a specific pattern a priori. One approach is to treat the parameter value as varying according to a random walk, as we now describe.

13.2 MODELING DIFFERENCES IN PARAMETER VALUES

Neither absolute change nor proportional change is particularly handy for describing variation in rates. A survival rate of 50% can change by 10% in absolute terms (to 40% or 60%) or in relative terms (to 45% or 55%). However, a survival rate of 95% cannot increase by 10%, either in absolute or relative terms; both result in rates greater than 100%. Changes in rates are more naturally described as consequences of changes in a transformed parameter. Here, we use the log odds transformation, also known as the logit: the *logit survival rate* for year j is $\theta_j = \text{logit}(S_j) = \log(S_j/(1 - S_j))$.

The *first differences* in logit survival rates are

$$\Delta_j = \theta_j - \theta_{j-1}, \quad j = 2, 3, \ldots, n. \tag{13.2}$$

Suppose that we model the first differences as independent and identically distributed (iid) mean zero random variables, independent of all previous values θ_j. Then the logit survival rates start with θ_1, and vary through time as a cumulative sum of changes of comparable

```
tau.theta.eps <- 0.0001*tau.theta
theta[1] ~ dnorm(0,tau.theta.eps)
logit(S[1]) <- theta[1]
for (i in 2:8){
    theta[i] ~ dunif(theta[i-1],tau.S)
    logit(S[i]) <- theta[i]
}
tau.theta ~ dgamma(0.001,0.001)
```

magnitude. The difference between θ_j and θ_k, for $j > k$, is

$$\theta_j - \theta_k = \Delta_{k+1} + \cdots + \Delta_j,$$

the sum of $j-k$ iid random variables. Thus writing σ_Δ^2 for the variance of one of the changes, the variance of $\theta_j - \theta_k$ is $(j-k)\sigma_\Delta^2$. For moderate values of σ_Δ^2, the model constrains adjacent survival rates to be similar, allowing for large differences to accumulate as time progresses.

The model implies that $[\theta_j|\theta_{j-1}] = N(\theta_{j-1}, \tau_\theta)$, for $j = 2, 3, \ldots, 8$. The distributions of θ_j are defined recursively after starting at θ_1. We implemented the model in BUGS, replacing the uniform prior specifications on S_i in Panel 13.1 with the code provided in Panel 13.2.

Note that θ_1 is given a vague prior distribution, $N(0, 0.0001\tau_\theta)$. It is natural to ask why we have used $0.0001\tau_\theta$ as the precision for θ_1, rather than simply 0.0001. After all, one might reasonably say, one anticipates (correctly) that τ_θ will not be too large, and that consequently $0.0001\tau_\theta$ will also be small; there should not be much difference in the priors, nor consequently in the resulting posteriors. The reason for including τ_θ is somewhat technical, but nonetheless intelligible: the slight difference is necessary for the joint prior distribution on θ_1 and τ_θ to be of the conjugate form. Choosing noninformative priors can be a tricky business, but the choice is often made easier when working in a conjugate family. In practical terms the difference between the two priors is the same as the difference between using n and $n-1$ as the divisor in estimating a variance: for small n (e.g., the 8 years in the present case) the difference is noticeable, for large n it is not.

13.3 RESULTS FOR DOVE ANALYSIS

We plot the posterior medians and 90% credible intervals for dove survival rates against year in Fig. 13.2, along with the raw estimates from our nonhierarchical preliminary model. Two features are apparent: first, that the pattern of variation has been substantially smoothed, second that the credible intervals have been made much shorter.

Both features reflect the limitations of the information provided by the data about the survival rates. Given noisy data, Bayesian inference is more heavily influenced by the prior specification. Here, we have noisy information about survival rates; the posterior means are close together, emphasizing the similarity of values indicated by the prior; they exhibit Bayesian shrinkage, having been shrunken toward each other. The effect of information content on such shrinkage is revealed in Fig. 13.3, where in addition to the raw estimates (red) and the posterior

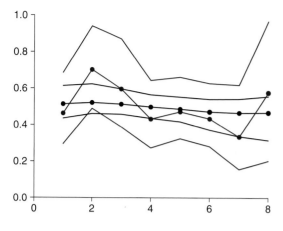

FIGURE 13.2   Posterior medians, 5th and 95th percentiles for dove annual survival plotted against year index. Blue curves are for preliminary analysis with independent $U(0,1)$ priors on survival rates. Black curves are for analysis based on first difference model.

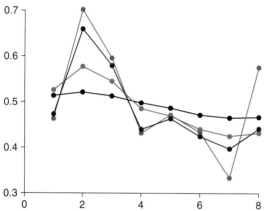

FIGURE 13.3   Shrinkage comparisons. Red curve gives posterior mean estimates of dove survival against year from unstructured analysis. Black, green, and blue curves are same, from first difference model with data of Table 13.1 frequencies multiplied by 1, 3, and 10.

means from our first difference model (black), we have included posterior means based on hypothetical data sets 3 and 10 times larger than given in Table 13.1, but with the same relative cell frequencies. The better the information provided by the data, the less the influence of the prior.

13.4 HIGHER ORDER DIFFERENCES

So if we can model the differences as mean zero random variables, and achieve some level of smoothing, why not model the *difference* of first differences as means zero random variables, and see what happens?

The second differences are defined using Eq. (13.2) as

$$\Delta_j^2 = \Delta_j - \Delta_{j-1} = (\theta_j - \theta_{j-1}) - (\theta_{j-1} - \theta_{j-2}) = (\theta_j - 2\theta_{j-1} + \theta_{j-2}), \tag{13.3}$$

for $j = 3, 4, \ldots, n$. If we assume that the second differences are iid mean zero normal random variables with precision τ, independent of θ_k for $k < j-1$, then the distribution of θ_j given θ_k,

PANEL 13.3   Changes to BUGS code of Panel 13.1 needed for first difference model.

```
tau.theta.eps <- 0.0001*tau.theta
for (i in 1:2){
    theta[i] ~ dnorm(0,tau.theta.eps)
    logit(S[i]) <- theta[i]
}
for (i in 3:8){
    theta[i] ~ dnorm(2*theta[i-1]-theta[i-2],tau.S)
    logit(S[i]) <- theta[i]
}
tau.theta ~ dgamma(0.001,0.001)
```

$k < j$ is normal with mean $2\theta_{j-1} - \theta_{j-2}$ and variance τ. The BUGS code of Panel 13.1 is modified using Panel 13.3 instead of Panel 13.2. Notice that instead of a vague prior on θ_1 alone, we must now have vague priors on θ_1 and θ_2. Breslow and Clayton (1993) smoothed breast cancer rates in Iceland by year of birth by treating second differences as iid random variables.

But why stop at second differences? We may calculate third differences using Eq. (13.3); then fourth differences, and fifth, and so on. A general pattern emerges: assuming the mth differences are iid normal, with mean 0 and precision τ, and that these are independent of θ_k for $k < j - m + 1$, then the conditional distribution of θ_j given θ_k, $k < j$ is normal with mean

$$\sum_{k=1}^{m} \binom{m}{k} (-1)^{k+1} \theta_{j-k} \tag{13.4}$$

and precision τ, for $j = m + 1, m + 2, \ldots, n$; we would place vague priors on the first m θ_i's. We could model the data this way, but does it make sense?

First, an observation about the mean (13.4). Suppose we were to construct a polynomial of degree $m - 1$, passing through the points $(1, \theta_{j-1})$, $(2, \theta_{j-2})$, $\ldots$, (m, θ_{j-m}); call the polynomial $f_j(x)$.[2] The polynomial does a great job of fitting the values θ_{j-k}, for $k = 1, 2, \ldots, m$. Indeed, for such k, $f_j(k) = \theta_{j-k}$. Extrapolating a bit, we might guess that $f_j(0)$ should do a reasonable job of fitting θ_{j-0}, i.e., θ_j. It turns out that $f(0)$ is exactly the same as (13.4). Thus assuming the mth difference model is equivalent to assuming that each next θ_j is equal to the extrapolation of an interpolating polynomial through the previous m values, plus noise. The first difference model can be thought of as a linear smoother, the second as a quadratic smoother, etc.

High-order polynomial interpolation and extrapolation is notoriously unstable. For example, the 6th degree polynomial passing through (1,7), (2,2), (3,6), (4,1), (5,5), (6,4), (7,3) is shown in Fig. 13.4. It oscillates fairly wildly between some of the 7 points, and shoots off to infinity as x moves beyond the interval [0,8]. Given the 7 data points alone, what would the reader

---

2. There is only one polynomial of degree $\leq m - 1$ passing through points $(x_1, y_1), (x_2, y_2), \ldots, (x_m, y_m)$. Known as the *Lagrange polynomial*, its exact form is

$$f(x) = \sum_{k=1}^{m} \left(\prod_{i=1, i \neq k}^{m} \frac{x - x_i}{x_k - x_i} \right) y_k.$$

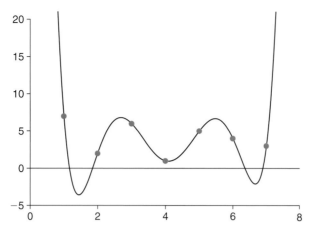

FIGURE 13.4   Sixth degree interpolating polynomial.

predict for an eighth value? Extrapolating to zero, the polynomial says 262. At $x = 8, 9$, and 10 the polynomial evaluates to 196, 1255, and 4738.

So, in addition to the obvious difficulty of interpretation, models based on high-order dif- ferences might be anticipated to be unstable in their estimates. Our experience is that MCMC performance in BUGS, implemented as described in this section, is rather poor, with very high autocorrelation requiring long runs.

We display results of two sets of four analyses in Fig. 13.5. Each set of four consists of an unstructured analysis of survival rates (i.e., iid $U(0,1)$ priors for survival rates), and analyses of first, second, and third difference models. For each set of analyses we plot posterior means (top row) and 95% credible intervals (bottom row). The left column gives results for the dove data of Table 13.1. As noted at the outset, the survival rates are rather poorly estimated, based on the data alone; as would be anticipated, the point estimates are rather substantially affected by the addition of the prior knowledge specified in the difference models, and the interval estimates are shortened. The right column gives results for a data set made by multiplying the original data frequencies by 10. Note that the point estimates are less affected by the prior information and the interval estimates are scarcely shortened, except for the very poorly estimated quantities S_6, S_7, and S_8.

13.5 AFTERWORD

The choice of a model for the signal in a collection of parameters needs to be made cautiously, and the implications of the choice reported advisedly.

The familiar mnemonic for posterior inference

$$\text{Inferential basis} = \text{Data} + \text{Prior knowledge}$$

applies in hierarchical modeling of process and signal in parameters. Here, the sense is (1) that given good data, estimates of parameters will be unaffected by the choice of modeled signal

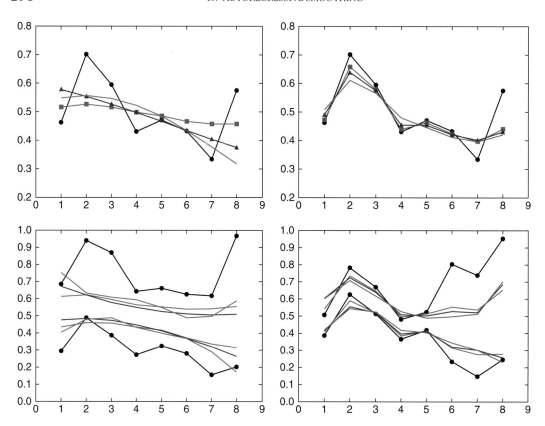

FIGURE 13.5   Posterior mean survival rates (top row) and 95% credible intervals (bottom row) for dove data set (Table 13.1) (left column) and for ten times the data of Table 13.1 (right column). Black curves for unstructured analysis (iid uniform priors on survival); red, blue and green curves for first, second and third difference models.

(a component of the prior) and (2) that given weaker data, the estimates of parameters will be influenced by the modeled signal. In using this mnemonic, it must be understood that the effect of the signal on estimates of individual parameters varies inversely to the strength of the modeled signal. For instance, if we insist that the parameters conform precisely to a specified pattern, or nearly so (through an informative prior on the variance of the noise component), then the effect of the prior will be more substantial.

Examining pattern in parameters is more challenging than examining pattern in data. There are fewer parameters than data, and parameters are imperfectly known. As a result, our hierarchical model may rest on essentially untestable assumptions, assumptions which have a strong inferential effect.

What happens if we get the pattern wrong? For example, what happens if we assume that parameter values change according to some particular pattern, when in fact some other pattern is correct, or no pattern prevails?

Consider a hierarchical model, with data X governed by parameters $\theta_1, \theta_2, \ldots, \theta_n$. Suppose that we have really good data, and the likelihood function $[X|\theta_1, \theta_2, \ldots, \theta_n]$ is such that all of the parameters θ_i are precisely estimated, except (say) for θ_5. We specify a pattern among the parameters, that $\theta_i = a + bi + \epsilon_i$, with values ϵ_i sampled independently from a mean zero normal distribution with variance σ^2. If n is reasonably large, we obtain precise estimates of the regression coefficients for the least squares line through the θ_i and a precise estimate of a residual variance. The posterior distribution for θ_5, being barely informed by the data, is strongly influenced by the prior specification, which in this case includes the linear model for the θ_i. Predictions for θ_5 will essentially amount to reading off a value from the least squares line through the other θ_i. If in reality the pattern of linear change were incorrect, we would have false confidence in our prediction for θ_5; it would probably be better to base our inference on whatever limited information the data actually provide for the parameter.

Misspecification of the signal may not matter too much, for those parameters which are strongly informed by the data. For such, lack of fit in the parameters to a misspecified model will be taken up as noise: the posterior distributions of individual parameters will be more or less correct. The difficulty with misspecification of signal will be in prediction of new parameters, or of parameters poorly informed by the data.

These caveats obviously apply when fitting straight lines, piecewise linear regressions, Fourier series, and in short to the entire ensemble of "parametric" models of signal and noise in parameters. They also apply when smoothing is the goal: it is important to have a clear notion of how strongly the data inform inference. For the dove data, under the first difference model, the posterior distribution of S_2 has mean of 0.527, and 95% credible interval (0.461, 0.623); this contrasts sharply with the posterior distribution obtained assuming iid uniform priors with mean of 0.702, and 95% credible interval (0.489, 0.940); the point estimate is substantially lower, and the interval estimate substantially shorter. The difference is the prior: it is what we assume, rather than what the data tell us.

PART IV

APPENDICES

A

Probability Rules

OUTLINE

A.1 Properties of Moments   301
A.2 Conditional Expectations   303
   A.2.1 Basic Rules for Conditional
     Expectations   303

A.2.2 Double Expectation Theorems   303
A.3 Independence and Correlation   304
   A.3.1 Independence and Prediction   304
   A.3.2 Correlation   305

In this appendix, we present 18 simple rules describing expected values of random variables. We include Chebyshev's inequality and the weak law of larger numbers as simple but interesting consequences of basic ideas.

None of the rules requires specific distributional assumptions; they are completely generic. The only assumptions required are that the moments under discussion exist.[1] We use μ for means, σ for standard deviations, and ρ for correlations, adding subscripts when necessary.

A.1 PROPERTIES OF MOMENTS

Let X and Y be random variables, and let a be a constant. Then the following rules are immediate consequences of the definition of expected values.

   1) $\mathrm{E}(aX) = a\mathrm{E}(X)$.
   2) $\mathrm{E}(X + Y) = \mathrm{E}(X) + \mathrm{E}(Y)$.

---

1. Not all random variables have finite moments. $f(x) = 1/x^2$ is a legitimate density function for $x \geq 1$, but its mean is $\int_1^\infty x f(x)\, dx = \infty$. The t-distribution with one degree of freedom, also known as the Cauchy distribution, has no finite moments. Remarkably, the sample mean of n observations from the Cauchy distribution has the same distribution as a single observation. This example dramatically demonstrates the necessity of finite variance for the convergence of sample means to population means (rule 9) and the Central Limit Theorem (rule 10).

3) $\text{Var}(X+Y)=\text{Var}(X)+\text{Var}(Y)+\text{Cov}(X,Y)$.

4) $\text{Var}(aX)=a^2\text{Var}(X)$.

5) If X and Y are independent, $\text{Cov}(X,Y)=0$.[2]

Let $X_1,X_2,\ldots,X_n$ be independent and identically distributed with mean μ and variance σ^2, and let $\bar{X}$ denote the sample mean. For rules 7, 9, and 10 we assume $\sigma^2<\infty$. It follows from 1) and 2) that

6) $\text{E}(\bar{X})=\mu$,

and from 3) to 5) that

7) $\text{Var}(\bar{X})=\frac{\sigma^2}{n}$.

Suppose that X has mean μ and variance $\sigma^2<\infty$, and let $k>0$. Then

8) $\Pr\left(|X-\mu|\geq k\sigma\right)\leq\frac{1}{k^2}$.

Rule 8, known as Chebyshev's inequality,[3] shows the appropriateness of the standard deviation, σ, as a measure of the spread of a distribution. Its proof is remarkably simple:

$$\Pr\left(|X-\mu|\geq k\sigma\right)=\text{E}\left\{\mathbf{I}\left(|X-\mu|\geq k\sigma\right)\right\}\leq\text{E}\left\{\frac{(X-\mu)^2}{k^2\sigma^2}\right\}$$

$$=\frac{1}{k^2\sigma^2}\text{E}(X-\mu)^2=\frac{1}{k^2}.$$

Note that the function $\mathbf{I}(\cdot)$ is the indicator function, equaling 1 if its argument is true, equaling 0 otherwise. The indicator of an event is a Bernoulli trial, hence its expected value is the probability of the event.

Chebyshev's inequality can be used along with rules 6) and 7) to prove the weak law of large numbers, that sample means "converge in probability" to population means:

9) For any $\epsilon>0$, $\Pr\left(|\bar{X}-\mu|\geq\epsilon\right)\to0$, as $n\to\infty$.

The proof: Replace X with $\bar{X}$ in rule 8, and set $k=\epsilon/\sigma(\bar{X})$. Then

$$\Pr\left(|\bar{X}-\mu|\geq\epsilon\right)\leq\frac{\sigma^2(\bar{X})}{\epsilon^2}=\frac{\sigma^2}{n\epsilon^2}\to0\quad\text{as }n\to\infty.$$

Rule 9 says that if we choose a tolerance $\epsilon>0$, be it ever so small, the probability that the sample mean will be more than ϵ away from the true mean goes to zero as sample size increases.

The Central Limit Theorem gives a more precise statement of the discrepancy between $\bar{X}$ and μ:

10) $\Pr\left(\frac{\bar{X}-\mu}{\sigma/\sqrt{n}}\leq z\right)\to\Phi(z)$ as $n\to\infty$.

2. But not the other way round. Suppose X takes values $-2,-1,1$, and 2 with equal probabilities, and let $Y=X^2$. Then $\text{E}(XY)=\text{E}(X^3)=0$ and $\text{E}(X)=0$, hence $\text{Cov}(X,Y)=0$, but X and Y are clearly not independent. More generally, any random variable X symmetrically distributed about zero will have $\text{E}(X^3)=\text{E}(X)=0$ (provided the third moment exists) and thus will satisfy $\text{Cov}(X,X^2)=0$.
3. This inequality is named for Russian mathematician Pafnuty L. Chebyshev (1821–1894).

Here, $\Phi(z)$ is the cumulative distribution function of a standard normal random variable (see Appendix 15 for details on the normal distribution).

A.2 CONDITIONAL EXPECTATIONS

Here, we present the "double expectation theorems," which relate features of conditional and marginal distributions, along with basic rules for manipulating conditional expectations.

A.2.1 Basic Rules for Conditional Expectations

The preceding rules can all be extended to describe conditional distributions and conditional expectations. Thus for instance, rule 2 becomes "$E(X+Y|Z)=E(X|Z)+E(Y|Z)$"; and rule 5 becomes "If X and Y are conditionally independent given Z, $Cov(X,Y|Z)=0$." Furthermore

11) $E(h(X,Y)|X=x_0)=E(h(x_0,Y)|X=x_0)$

and

12) $E(h(X)g(X,Y)|X)=h(X)E(g(X,Y)|X)$.

Rule 11 is called the "substitution rule": values conditioned upon can be substituted into the expressions on the left-hand side of the solidus.

A.2.2 Double Expectation Theorems

The conditional distribution of Y given $X=x$, denoted $f_{Y|X}(y|x)$, is a specific probability distribution, defined in terms of a fixed value x; its features depend on x. For instance, suppose that given $X=x$, Y is a Poisson random variable with parameter x. Then

$$f_{Y|X}(y|x)=\frac{\exp(-x)x^y}{y!}, \quad y=0,1,\ldots,\infty.$$

The mean and variance of this distribution are denoted by $E(Y|X=x)$ and $Var(Y|X=x)$; because the conditional distribution is Poisson, both of these quantities are equal to the parameter x.

If asked "how large do we expect Y to be?," a perfectly legitimate answer would be "Well, that depends on the value of X." In so doing, we think of the mean and variance of Y as random variables, $E(Y|X)=Var(Y|X)=X$.

Alternatively, we might choose to answer the question about Y without reference to X, basing our response on the marginal distribution of Y. Suppose that X is an exponential random variable with parameter λ. Then the joint distribution of Y and X is

$$f_{Y,X}(y,x)=f_{Y|X}(y|x)f_X(x)=\frac{\exp(-x)x^y}{y!}\lambda\exp(-\lambda x),$$

from which it follows that the marginal distribution of Y is

$$f_Y(y)=\int_0^\infty f_{Y,X}(y,x)dx=\left(\frac{\lambda}{\lambda+1}\right)\left(\frac{1}{\lambda+1}\right)^y. \tag{A.1}$$

Thus, the expected value of Y is

$$E(Y) = \sum_{y=0}^{\infty} y f_Y(y) = \frac{1}{\lambda},$$

and the variance of Y is

$$Var(Y) = \sum_{y=0}^{\infty} (y - E(Y))^2 f_Y(y) = \frac{1+\lambda}{\lambda^2}.$$

So, we have a second answer to the question "how large do we expect Y to be?" The first, that $E(Y|X) = X$, is a statement about the conditional distribution; the second, that $E(Y) = 1/\lambda$, is based on the marginal distribution. The first is the conditional expectation, and the second is the marginal expectation.

As it turns out, we need not have computed the marginal distribution (A.1) in order to obtain the marginal mean and variance. The conditional and marginal expectations are related by the "Double expectation theorems," which we add to our list of simple rules:

13) $E(Y) = E\{E(Y|X)\}$.

14) $Var(Y) = E\{Var(Y|X)\} + Var\{E(Y|X)\}$.

So, given that $E(Y|X) = X$, and that X was exponential with mean $1/\lambda$, it follows as an immediate consequence of rule 13 that $E(Y) = 1/\lambda$; the mean of the conditional means is the marginal mean. Rule 14 is applied as follows:

$$\begin{aligned} Var(Y) &= E\{Var(Y|X)\} + Var\{E(Y|X)\} \\ &= E\{X\} + Var\{X\} \\ &= \frac{1}{\lambda} + \frac{1}{\lambda^2}; \end{aligned}$$

note here that we have used the fact that the variance of an exponential random variable is the square of its mean.

Another double expectation theorem, for covariances, is as follows:

15) $Cov(X,Y) = E\{Cov(X,Y|Z)\} + Cov\{E(X|Z), E(Y|Z)\}$.

A.3 INDEPENDENCE AND CORRELATION

A.3.1 Independence and Prediction

Informally, independence can be understood in terms of information: Y is independent of X means that knowledge of X provides no insights into the value of Y. Whether or not it will snow or not on January 30th in Buffalo, New York is independent of the outcome of a coin flip on June 5th in Sao Paulo, Brazil; if you were trying to set odds on whether it will snow in Buffalo, a phone call from a coin-flipping friend in Brazil might be welcome on personal grounds, but would be worthless on inferential grounds.

In terms of probability distributions, independence is expressed by whether or not a joint distribution factors into marginal distributions, that is, $[Y,X] = [Y][X]$. It is always true that $[Y,X] = [Y|X][X]$, hence independence means that $[Y|X] = [Y]$, and consequently that $E(Y|X) = E(Y)$.

If X and Y are not independent, knowledge of X should help in predicting Y.

16) $E((Y-a)^2|X)$ is minimized by setting $a = E(Y|X)$.

17) $E(|Y-a| \, |X)$ is minimized when a is a median of $[Y|X]$.

Rule 16 says that if we wish to predict Y using X, the mean squared error (MSE) of our prediction is minimized by predicting $Y = E(Y|X)$; rule 17 says that the median minimizes the mean absolute deviation.

A.3.2 Correlation

Correlation is a measure of linear association. Here's why: suppose that we wish to approximate Y by a linear function of X, $a + bX$; the problem is to choose the values a and b so as to make a good approximation. One way to do so is to minimize the average squared difference between Y and $a + bX$. Using rules 1) to 4), and a bit of calculus, it can be demonstrated that

18) $E\left(\left[Y - (a+bX)\right]^2\right)$ is minimized by setting $b = \frac{\rho\sigma_Y}{\sigma_X}$ and $a = \mu_Y - b\mu_X$; the MSE of the best linear approximation is $(1 - \rho^2)\sigma_Y^2$.

This result does not rely on any specific distributional assumptions, such as normality; the only assumption is that the variances in question are finite. Thus, the equation

$$y = \left(\mu_Y - \frac{\rho\sigma_Y}{\sigma_X}\mu_X\right) + \frac{\rho\sigma_Y}{\sigma_X}x$$
$$= \mu_Y + \frac{\rho\sigma_y}{\sigma_x}(x - \mu_X), \tag{A.2}$$

is the *least squares line*. Note that the sign of ρ determines whether the least squares line is increasing or decreasing; also, the minimum MSE is a decreasing function of $|\rho|$. Thus, the magnitude of the correlation determines the strength of the linear relation.

If the joint distribution of $(X,Y)'$ is bivariate normal, then the conditional distribution of Y given X is univariate normal, with mean given by the least squares line (A.2), and variance equal to the MSE, $(1 - \rho^2)\sigma_Y^2$.

Probability Distributions

OUTLINE

B.1   Uniform Distribution   307
B.2   Discrete Uniform Distribution   308
B.3   Normal Distribution   308
B.4   Multivariate Normal Distribution   309
B.5   Bernoulli Trials and the Binomial Distribution   311
B.6   Poisson Distribution   312

B.7   Multinomial Distribution   313
B.8   Exponential Distribution   315
B.9   Gamma Distribution   317
B.10   Beta Distribution   319
B.11   t-Distribution   320
B.12   Negative Binomial Distribution   321

In this appendix, we list a number of the most widely used probability distributions. We summarize characteristics of the distributions, including moments, alternative parameterizations, relations to other distributions, and methods of generating samples.

B.1   UNIFORM DISTRIBUTION

We say that X has a uniform distribution on the interval $[a,b]$, and write $X \sim U(a,b)$, if X has a constant density function $f(t) = 1/(b-a)$ on $[a,b]$. The *support* of X is the interval $[a,b]$, meaning that $f(t) = 0$ for $t \notin [a,b]$.

The mean and variance of the uniform distribution are

$$\mu = \frac{a+b}{2}$$

and

$$\sigma^2 = \frac{(b-a)^2}{12}.$$

Most simple random number generators produce data that, while not truly random, cannot be easily distinguished from the samples of the uniform distribution on [0,1]. These can be transformed into samples from other distributions. In particular, if $X \sim U(0,1)$, and $b > a$, then $Y = a + (b-a)X \sim U(a,b)$.

B.2 DISCRETE UNIFORM DISTRIBUTION

Suppose that $S = \{s_1, s_2, \ldots, s_n\}$ is a finite set and that $\Pr(X = s_i) = 1/n$, for $i = 1, 2, \ldots, n$. Then X is said to have a discrete uniform distribution on the set S, denoted $X \sim DU(S)$. One might think of X as a draw from a well-mixed urn of balls, indistinguishable except for labels $s_1, s_2, \ldots, s_n$, which cannot be read while the balls are in the urn.

The discrete uniform distribution on the integers $x = a, a+1, \ldots, b$ has mean $(a+b)/2$ and variance

$$\frac{(b-a+1)^2 - 1}{12}.$$

For integers $a < b$, one can sample $X \sim DU(\{a, a+1, \ldots, b\})$ as $X = \lceil Y \rceil$ where $Y \sim U(a-1,b)$ and $\lceil \cdot \rceil$ is the "least integer greater than" function: X is obtained by rounding Y upward.

B.3 NORMAL DISTRIBUTION

The normal distribution is probably the most familiar of probability distributions. X is a normal (or *Gaussian*[1]) random variable if it has density function

$$\phi(x; \mu, \sigma^2) = \frac{1}{\sqrt{2\pi}\sigma} \exp\left\{ -\frac{1}{2\sigma^2}(x-\mu)^2 \right\}, \quad -\infty < x < \infty. \tag{B.1}$$

The parameters $\mu \in (-\infty, \infty)$ and $\sigma^2 > 0$ are the mean and variance of the distribution; we write $X \sim N(\mu, \sigma^2)$. The median and mode of the normal distribution are also equal to μ. It is sometimes convenient to parameterize the normal distribution using the *precision* $\tau = 1/\sigma^2$.

The standard normal distribution is the version of (B.1) with $\mu = 0$ and $\sigma = 1$ (Fig. B.1). The capital letter Z is often used to designate a random variable with the standard normal distribution. The *cdf* of Z is usually designated by $\Phi(z) = \Pr(Z \leq z)$.

If $Z \sim N(0,1)$ and $X = \mu + \sigma Z$, then $X \sim N(\mu, \sigma^2)$. Consequently,

$$\Pr(X \leq x) = \Phi\left(\frac{x-\mu}{\sigma} \right). \tag{B.2}$$

Given U_1 and U_2, a pair of independent $U(0,1)$ random variables, the Box–Muller transformation produces independent standard normal random variables X_1 and X_2 as

$$X_1 = \cos(2\pi U_2)\sqrt{-2\ln(U_1)} \tag{B.3}$$

---

1. For Karl Friedrich Gauss (1777–1855), possibly the greatest mathematician of all time.

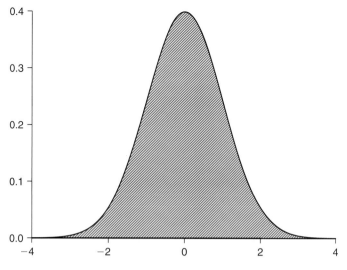

FIGURE B.1   The standard normal density, $\phi(x;0,1)$.

and

$$X_2 = \sin(2\pi U_2)\sqrt{-2\ln(U_1)}. \tag{B.4}$$

B.4 MULTIVARIATE NORMAL DISTRIBUTION

A vector $X = (X_1, X_2, \ldots, X_k)'$ has a multivariate normal (MVN) distribution if its joint distribution function is

$$\phi_k(x; \mu, \Sigma) = \frac{1}{(2\pi)^{(k/2)}|\Sigma|^{(1/2)}} \exp\left(-\frac{1}{2}(x-\mu)'\Sigma^{-1}(x-\mu)\right);$$

we write $X \sim N_k(\mu, \Sigma)$. Here, Σ is a $k \times k$ symmetric matrix of rank $q \leq k$, the variance–covariance matrix of X; μ is the vector of means, with ith component equal to $E(X_i)$. The bivariate normal distribution is the MVN with $k = 2$ (Fig. B.2).

Suppose $X \sim N_k(\mu, \Sigma)$, and Σ is of full rank. Let A denote a $p \times k$ matrix of constants, of rank $p \leq k$, and let c be a $p \times 1$ vector of constants. Then,

$$Y = AX + c \sim N_p\left(A\mu + c, A\Sigma A'\right).$$

This says that linear combinations of MVN random variables have a MVN distribution. In particular, if $X_1 \sim N(\mu_1, \sigma_1^2)$ is independent of $X_2 \sim N(\mu_2, \sigma_2^2)$, and $Y = aX_1 + bX_2$, where a and b are constants, then

$$Y \sim N\left(a\mu_1 + b\mu_2, a^2\sigma_1^2 + b^2\sigma_2^2\right).$$

Suppose that (X, Y) has a bivariate normal distribution, with means μ_X and μ_Y, variances σ_X^2 and σ_Y^2, and correlation ρ. Then, the conditional distribution of Y given $X = x$ is normal

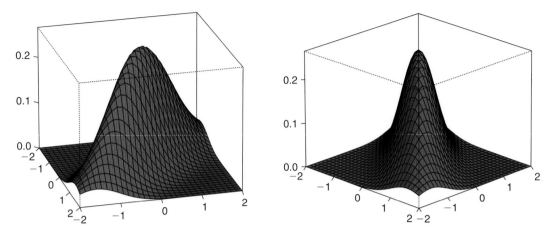

FIGURE B.2   Two views of the bivariate normal distribution with standard normal marginal distributions and correlation $\rho=0.8$.

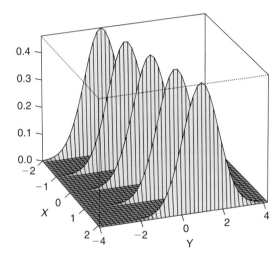

FIGURE B.3   Conditional distributions of Y given $X=-2,-1,0,1,$ and 2, given that X and Y have marginal standard normal distributions and correlation $\rho=0.5$.

with mean

$$\mu_{Y|X=x} = E(Y|X=x) = \mu_Y + \frac{\rho\sigma_Y}{\sigma_X}(x-\mu_X), \tag{B.5}$$

and variance

$$\sigma^2_{Y|X=x} = \text{Var}(Y|X=x) = (1-\rho^2)\sigma_Y^2.$$

An example is presented in Fig. B.3. It can be seen from (B.5) and material presented in Appendix A.3.2 that the conditional expectation of Y given X is the least squares line.

B.5 BERNOULLI TRIALS AND THE BINOMIAL DISTRIBUTION

A random variable with two categorical outcomes is called a Bernoulli trial. One of the outcomes is generically labeled "success"; the probability of success is usually designated as p.

Let X denote the total number of successes in a fixed number n of independent Bernoulli trials with common success parameter $p \in [0, 1]$. Then X has a binomial distribution with index n and success parameter p; the distribution of X is

$$\Pr(X = k) = \binom{n}{k} p^k (1 - p)^{n-k}, \tag{B.6}$$

$k = 0, 1, 2, \ldots, n$, where

$$\binom{n}{k} = \frac{n!}{k!(n-k)!}.$$

We use $B(k; n, p)$ as a shorthand for the probability mass function (B.6), and write $X \sim B(n, p)$. The mean and variance of the binomial distribution are $\mu = np$ and $\sigma^2 = np(1 - p)$, respectively. The mode of X is the greatest integer less than or equal to $(n+1)p$. If $m = (n+1)p$ is an integer, then both $m - 1$ and m are modes.

Calculation of binomial probabilities could be a headache. For example, the probability of exactly 50 "Heads" in 100 independent tosses of a fair coin is

$$B(50; 100, p) = \frac{100!}{50!50!} \times \left(\frac{1}{2}\right)^{100};$$

$100! = 93326215443944152681699238856266700490715968264381621468592963895217599993229$ $91560894146397615651828625369792082722375825118521091686400000000000000000000000000$, not the sort of number amenable to back-of-the-envelope computations. Calculation of the cumulative binomial distribution would be even worse: to compute $\Pr(X \leq 47)$, for instance, we would need to calculate $B(k; 100, p)$ for all values $k = 0, 1, \ldots, 47$, and then add them up.

Instead, we may approximate the cumulative distribution of a binomial random variable using $\Phi(\cdot)$, the standard normal *cdf*. Substituting $\mu = np$ and $\sigma^2 = np(1 - p)$ in (B.2), we have

$$\Pr(X \leq k) \approx \Phi\left(\frac{k - np}{\sqrt{np(1-p)}}\right). \tag{B.7}$$

This approximation can be improved upon, by adding a "continuity correction." Binomial random variables are integer valued, whereas the normal distribution is continuous. It seems reasonable, therefore, to associate all values in the interval $(k - 1/2, k + 1/2)$ with the outcome $X = k$, and to use $k + 1/2$ in the right-hand side of (B.7), namely,

$$\Pr(X \leq k) \approx \Phi\left(\frac{k + \frac{1}{2} - np}{\sqrt{np(1-p)}}\right). \tag{B.8}$$

The performance of this normal approximation with continuity correction is typically deemed satisfactory if np and $n(1 - p)$ are both greater than 5.

For example, the probability of exactly 50 "Heads" in 100 independent tosses of a fair coin is $B(50;100,p) = 0.07959$. Using (B.8), we have

$$\Pr(X = 50) = \Pr(X \le 50) - \Pr(X \le 49)$$

$$\approx \Phi\left(\frac{50.5 - 50}{5}\right) - \Phi\left(\frac{49.5 - 50}{5}\right) = 0.07966,$$

an error of less than 0.1%. The probability of 47 or fewer "Heads" in 100 flips of a fair coin is, to five decimal place accuracy, 0.30865. Using (B.8), we obtain an estimate of 0.30854, an error of less than 0.04%.

B.6 POISSON DISTRIBUTION

The Poisson distribution is one of the most important distributions in statistics. It is derived as a limit of binomial distributions and is useful for describing many types of count data.

A nonnegative integer-valued random variable X has a Poisson distribution with parameter $\lambda > 0$ if

$$\Pr(X = k) = \frac{\exp(-\lambda)\lambda^k}{k!} \tag{B.9}$$

for $k = 0, 1, 2, \dots$; we write $X \sim P(\lambda)$. The parameter λ is referred to as the rate parameter, for reasons that will become clear subsequently. The mean and variance of the Poisson distribution are equal: $E(X) = \text{Var}(X) = \lambda$. The mode of X is the greatest integer less than or equal to λ; if λ is an integer, then both $\lambda - 1$ and λ are modes.

The Poisson distribution is related to the binomial as follows: define $\lambda = np$, and substitute $p = \lambda/n$ in (B.6). On rearrangement, we have

$$\Pr(X = k) = \frac{n!}{k!(n-k)!}\left(\frac{p}{1-p}\right)^k (1-p)^n$$

$$= \frac{1}{k!}\frac{n!}{(n-k)!}\left(\frac{\lambda}{n-\lambda}\right)^k\left(1 - \frac{\lambda}{n}\right)^n$$

$$= \frac{1}{k!}\left[\frac{n!}{(n-k)!\,(n-\lambda)^k}\right]\lambda^k\left(1 - \frac{\lambda}{n}\right)^n. \tag{B.10}$$

As $n \to \infty$, the term in square brackets approaches 1, and $(1 - \lambda/n)^n \to \exp(-\lambda)$. Thus, (B.10) approaches (B.9) as $n \to \infty$.

This means that if $X \sim B(n,p)$ with n large and p small, then the distribution of X is almost the same as the distribution of a Poisson random variable with rate parameter $\lambda = np$. This approximation is illustrated in Table B.1. Note that since $np = 2$, the normal approximation to the binomial would not be deemed satisfactory.

The Poisson approximation to the binomial distributions leads to the description of Poisson random variables as frequencies of rare events. The Poisson property of having equal mean

TABLE B.1 Binomial probability with $np=2$ converges to P(2) probability as $n \to \infty$.

k	$B(5, 0.4)$	$B(10, 0.2)$	$B(40, 0.05)$	$B(100, 0.02)$	$P(2)$
0	0.078	0.107	0.129	0.133	0.135
1	0.259	0.268	0.271	0.271	0.271
2	0.346	0.302	0.278	0.273	0.271
3	0.230	0.201	0.185	0.182	0.180
4	0.077	0.088	0.090	0.090	0.090
5	0.010	0.026	0.034	0.035	0.036
6	0	0.006	0.010	0.011	0.012
7	0	0.001	0.003	0.003	0.003
8	0	0.000	0.001	0.001	0.001

and variance is seen as a consequence of the approximation: if $np \to \lambda$, and $n \to \infty$, p must go to zero, $1-p$ to 1, and the binomial variance $np(1-p) \to \lambda$. The approximation gives sense to another characteristic of Poisson random variables, that if $X \sim P(\lambda)$ is independent of $Y \sim P(\mu)$, then $Z = X + Y \sim P(\lambda + \mu)$: random variable Z can be thought of as the total number of two sorts of rare events. Finally, this concept provides the background for understanding the Poisson process, a time-indexed stochastic process $\{N(t); t > 0\}$ with the property that increments over disjoint time intervals are independent Poisson random variables, with rates proportional to the lengths of the intervals.

It would be a mistake to think that the Poisson distribution is *only* useful as a description of the frequencies of rare events, or to think that its usefulness was limited by the equality of mean and variance. Poisson distributions are useful for modeling counts; they are the starting point of loglinear modeling, and consequently are as important to discrete data analysis as normal distributions are to traditional linear models. Random effects loglinear models allowing the variance to exceed the mean are increasingly commonplace and easily fit using Bayesian methods.

B.7 MULTINOMIAL DISTRIBUTION

The binomial distribution describes events with two possible outcomes; events with $k \geq 2$ possible outcomes are described by multinomial distributions. These distributions are important in ecological applications, having a central role in band-recovery and other mark-recapture models. In most mark-recapture models, animals are caught, marked, and released at discrete sampling occasions. Each animal in the study has associated with it an encounter history that summarizes capture and recovery events, and there are a finite number of such histories possible during the study. The vector describing the numbers of animals with each distinct history can often be modeled as a multinomial random variable with the parameter vector as a function of demographic parameters of interest and nuisance parameters governing the capture and recovery process.

Suppose that $e_1, e_2, \ldots, e_k$ are mutually exclusive and exhaustive outcomes for a random variable Y, and that $\Pr(Y = e_j) = \pi_j$; the values π_j are nonnegative and sum to 1. The random variable Y is said to have a categorical distribution. A sample $Y_1, Y_2, \ldots, Y_N$ can be summarized by a vector $X = (X_1, X_2, \ldots, X_k)'$, with X_j being the count of $Y_i = j$.

We say then that X has a k-cell multinomial distribution with index N and parameter vector $\boldsymbol{\pi} = (\pi_1, \pi_2, \ldots, \pi_k)'$, and write $X \sim M_k(N, \boldsymbol{\pi})$. The probability of an outcome $X = x = (x_1, x_2, \ldots, x_k)'$ is

$$\Pr(X = x) = \frac{N!}{\prod_{j=1}^{k} x_j!} \prod_{j=1}^{k} \pi_j^{x_j}.$$

The mean value of X is, not surprisingly, $N\boldsymbol{\pi}$. The variance–covariance matrix for X has $-N\pi_i\pi_j$ in row i, column j, for $i \neq j$, and $N\pi_i(1 - \pi_i)$ as its ith diagonal element.

Multinomial modeling involves specification of a parametric model for $\boldsymbol{\pi}$ in terms of observed covariates z; we assume that $\boldsymbol{\pi} = \boldsymbol{\pi}(\boldsymbol{\theta}; z)$.

Multinomial random variables have an important association with Poisson random variables. Suppose that vector $A = (A_1, A_2, \ldots, A_n)'$ is made up of independent Poisson random variables, the ith of which has mean μ_i. Conditioning on their sum $T = \sum_i A_i$, the vector A has a multinomial distribution, $A|T \sim M_n(T, \boldsymbol{\pi})$, with

$$\pi_i = \frac{\mu_i}{\sum_j \mu_j}. \tag{B.11}$$

In particular, suppose that $\mu_i = \lambda q(\boldsymbol{\theta}; z_i)$; here, λ is a baseline rate, and $q(\boldsymbol{\theta}; z)$ is a function describing the effects of covariates on the rate. Index counts of animal populations are often described this way, with λ reflecting baseline abundance and detection rates, $q(\boldsymbol{\theta}; z)$ describing population trend and observer effects. Conditioning on the total has the effect of removing parameter λ from the model: it cancels out of the numerator and denominator of the right-hand side of Eq. (B.11).

Multinomial Factorization Theorem

Suppose that of the outcomes $e_1, \ldots, e_k$, our interest is restricted to a subset $\mathcal{S} = \{e_j\}_{j \in \mathcal{K}}$ where $\mathcal{K} \subseteq \{1, 2, \ldots, k\}$. Let $X_{\mathcal{S}}$ denote the corresponding subset of X. The joint probability of the elements of $X_{\mathcal{S}}$, conditional on their total, is also multinomial:

$$\Pr(X_{\mathcal{S}} = x_{\mathcal{S}} \mid N_{\mathcal{S}}) = \frac{N_{\mathcal{S}}!}{\prod_{j \in \mathcal{K}} x_j!} \prod_{j \in \mathcal{K}} \gamma_j^{x_j},$$

where

$$N_{\mathcal{S}} = \sum_{j \in \mathcal{K}} X_j$$

and

$$\gamma_j = \frac{\pi_j}{\sum_{h \in \mathcal{K}} \pi_h}, \quad j \in \mathcal{K}.$$

Example

Suppose we have observations of a categorical random variable; with outcomes $\{1,2,3,4,5\}$ and outcome probabilities $\pi_1,\ldots,\pi_5$, which we roll N times. If we restrict our interest only to those outcomes that were less than 4, the joint probability for $X_S = (X_1, X_2, X_3)'$, given $X_1 + X_2 + X_3 = T$, is

$$\Pr(X_S = (x_1, x_2, x_3)' \mid T) = \frac{T!}{x_1! x_2! x_3!} \prod_{j=1}^{3} \left(\frac{\pi_j}{\pi_1 + \pi_2 + \pi_3} \right)^{x_j}.$$

B.8 EXPONENTIAL DISTRIBUTION

We say that X is exponential with parameter $\lambda > 0$, and write $X \sim E(\lambda)$, if X has density function

$$f(t) = \lambda \exp(-\lambda t), \quad t > 0. \tag{B.12}$$

The mean and variance of the distribution are $1/\lambda$ and $1/\lambda^2$, respectively. The mode of an exponential random variable X is 0 and the median is $\ln(2)/\lambda$.

The exponential distribution is often described as the waiting time between the rare events of a Poisson process. Suppose that $\{N(t); t > 0\}$ is a Poisson process with rate λ, so that $N(t) \sim P(\lambda t)$. Letting X denote the time until the first event, and $F(t)$ denote the *cdf* of X,

$$\Pr(X > t) = 1 - F(t) = \Pr(N(t) = 0) = \frac{(\lambda t)^0 \exp(-\lambda t)}{0!} = \exp(-\lambda t)$$

so that

$$F(t) = 1 - \exp(-\lambda t). \tag{B.13}$$

Differentiating both sides of (B.13) with respect to t yields (B.12).

Exponential random variables are easily generated, given a uniform random number generator: if $U \sim U(0,1)$, then $X = -\log(U)/\lambda \sim E(\lambda)$.

The minimum of several independent exponential random variables itself is an exponential random variable. Suppose that $X_i \sim E(\lambda_i)$, $i = 1, 2, \ldots, n$ are independent. Then,

$$Y = \min\{X_1, X_2, \ldots, X_n\} \sim E(\lambda_1 + \lambda_2 + \cdots + \lambda_n).$$

Furthermore, $\Pr(X_i = \min\{X_1, X_2, \ldots, X_n\}) = \lambda_i / \sum_j \lambda_j$.

Lack of Aging Property

The exponential distribution plays a key role in survival analysis, having a number of remarkable distinguishing features, chief of which is the lack of aging property. Let t and h be any nonnegative numbers, and suppose that X is a random variable with the property that

$$\Pr(X > t + h \mid X > t) = \Pr(X > h). \tag{B.14}$$

The probability of surviving an additional h units of time, given survival through time t, is the same as the probability of surviving the initial h units of time. If human lifetimes followed an exponential distribution, the proportion of 50 year olds surviving to age 55, the proportion of 10 year olds surviving to age 15, and the proportion of newborns surviving to age 5 would all be the same.

It is easily demonstrated that (B.14) is true if and only if $X \sim E(\lambda)$. In terms of the survival function $S(t) = 1 - F(t)$, (B.14) can be written as

$$\frac{S(t+h)}{S(t)} = S(h). \tag{B.15}$$

Multiplying both sides of (B.15) by $S(t)$, then subtracting $S(t)$ from both sides, then dividing by h, we have

$$\frac{S(t+h) - S(t)}{h} = S(t) \left(\frac{S(h) - 1}{h} \right). \tag{B.16}$$

For a nonnegative random variable $S(0) = 1$, so we replace the 1 on the right-hand side of (B.16) with $S(0)$. Then, taking the limits as h approaches zero on both sides of the equation, we obtain

$$S'(t) = S(t)S'(0). \tag{B.17}$$

$S'(0)$ is a constant; call it λ. Thus, (B.17) is seen as a simple differential equation, the unique solution of which is

$$S(t) = \exp(-\lambda t).$$

Thus, the lack of aging property (B.14) corresponds exclusively to the exponential distribution.

Hazard Functions

The lack of aging property is often described in terms of the hazard function or force of mortality function. Consider the probability that an individual, having survived to age t, dies in the next brief interval $(t, t+h]$. Assuming that survival is a continuous random variable, the probability of death in the interval $(t, t+h]$ approaches zero as $h \to 0$; however, we may scale the probability by the length of the interval to obtain an instantaneous rate, which may vary as a function of t. Thus, the hazard function is defined as

$$\lambda(t) = \lim_{h \to 0} \frac{1}{h} \Pr \left(X \in (t, t+h] | X > t \right)$$

$$= \lim_{h \to 0} \frac{F(t+h) - F(t)}{h \, S(t)} = \frac{f(t)}{S(t)}.$$

For the exponential distribution, and the exponential distribution alone, the hazard function is constant, $\lambda(t) \equiv \lambda$: the lack of aging property is equivalent to a constant force of mortality.

The hazard function can be computed from the distribution function and vice versa. Since the hazard function is defined as $\lambda(t) = f(t)/S(t)$,

$$\int_0^x \lambda(t)dt = \int_0^x \frac{f(t)\,dt}{S(t)} = -\log(S(x)).$$

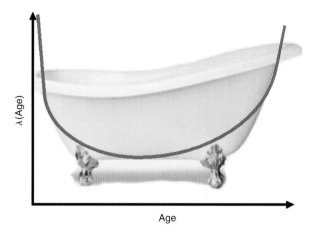

FIGURE B.4   A hazard function.

Thus,

$$S(x) = \exp(-x\Lambda(x)),$$

where

$$\Lambda(x) = \frac{1}{x} \int_0^x \lambda(t) dt$$

is the interval averaged hazard function. The one-to-one correspondence of hazard functions and distribution functions means that characteristics of the survival distribution may be expressed in terms of the hazard function and vice versa. The exponential distribution, with its constant hazard function, is thus a natural baseline for investigating patterns of survival. For instance, human survival is often described as having a "bathtub shaped" hazard function (Fig. B.4): the force of mortality is highest for newborns and elderly, declining in childhood, and increasing in old age.

Weibull Distribution

If $X \sim E(\lambda)$ and $\alpha > 0$ is a constant, then $Y = X^{1/\alpha}$ has a Weibull distribution, $Y \sim W(\lambda, \alpha)$. The hazard function of the Weibull distribution is $\lambda_W(t) = (\alpha\lambda)t^{(\alpha-1)}$, a strictly decreasing function for $\alpha < 1$, and a strictly increasing function for $\alpha > 1$.

B.9 GAMMA DISTRIBUTION

We say that X has a gamma distribution with parameters α and β and write $X \sim Ga(\alpha, \beta)$, if X has density function

$$g(x; \alpha, \beta) = \frac{\beta^\alpha}{\Gamma(\alpha)} x^{\alpha-1} \exp(-\beta x), \quad x > 0,$$

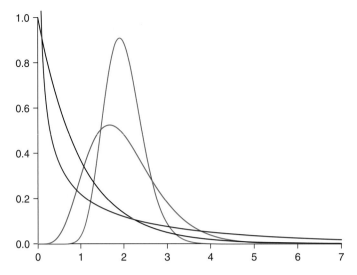

FIGURE B.5   Gamma distributions with mean = 2. Shape parameters $\alpha = 20$ (red), $\alpha = 6$ (green), $\alpha = 1$ (black; exponential distribution), $\alpha = 0.5$ (blue).

for values $\alpha, \beta > 0$. Here, $\Gamma(z)$ is the gamma function, defined for $z > 0$ by

$$\Gamma(z) = \int_0^\infty t^{z-1} \exp(-t)dt.$$

Parameter α is referred to as the *shape* parameter, and β as the *scale* parameter. The mean and variance of a gamma distribution are α/β and α/β^2, respectively. The mode is $(\alpha - 1)/\beta$ for $\alpha \geq 1$, and zero for $\alpha < 1$; the median does not have a simple closed form expression.

If $X \sim Ga(\alpha, \beta)$, then $E(X) = \alpha/\beta$ and $Var(X) = \alpha/\beta^2$. Several gamma distributions are illustrated in Fig. B.5.

For positive integer k, the distribution obtained setting $\alpha = k/2$ and $\beta = 1/2$ is referred to as the chi-squared distribution with k degrees of freedom (*df*), denoted χ_k^2. If $Z_1, Z_2, \ldots, Z_k$ are independent and identically distributed (iid) standard normal random variables, then

$$\sum_{i=1}^k Z_i^2 \sim \chi_k^2. \tag{B.18}$$

The exponential distributions are members of the gamma family, obtained by setting $\alpha = 1$. In particular, the exponential distribution with mean 2 is the same as the chi-squared distribution with 2 *df*.

Suppose that $X \sim Ga(\alpha, \beta)$ is independent of $Y \sim Ga(\gamma, \beta)$, and $a > 0$. Then,

1. if $Z = aX$, then $Z \sim Ga(\alpha, \beta/a)$;
2. if $Z = X + Y$, then $Z \sim Ga(\alpha + \gamma, \beta)$.

The gamma family of distributions is seldom used to model biological data per se. Its primary importance (under the Bayesian paradigm) is as an expression of uncertainty about unknown parameters.

The gamma distribution is also of interest as the distribution of the sample variance. Suppose that $X_1, X_2, \ldots, X_n$ are random variables. The sample mean is defined as

$$\bar{X} = \frac{1}{n} \sum_{i=1}^{n} X_i,$$

and the sample variance as

$$S^2 = \frac{1}{n-1} \sum_{i=1}^{n} (X_i - \bar{X})^2.$$

If the X_i are iid with a $N(\mu, \sigma^2)$ distribution, then $(n-1)S^2/\sigma^2 \sim \chi^2_{n-1}$; this is equivalent to saying that $S^2 \sim Ga\left(\frac{n-1}{2}, \frac{n-1}{2\sigma^2}\right)$.

Wishart Distribution

Equation (B.18), identifying the chi-squared distribution as that of the sum of squared standard normal random variables, has a multivariate extension. Suppose that $X_i, i = 1, 2, \ldots, k$ are iid p-dimensional multivariate normal random variables, $X_i \sim N_p(0, \Sigma)$. Then,

$$W = \sum_{i=1}^{k} X_i X_i'$$

has a *Wishart* distribution with k degrees of freedom and scale matrix Σ. The p-dimensional Wishart distribution has support on the set of p dimensional positive definite matrices, for $p = 1, 2, \ldots$. The expected value of W is $k\Sigma$, the mode of W is $(k-p-1)\Sigma$, and the density function is

$$f(W) \propto |W|^{(k-p-1)/2} \exp\left(-\frac{1}{2} \mathrm{Tr}(\Sigma^{-1} W)\right),$$

where $\mathrm{Tr}(A)$ is the *trace* operator, returning the sum of the diagonal elements of its argument.

B.10 BETA DISTRIBUTION

We say that X has a beta distribution with parameters $a, b > 0$ and write $X \sim Be(a, b)$, if X has density function

$$Be(x; a, b) = \frac{\Gamma(a+b)}{\Gamma(a)\Gamma(b)} x^{a-1} (1-x)^{b-1}, \tag{B.19}$$

$0 < x < 1$. The mean and variance of the beta distribution are

$$\mu = \frac{a}{a+b},$$

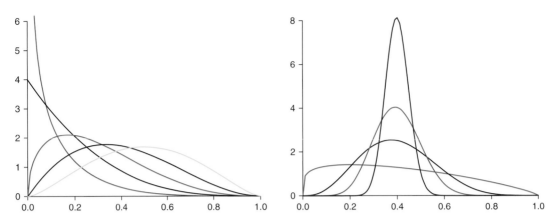

FIGURE B.6   Beta distributions: left panel has $\theta=5, \mu=0.1, 0.2, 0.3, 0.4$, and 0.5 (red, black, green, blue, yellow); right panel has $\mu=0.4, \theta=3, 10, 25, 100$ (green, black, red, blue).

and

$$\sigma^2 = \frac{ab}{(a+b)^2(a+b+1)} = \frac{\mu(1-\mu)}{\theta+1},$$

where $\theta=a+b$. The mode of a beta distribution with $\alpha, \beta \geq 1$ is

$$\frac{\alpha-1}{\alpha+\beta-2}.$$

If $\alpha < 1$ there is a mode at 0, and if $\beta < 1$ there is a mode at 1.

The beta density is usually parameterized as in Eq. (B.19), but the (μ, θ) parameterization is often easier to work with. Note that the variance is a function of the mean; the concentration parameter θ provides a more natural measure of the relative spread in the distribution. The greater the value of $\theta > 0$, the greater the concentration of mass around the mean (Fig. B.6).

The uniform distribution on [0,1] is a member of the beta family, namely $\beta(1,1)$.

Beta random variables can be obtained from gamma random variables: if $X \sim Ga(p, \lambda)$ is independent of $Y \sim Ga(q, \lambda)$, then $Z = X/(X+Y) \sim Be(p,q)$.

B.11 t-DISTRIBUTION

The Student t-distribution with k degrees of freedom (df) is defined as the distribution of the ratio

$$T = \frac{Z}{\sqrt{V/k}},$$

where $Z \sim N(0,1)$ and $V \sim \chi_k^2$ are independent. Its density function is

$$f(t) = \frac{\Gamma((k+1)/2)}{\sqrt{k\pi}\,\Gamma(k/2)}(1+t^2/k)^{-(k+1)/2}, \quad -\infty < t < \infty.$$

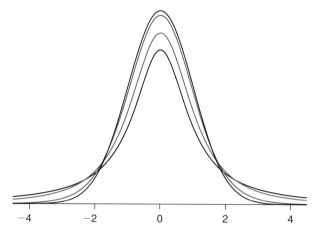

FIGURE B.7 t-distributions with $k = 1, 2$, and 10 degrees of freedom (blue, green, red) and standard normal distribution (black).

The t-distribution has mode and median at zero; its mean is zero if $k > 1$ and undefined otherwise. Its variance is $k/(k-2)$ for $k > 2$, and infinite otherwise. As evident from Fig. B.7, the t-distribution approximates the standard normal for large values of k; for $k > 30$ they are virtually indistinguishable.

The t-distribution is best known for its role in frequentist analysis, in constructing hypothesis tests, and in interval estimates for the mean of a normal distribution with unknown variance. Given a sample $X_i, i = 1, 2, \ldots, n$ with $X_i \sim N(\mu, \sigma^2)$,

$$t = \frac{\bar{X} - \mu}{s/\sqrt{n}}$$

has a t-distribution with $n - 1$ df.

In Bayesian analysis, the t-distribution is the posterior distribution of the mean assuming flat normal and vague gamma priors on μ and σ^2. Although there is usually no a priori reason for choosing a t-distribution in modeling ecological data, it may sometimes be useful as a heavy-tailed alternative to the normal.

B.12 NEGATIVE BINOMIAL DISTRIBUTION

We have noted that the Poisson distribution is often chosen as a starting point for the analysis of count data. The usefulness of the Poisson distribution is limited by the fact that if $X \sim P(\lambda)$ then $E(X) = Var(X)$. Count data are typically overdispersed relative to the Poisson distribution: their variance is larger than their mean.

A solution is to model the mean λ of the Poisson distribution as a random variable. It follows from the double expectation theorems (Appendix A.2.2) that if $[X|\lambda] \sim P(\lambda)$ and $Var(\lambda) > 0$, then $Var(X) > E(X)$.

The negative binomial distribution is a useful alternative to the Poisson distribution, obtained by supposing that $[X|\lambda] \sim P(\lambda)$ with $\lambda \sim Ga(\alpha, \beta)$. We have

$$
\begin{aligned}
\Pr(X=k) &= \int [X|\lambda][\lambda] \, d\lambda \\
&= \int \frac{e^{-\lambda} \lambda^k}{k!} \times \frac{\beta^\alpha}{\Gamma(\alpha)} \lambda^{\alpha-1} e^{-\beta\lambda} \, d\lambda \\
&= \frac{\beta^\alpha}{k! \, \Gamma(\alpha)} \int \lambda^{k+\alpha-1} e^{-(\beta+1)\lambda} \, d\lambda,
\end{aligned}
\tag{B.20}
$$

for $k = 0, 1, \ldots$, $\alpha > 0$; and $\beta > 0$.

Now we "integrate like a statistician." Noting that $a = k + \alpha > 0$ and that $b = \beta + 1 > 0$, we see that the integrand is nearly in the form of a gamma density function; the only problem is that the normalizing constant is missing. So, we multiply inside the integral by the necessary normalizing constant and outside by its inverse. Thus,

$$
\begin{aligned}
\Pr(X=k) &= \frac{\beta^\alpha}{k! \, \Gamma(\alpha)} \frac{\Gamma(a)}{b^a} \int_\lambda \frac{b^a}{\Gamma(a)} \lambda^{a-1} e^{-b\lambda} \, d\lambda \\
&= \frac{\beta^\alpha}{k! \, \Gamma(\alpha)} \frac{\Gamma(a)}{b^a} \times 1.
\end{aligned}
$$

Substituting $k + \alpha$ for a and $1 + \beta$ for b, we obtain the negative binomial distribution function:

$$
\Pr(X=k) = \frac{\Gamma(k+\alpha)}{k! \, \Gamma(\alpha)} \left(\frac{\beta}{1+\beta} \right)^\alpha \left(\frac{1}{1+\beta} \right)^k,
\tag{B.21}
$$

for $k = 0, 1, \ldots$.

The negative binomial distribution has mean $E(X) = \alpha/\beta$ and variance

$$
\mathrm{Var}(X) = \frac{\alpha(1+\beta)}{\beta^2}.
$$

If $\alpha \le 1$ the mode is 0. Otherwise, the mode of X is the greatest integer less than or equal to

$$
m = \frac{\alpha - 1}{\beta}.
$$

If m is an integer, then both $m - 1$ and m are modes. The median does not have a closed form expression.

The negative binomial distribution also arises in describing sequences of Bernoulli trials. Let α be an integer, and let $p = \beta/(1+\beta) \in (0,1)$. Then, the probability of X failures before success number α is given by Eq. (B.21). The geometric distribution is obtained by setting $\alpha = 1$.

Bibliography

Agresti, A. (1994). Simple capture-recapture models permitting unequal catchability and variable sampling effort. *Biometrics* **50**, 494–500.

Agresti, A. (2002). "Categorical Data Analysis." Wiley, Hoboken.

Aitkin, M. (1991). Posterior Bayes factors (with discussion). *J. R. Stat. Soc., Ser. B* **53**, 111–142.

Albert, J. (2007). "Bayesian Computation with R." Springer, New York.

Arnason, A. N. (1972). Parameter estimates from mark-recapture experiments on two populations subject to migration and death. *Res. Popul. Eco.* **13**, 97–113.

Barker, R. J. (1997). Joint modeling of live-recapture, tag-resight, and tag-recovery data. *Biometrics* **53**, 666–677.

Barker, R. J. (2008). Theory and application of mark-recapture and related techniques to aerial surveys of wildlife. *Wildl. Res.* **35**, 268–274.

Barker, R. J., Burnham, K. P., and White, G. C. (2004). Encounter history modeling of joint mark-recapture, tag-resighting and tag-recovery data under temporary emigration. *Stat. Sin.* **14**, 1037–1055.

Berger, J. O., and Pericchi, L. R. (1996). The intrinsic Bayes factor for model selection and prediction. *J. Am. Stat. Assoc.* **91**, 109–122.

Bernardo, J. M. (1979). Reference posterior distributions for Bayesian inference. *J. R. Stat. Soc., Ser. B* **41**, 113–147.

Besag, J. (1986). On the statistical analysis of dirty pictures. *J. R. Stat. Soc., Ser. B* **48**, 259–302.

Bishop, J. A., Cook, L. M., and Muggleton, J. (1978). The response of two species of moths to industrialization in northwest England. II. Relative fitness of morphs and populations size. *Philos. Trans. R. Soc., Lond., Proc. B* **281**, 517–540.

Bonner, S. J., and Schwarz, C. J. (2006). An extension of the Cormack-Jolly-Seber model for continuous covariates with application to *Microtus pennsylvanicus*. *Biometrics* **62**, 142–149.

Borchers, D. L., Buckland, S. T., and Zucchini, W. (2002). "Estimating Animal Abundance: Closed Populations." Springer, London.

Box, G. E. P., and Tiao, G. C. (1973). "Bayesian Inference in Statistical Analysis." Wiley, New York.

Breslow, N. E., and Clayton, D. G. (1993). Approximate inference in generalized linear mixed models. *J. Am. Stat. Assoc.* **88**, 925.

Brewer, K. R. W. (1981). Estimating marihuana usage using randomized response: some paradoxical findings. *Aust. J. Stat.* **23**, 139–148.

Brooks, S. P., and Gelman, A. (1998). General methods for monitoring convergence of iterative simulations. *J. Comput. Graph. Stat.* **7**, 434–455.

Brownie, C., Anderson, D. R., Burnham, K. P., and Robson, D. A. (1985). Statistical inference from band recovery data – a handbook. Technical report, U.S. Department of the Interior, Fish and Wildlife Service Resource Publication No. 156, Washington, DC.

Brownie, C., Hines, J. E., Nichols, J. D., Pollock, K. H., and Hestbeck, J. B. (1993). Capture-recapture studies for multiple strata including non-Markovian transitions. *Biometrics* **49**, 1173–1187.

Buckland, S. T., Anderson, D. R., Burnham, K. P., and Laake, J. L. (1993). "Distance Sampling: Estimating Abundance of Biological Populations." Chapman and Hall, London.

Buckland, S. T., Burnham, K. P., and Augustin, N. H. (1997). Model selection: an integral part of inference. *Biometrics* **53**, 603–618.

Burnham, K. P. (1991). On a unified theory for release-resampling of animal populations. In "Taipei Symposium in Statistics," 11–35. Chao, M. T., and Cheng, P. E. (eds.). Institute of Statistical Science, Academia Sinica. Taipei, Taiwan, R.O.C.

Burnham, K. P. (1993). A theory for combined analysis of ring recovery and recapture data. In "Marked Individuals in Bird Population Studies," 199–213. Lebreton, J.-D., and North, P. (eds.). Birkhauser Verlag, Basel.

Burnham, K. P., and Anderson, D. R. (2002). "Model Selection and Multimodel Inference. A Practical Information-Theoretic Approach," 2nd edition. Springer, New York.

Burnham, K. P., and Anderson, D. R. (2004). Multimodel inference. Understanding AIC and BIC in model selection. *Sociol. Methods Res.* **33**, 261–304.

Cam, E., Link, W. A., Cooch, E. G., Monnat, J.-Y., and Danchin, E. (2002). Individual covariation in life-history traits: seeing the trees despite the forest. *Am. Naturalist* **159**, 96–105.

Caswell, H. (2001). "Matrix Population Models," 2nd edition. Sunderland, Massachusetts, Sinauer Associates.

Chatfield, C. (1995). Model uncertainty, data mining and statistical inference. *J. R. Stat. Soc., Ser. A* **158**, 419–466.

Chen, M.-H., Ibrahim, J. G., and Shao, Q.-M. (2006). Posterior propriety and computation for the Cox regression model with applications to missing covariates. *Biometrika* **93**, 791–807.

Cormack, R. M. (1964). Estimates of survival from the sighting of marked animals. *Biometrika* **51**, 429–438.

Cormack, R. M. (1989). Log-linear models for capture-recapture. *Biometrics* **45**, 395–413.

Coull, B. A., and Agresti, A. (1999). The use of mixed logit models to reflect heterogeneity in capture-recapture studies. *Biometrics* **55**, 294–301.

Cox, D. (1972). Regression models and life-tables. *J. R. Stat. Soc., Ser. B* **34**, 187–220.

Cox, D. R. (1975). Partial likelihood. *Biometrika* **62**, 269–276.

Crosbie, S. F. (1979). The mathematical modelling of capture-mark-recapture experiments on animal populations. Ph.D. thesis, University of Otago, Dunedin, New Zealand.

Crosbie, S. F., and Manly, B. F. J. (1985). Parsimonious modelling of capture-mark-recapture studies. *Biometrics* **41**, 385–398.

Dale, A. I. (1991). "A History of Inverse Probability: From Thomas Bayes to Karl Pearson." Springer, New York.

Darroch, J. N. (1958). The multiple-recapture census. I. Estimation of a closed population. *Biometrika* **45**, 343–359.

Dempster, A. P., Laird, N. M., and Rubin, D. B. (1977). Maximum likelihood from incomplete data via the EM algorithm. *J. R. Stat. Soc., Ser. B* **39**, 1–38.

Dennis, B. (1996). Discussion: should ecologists become Bayesians? *Ecol. Appl.* **6**, 1095–1103.

Dorazio, R. M., and Royle, J. A. (2003). Mixture models for estimating the size of a closed population when capture rates vary among individuals. *Biometrics* **59**, 351–364.

Draper, D. (1995). Inference and hierarchical modeling in the social sciences. *J. Educ. Behav. Stat.* **20**, 115–147.

Dupuis, J. A. (1995). Bayesian estimation of movement and survival probabilities from capture-recapture data. *Biometrika* **82**, 761–772.

Dupuis, J. A., and Schwarz, C. J. (2007). A Bayesian approach to the multi-state Jolly-Seber capture recapture model. *Biometrics* **63**, 1015–1022.

Fienberg, S. E. (1972). The multiple recapture census for closed populations and incomplete 2^k contingency tables. *Biometrika* **59**, 591–603.

Fisher, R. A. (1925a). "Statistical Methods for Research Workers." Oliver and Boyd, Edinburgh.

Fisher, R. A. (1925b). Theory of statistical information. *Proc. Cambridge Philos. Soc.* **22**, 700–725.

Fisher, R. A. (1935). "The Design of Experiments." Oliver and Boyd, Edinburgh.

Flegal, J. M., Haran, M., and Jones, G. L. (2008). Markov chain Monte Carlo: can we trust the third significant figure? *Stat. Sci.* **23**, 250–260.

Galambos, J., and Kotz, S. (1978). "Lecture Notes in Mathematics, volume 675: Characterizations of Probability Distributions." Springer-Verlag, Berlin-Heidelberg-New York.

Gelman, A., and Hill, J. (2007). "Data Analysis Using Regression and Multilevel/Hierarchical Models." Cambridge University Press, Cambridge.

Gelman, A. B., Carlin, J. S., Stern, H. S., and Rubin, D. B. (1995). "Bayesian Data Analysis." Chapman and Hall, Boca Raton.

Gelman, A. B., Carlin, J. S., Stern, H. S., and Rubin, D. B. (2004). "Bayesian Data Analysis," 2nd edition. Chapman and Hall, Boca Raton.

Gelman, A. R. (2006). Prior distributions for variance parameters in hierarchical models (comment on article by Browne and Draper). *Bayesian Anal.* **1**, 515–534.

Geman, S., and Geman, D. (1984). Stochastic relaxation, Gibbs distributions and the Bayesian restoration of images. *IEEE Trans. Pattern Anal. Mach. Intel.* **6**, 721–741.

Good, I. J. (1983). "Good Thinking: The Foundations of Probability and its Applications." University of Minnesota Press, Minneapolis.

Green, P. J. (1995). Reversible jump Markov chain Monte Carlo computation and Bayesian model determination. *Biometrika* **82**, 711–732.

Haldane, J. B. S. (1931). A note on inverse probability. *Proc. Camb. Philos. Soc.* **28**, 55–61.

Haramis, G. M., Link, W. A., Ostenton, P. C., Carter, D. B., Weber, R. G., Teece, M. A., and Mizrahi, D. S. (2007). Stable isotope and pen feeding trial studies confirm the values of horseshoe crab *Limulus polyphemus* eggs to spring migrant shorebirds in Delaware Bay. *J. Avian Biol.* **38**, 367–376.

Hoeting, J. A., Madigan, D., Raftery, A. E., and Volinsky, C. T. (1999). Bayesian model averaging: a tutorial. *Stat. Sci.* **14**, 382–417.

Huggins, R. M. (1989). On the statistical analysis of capture experiments. *Biometrika* **76**, 133–140.

Huggins, R. M. (1991). Some practical aspects of a conditional likelihood approach to capture experiments. *Biometrics* **47**, 725–732.

Huggins, R. M. (2002). A parametric empirical Bayes approach to the analysis of capture-recapture experiments. *Aust. New Zealand J. Stat.* **44**, 55–62.

Jeffreys, H. (1946). An invariant form for the prior probability in estimation problems. *Proc. R. Soc. Lond. A* **186**, 453–461.

Jeffreys, H. (1961). "Theory of Probability," 3rd edition. Oxford University Press, Oxford.

Johnson, D. H. (2008). In defense of indices: the case of bird surveys. *J. Wildlife Manag.* **72**, 857–868.

Johnson, N. L., Kemp, A. W., and Kotz, S. (2005). "Univariate Discrete Distributions," 3rd edition. John Wiley and Sons.

Jolly, G. (1965). Explicit estimates from capture-recapture data with both death and immigration-stochastic model. *Biometrika* **52**, 225–247.

Kass, R. E., and Raftery, A. E. (1995). Bayes factors. *J. Am. Stat. Assoc.* **90**, 773–795.

Kass, R. E., and Wasserman, L. (1996). The selection of prior distributions by formal rules. *J. Am. Stat. Assoc.* **91**, 1343–1370.

Kendall, W., Nichols, J., and Hines, J. (1997). Estimating temporary emigration and breeding proportions using capture-recapture data with pollocks robust design. *Ecology* **78**, 563–578.

Kendall, W. L., and Nichols, J. D. (1995). On the use of secondary capture-recapture samples to estimate temporary emigration and breeding proportions. *J. Appl. Stat.* **22**, 751–762.

King, R., and Brooks, S. P. (2002). Bayesian model discrimination of multiple strata capture-recapture data. *Biometrika* **89**, 785–806.

King, R., and Brooks, S. P. (2008). On the Bayesian estimation of a closed population size in the presence of heterogeneity and model uncertainty. *Biometrics* **64**, 816–824.

Kotz, S., Balakrishnan, N., and Johnson, N. L. (1994). "Continuous Univariate Distributions," Volume 1, 2nd edition. Wiley, New York.

Kotz, S., Balakrishnan, N., and Johnson, N. L. (1995). "Continuous Univariate Distributions," Volume 2, 2nd edition. Wiley, New York.

Kotz, S., Balakrishnan, N., and Johnson, N. L. (1997). "Discrete Multivariate Distributions." Wiley, New York.

Kotz, S., Balakrishnan, N., and Johnson, N. L. (2000). "Continuous Multivariate Distributions, Volume 1: Models and Applications," 2nd edition. Wiley, New York.

Laird, N. M., and Louis, T. A. (1987). Empirical Bayes confidence intervals based on bootstrap samples (with discussion). *J. Am. Stat. Assoc.* **82**, 739–757.

Leamer, E. E. (1978). "Specification Searches." Wiley, New York.

Lewis, J. C., and Morrison, J. A. (1978). Some demographic characteristics of mourning dove populations in western oklahoma. *Proc. Oklahoma Acad. Sci.* **58**, 27–31.

Link, W. A. (2003). Nonidentifiability of population size from capture-recapture data with heterogeneous detection probabilities. *Biometrics* **59**, 1123–1130.

Link, W. A. (2006). Reply to: "On identifiability in capture-recapture models." *Biometrics* **62**, 936–939.

Link, W. A., and Barker, R. J. (2005). Modeling association among demographic parameters in analysis of open population capture-recapture data. *Biometrics* **61**, 46–54.

Link, W. A., and Barker, R. J. (2006). Model weights and the foundations of multi-model inference. *Ecology* **87**, 2626–2635.

Link, W. A., and Barker, R. J. (2008). Efficient implementation of the Metropolis-Hastings algorithm, with application to the Cormack-Jolly-Seber model. *Environ. Ecol. Stat.* **15**, 79–87.

Link, W. A., Cooch, E. G., and Cam, E. (2002). Model-based estimation of individual fitness. *J. Appl. Stat.* **29**, 207–224.

Little, R. (2006). Calibrated Bayes: a Bayes/frequentist roadmap. *Am. Stat.* **60**, 213–223.

Lukacs, P. M., and Burnham, K. P. (2005). Review of capture-recapture methods applicable to noninvasive genetic sampling. *Mol. Ecol.* **14**, 3909–3919.

MacKenzie, D. I., Nichols, J. D., Lachman, G. B., Droege, S., Royle, J. A., and Langtimm, C. (2002). Estimating site occupancy rates when detection probabilities are less than one. *Ecology* **83**, 2248–2255.

MacKenzie, D. I., Nichols, J. D., Pollock, K. H., Bailey, L. L., and Hines, J. E. (2006). "Occupancy Estimation and Model." Academic Press, New York.

Manly, B. F. J., Seyb, A., and Fletcher, D. (1997). Fisheries bycatch monitoring and control. In "Statistics in Ecology and Environmental Monitoring 2: Decision Making and Risk Assessment in Biology," 121–130. Fletcher, D., Kavalieris, L., and Manly, B. F. J. (eds.). Otago University Press.

McCullagh, P., and Nelder, J. A. (1989). "Generalized Linear Models," 2nd edition. Chapman and Hall, London.

McGraw, J. B., and Caswell, H. (1996). Estimation of individual fitness from life-history data. *Am. Naturalist* **147**, 47–64.

Metropolis, N., Rosenbluth, A. W., Rosenbluth, M. N., and Teller, A. H. (1953). Equation of state calculations by fast computing machines. *J. Chem. Phys.* **21**, 1087–1092.

Nichols, J. D. (2005). Modern open population capture-recapture models. In "Handbook of Capture-Recapture Analysis," 88–123. Amstrup, S. C., McDonald, T. L., and Manly, B. F. J. (eds.). Princeton University Press, Princeton.

Nichols, J. D., Kendall, W. L., Hines, J. E., and Spendelow, J. (2004). Estimation of sex-specific survival from capture-recapture data when sex is not always known. *Ecology* **85**, 3192–3201.

Norris, J. L., and Pollock, K. H. (1996). Nonparametric mle under two closed capture-recapture models with heterogeneity. *Biometrics* **52**, 639–649.

Otis, D. L., Burnham, K. P., White, G. C., and Anderson, D. R. (1978). Statistical inference from capture data on closed animal populations. *Wildl. Monogr.* **62**, 1–135.

Pledger, S. (2000). Unified maximum likelihood estimates for closed capture-recapture models using mixtures. *Biometrics* **56**, 434–442.

Pollock, K. H. (1982). A capture-recapture design robust to unequal probability of capture. *J. Wildl. Manag.* **46**, 752–757.

Pollock, K. H., Nichols, J. D., Brownie, C., and Hines, J. E. (1990). Statistical inference for capture recapture experiments. *Wildl. Monogr.* **62**, 1–135.

Pradel, R. (1996). Utilization of capture-mark-recapture for the study of recruitment and population growth rate. *Biometrics* **52**, 703–709.

Pradel, R. (2005). Multievent: an extension of multistate capture-recapture models to uncertain states. *Biometrics* **61**, 442–447.

Raftery, A. E. (1999). Bayes factors and BIC: comment on "A critique of the Bayesian information criterion for model selection." *Sociol. Methods Res.* **27**, 411–427.

Rasch, G. (1961). On general laws and the meaning of measurement in psychology. In *Proceedings of the fourth Berkeley Symposium on Mathematics, Statistics, and Probability, IV.* 321–334. Neyman, J. (ed). University of California Press, Berkeley, CA.

Rattner, B. A., McKernan, M. A., Eisenreich, K. M., Link, W. A., Olsen, G., Hoffman, D. J., Knowles, K. A., and McGowan, P. C. (2006). Toxicity and hazard of vanadium to mallard ducks (*Anas platyrhynchos*) and Canada geese (*Branta canadensis*). *J. Toxicol. Environ. Health Part A* **69**, 331–351.

Robbins, C. S., Bystrak, D., and Geissler, P. (1986). The breeding bird survey: its first fifteen years, 1965–1979. Technical report, U.S. Fish. *Wildl. Serv., Resour. Publ.* 157.

Roberts, G. O. (1996). Markov chain concepts related to sampling algorithms. In "Markov Chain Monte Carlo in Practice," 45–57. Gilks, W. R., Richardson, S., and Spiegelhalter, D. J. (eds.). Chapman and Hall, London.

Royle, J. A., Dorazio, R. M., and Link, W. A. (2007). Analysis of multinomial models with unknown index using data augmentation. *J. Comput. Graph. Stat.* **16**, 1–19.

Rubin, D. B. (1976). Inference and missing data. *Biometrika* **63**, 581–592.

Sauer, J. R., and Link, W. A. (2002). Hierarchical modeling of population stability and species group attributes from survey data. *Ecology* **86(6)**, 1743–1751.

Scheaffer, R. L., Mendenhall, W. I., and Ott, R. L. (1996). "Elementary Survey Sampling," 5th edition. Duxbury Press, Boston.

Schofield, M. R. (2007). Hierarchical capture-recapture models. Ph.D. thesis, University of Otago, Dunedin, New Zealand.

Schofield, M. R., Barker, R. J., and MacKenzie, D. I. (2008). Flexible hierarchical mark-recapture model for open populations using WinBUGS. *Environ. Ecol. Stat.* (In press.)

Schwarz, C. J., and Arnason, A. N. (1996). A general methodology for the analysis of capture-recapture experiments in open populations. *Biometrics* **52**, 860–873.

Schwarz, C. J., Schweigert, J. F., and Arnason, A. (1993). Estimating migration rates using tag-recovery data. *Biometrics* **49**, 177–193.

Schwarz, G. (1978). Estimating the dimension of a model. *Ann. Stat.* **6**, 461–464.

Seber, G. A. F. (1965). A note on the multiple-recapture census. *Biometrika* **52**, 249–259.

Seber, G. A. F. (1970). Estimating time-specific survival and reporting rates for adult birds from band returns. *Biometrika* **57**, 313–318.

Seber, G. A. F. (1982). "Estimating Animal Abundance and related Parameters," 2nd edition. Charles Griffin and Co, London.

Shealer, D. A., and Spendelow, J. A. (2002). Individual foraging strategies of kleptoparasitic roseate terns. *Waterbirds* **25**, 436–441.

Siniff, D. B., and Skoog, R. O. (1964). Aerial censusing of caribou using stratified random sampling. *J. Wildlife Manag.* **28**, 391–401.

Spiegelhalter, D. J., Best, N. G., Carlin, B. P., and van der Linde, A. (2002). Bayesian measures of model complexity and fit. *J. R. Stat. Soc., Ser. B* **64**, 583–639.

Stigler, S. M. (1983). Who discovered Bayes's theorem? *Am. Stat.* **37**, 290–296.

Tjur, T. (1982). A connection between Rasch's item analysis model and a multiplicative Poisson model. *Scand. J. Stat.* **9**, 23–30.

Vaida, F., and Blanchard, S. (2005). Conditional Akaike information for mixed-effects models. *Biometrika* **92**, 351–370.

Walters, M. J. (1992). "A Shadow and a Song." Chelsea Green Publishing Co. Post Mills, Vermont.

Warner, S. (1965). Randomized response: a survey technique for eliminating evasive answer bias. j. *J. Am. Stat. Soc.* **60**, 63–69.

Weakliem, D. L. (1999). A critique of the Bayesian information criterion for model selection. *Sociol. Methods Res.* **27**, 359–397.

White, G. C., and Burnham, K. P. (1999). Program MARK, survival estimation from populations of marked animals. *Bird Study* **46**, S120–S139.

Williams, B. K., Nichols, J. D., and Conroy, M. J. (2002). "Analysis and Management of Animal Populations." Academic Press, New York.

Yang, R., and Berger, J. (1996). A catalog of noninformative priors. Technical report, Institute of Statistics and Decision Science, Duke University, Durham.

Yoshizaki, J. (2007). Use of natural tags in closed population capture-recapture studies: modeling misidentification. Ph.D. thesis, North Carolina State University, Raleigh.

Index of Examples

Deformed New Zealand 50¢ coin 5

Bycatch of New Zealand sealions in the southern squid fishery 8

Roseate tern kleptoparasitism 23, 37, 122

Drift dive counts of trout 49

Fisher's tick data 51, 105

Horseshoe crab isotope study 73

Grassland nesting species in the Breeding Bird Survey 87

Vanadium in ducks and geese 93

Gonodontis mark-recapture 101, 256

Dusky seaside sparrows 111

Bird band returns 141, 244

Returns of tagged Lake Brunner trout 154

Randomized response for Behavior X 165

American toads in Maryland 178

Line transect study of wooden stakes 184

Muskrat house surveys 188, 197

Aerial counts of Alaskan caribou 192

Shearwater burrow occupancy 194

Meadow voles – closed population mark-recapture 206

Snowshoe hare mark-recapture 210

Reid's deer mice study 211

Koala counts modeled using mark-recapture 218

Meadow voles 261, 262, 266

Kittiwake individual fitness simulation 276

Mourning dove band returns 288

Index

A

Absorbing state, 59
AIC, *see* Akaike's information
    criterion
Akaike's information criterion
    (AIC), 129, 150–153, 159–160
Aperiodic Markov chain, 60
Arnason-Schwarz model, 268
Asymptotic normality, finite
    population sampling,
    191–192
Autocorrelation, Markov chains,
    62–64, 70, 72–74
Autoregressive smoothing
    caveats, 296–297
    mourning dove banding analysis
      higher order differences,
        294–296
      modeling differences in
        parameter values, 291–292
      posterior medians, 292–293
      preliminary analysis, 288–291
      shrinkage comparisons,
        292–293
    rationale, 287–288

B

Bayes, Thomas, 4–5, 8
Bayes factor, *see also* Multimodel
    inference
    estimation, 156–157
    likelihood ratio statistics, 134
    multiplier of odds, 135
    relative support measures, 136
    updating, 135–136
    vague prior problems, 137–139

Bayes' theorem
    conditional probability, 8–9
    inference mechanism, 11–12
    joint probability, 10
Bayesian inference
    basics, 36–37
    binomial success rate interval
      estimate
        credible interval, 40–41
        overview, 37–40
        prior knowledge
          incorporation, 41–42
    observed versus unobserved
      quantities, 78
Bayesian information criterion
    (BIC), 129, 148–149
Bayesian māramatanga
    derived parameters, 78–79
    overview, 78
    posterior predictive distribution,
      79–81
Bayesian multimodel inference
    (BMI)
    objections, 130–131
    principles, 129–130
    specified model example,
      131–132
    unknown parameter example,
      132–134
Bayesian *p*-value, 106–107
BBS, *see* North American Breeding
    Bird Survey
Bernoulli trial, 5, 23, 311–312
Beta probability distribution, 21,
    319–320

BIC, *see* Bayesian information
    criterion
Binomial probability distribution
    Bernoulli trials, 311–312
    negative binomial distribution,
      321–322
    overview, 21, 25
    reversible jump Markov chain
      Monte Carlo for model
      comparison, 147–148
BMI, *see* Bayesian multimodel
    inference
Bracket notation, distribution
    functions, 18
Breeding bird survey, *see* North
    American Breeding Bird
    Survey
Brooks–Gelman–Rubin diagnostic,
    Markov chain Monte Carlo
    diagnostics, 71
Brownie models, BUGS code for
    comparison, 142–143
BUGS code
    autoregressive smoothing, 290,
      292, 294
    comparison of models
      Brownie models, 142
      geometric and Poisson
        models, 140
    hidden data models, 170,
      172–173, 175, 180, 185, 187,
      190–191, 193, 195, 199
    individual fitness, 279
    line-transect model, 185, 187
    logit normal model, 70
    mark-recapture models

BUGS code (*continued*)
  closed-population models data input file, 220
    koala data model, 221
    model M_h, 210, 213
    model M_t, 227
  open-population models
    CMSA model, 257–258, 261, 263–264
    Cormack–Jolly–Seber model, 252
    exponential survival model, 242
    North American Breeding Bird Survey summary analysis, 90
  posterior predictive analysis of Fisher's tick data, 105
  trout return data analysis, 156
  vanadium dose-response study, 95–97

C

Canada goose, vanadium poisoning, 93–94
Caribou, finite population sampling, 193
CDF inversion, *see* Cumulative distribution function inversion
CDL, *see* Complete data likelihood
Censoring, *see* Continuous-time survival models
Change of variables theorem, 21, 82
CI, *see* Confidence interval (if you're a frequentist) or Credible interval (if you're a Bayesian)
CJS model, *see* Cormack–Jolly–Seber model
Cluster sampling, finite population sampling, 194–197
CMSA model, *see* Crosbie-Manly-Schwarz-Arnason model
Coherency, Bayesian approach, 8
Coin flip
  bent coin flip outcomes, 5–7, 11–12
  probability space, 15–16
Complete data likelihood (CDL)
  finite population sampling, 188–189

observed data likelihood comparison, 164–165
open population models
  CMSA model, 255–256
  full models, 253, 269–270
posterior distribution calculation in hidden data models with Markov chain Monte Carlo using BUGS
  complete data likelihood, 169–171
  observed data likelihood, 171–172
  partially integrated complete data likelihood, 172–173
Conditional distribution function, 17–18
Conditional expectations
  basic rules, 303
  double expectation theorems, 303–304
Conditional independence, variables, 20
Conditional probability, fishing bycatch rate example, 8–9
Confidence interval (CI)
  binomial success rate
    approximate confidence intervals, 30–32
    exact confidence interval, 32–35
  complications, 35
  coverage properties, 35–36
  overview, 29–30
Conjugacy, *see* Posterior distribution
Conjugate prior, 48
Continuous-time survival models
  hazard function, 240–241
  interval censoring, staggered entry, and known rates, 243–244
  observed data likelihood, 242–243
  right censoring and complete data likelihood, 241–242
Cormack–Jolly–Seber (CJS) model
  moth capture and release study, 101–104
  open model
    applications, 248
    Gibbs sampling algorithm, 250–252

observed data likelihood, 248–250
  parameters and statistics, 249
  overview, 98–99, 104
  uniform prior on γ and τ, 100
  uniform prior on p and ϕ, 101
Correlation, 305
Credible interval, 40–41
Crosbie-Manly-Schwarz-Arnason model, 268
  complete data likelihood, 255–256
  moth coloration study, 256–260
  multiple groups, 260–261
  multistate models, 268–269
  robust design
    overview, 261–262
    temporary emigration, 264–268
    vole examples, 262–264, 266–268
Cumulative distribution function (CDF) inversion, posterior distribution calculation, 55

D

DAG, *see* Directed acyclic graph
Data augmentation, 165
Deer mouse, mark-recapture model example, 211–214
Derived parameters
  Bayesian maramatanga, 78–79
  dose–response study
    derived parameters, 95
    laboratory exposure, 94–95
    mallard drake exposure to sodium metavanadate, 93–94
    out of sample prediction, 95–98
  mean residual lifetime estimation with derived parameter ψ frequentist analysis, 81–82
  posterior distribution calculation via change of variables theorem, 82
  posterior features based on posterior of p, 82–83
  North American Breeding Bird Survey summary analysis

derived parameters, 87–89
grassland bird trend
summaries, 89–92
overview, 85–86, 93
Deviance information criterion
(DIC), 129, 153–154
DIC, *see* Deviance information
criterion
Directed acyclic graph (DAG)
finite population sampling, 189
line-transect model, 184
mark-recapture models, 209
randomized response survey
data, 167, 175
Discrete uniform distribution,
307–308
Distance sampling
BUGS code, 185, 187
directed acyclic graph, 184
line-transect model, 182–186
overview, 181–182
Distribution functions
bracket notation, 18
conditional distribution
function, 17–18
marginal distribution function,
17
overview, 17
Doctrine of chances, 5–8
Double expectation theorems,
303–304
Dove, *see* Mourning dove
Dusky seaside sparrow
extinction, 110–111
prior distribution example
hypergeometric distribution,
111–113
posterior distribution, 113–115

E

Event, probability theory, 15
Exactness, Bayesian approach, 7
Exponential probability
distribution
hazard function, 316–317
lack of aging property, 315–316
overview, 21, 315
Weibull distribution, 317

F

Finite population sampling
auxiliary variables, 197–199
cluster sampling, 194–197

muskrat survey data
asymptotic normality, 191–192
Bayesian analysis, 189
complete data likelihood,
188–189
directed acyclic graph, 189
infinite population inference
comparison, 189–190
overview, 186–187
stratification, 192–194
Fitness, *see* Individual fitness;
Population fitness

G

Gamma probability distribution
overview, 21, 317–319
Wishart distribution, 319
Gibbs sampling algorithm
Cormack–Jolly–Seber model,
250–252
hidden data model coding,
173–174
mark-recapture models
Gibbs sampling
overview, 229–230, 232–233
priors and full conditionals
for p_t and α, 233
priors and full conditionals
for X and N, 233–234
model M_t, 204–206
Markov chain Monte Carlo,
67–70
occupancy models, 177–178
reversible jump Markov chain
Monte Carlo, 146–147

H

Hazard function
continuous-time survival
models, 240–241
exponential probability
distribution, 316–317
Heterogeneity models, *see*
Mark-recapture models
Hidden data models
comparison of models, 199–200
complete data likelihood versus
observed data likelihood,
164–165
distance sampling
BUGS code, 185, 187
directed acyclic graph, 184

line-transect model, 182–186
overview, 181–182
finite population sampling
auxiliary variables, 197–199
cluster sampling, 194–197
muskrat survey data
asymptotic normality,
191–192
Bayesian analysis, 189
complete data likelihood,
188–189
directed acyclic graph, 189
infinite population inference
comparison, 189–190
overview, 186–187
stratification, 192–194
occupancy models
complex models, 180–181
Gibbs sampler, 177–178
overview, 176–177
toad data example, 178–180
randomized response survey
data
directed acyclic graph, 167,
175
overview, 165–168
posterior distribution
calculation with Markov
chain Monte Carlo using
BUGS
complete data likelihood,
169–171
Gibbs sampler coding,
173–174
observed data likelihood,
171–172
partially integrated
complete data likelihood,
172–173
Hierarchical models
overview of features, 163–164
Highest posterior density interval
(HPDI), 40
History plot, Markov chain, 61–62,
71–72
Horseshoe crab, egg consumption
by migratory shorebirds,
73–74
HPDI, *see* Highest posterior
density interval
Hypergeometric distribution,
111–113

I

Improper priors
  binomial success rate, 118–119
  definition, 117–118
  normal mean, 119–120
Independence, predictions,
    304–305
Individual fitness
  definition, 273–274
  example of analysis, 276–281
  group context, 275–276
  known parameters, 281–282
  population distribution
      estimation, 282–285
  realized fitness, 274–275
Inference
  Bayesian interval estimation,
      36–42
  confidence interval, 29–36
  likelihood, 25–29
  point estimation, 43–45
Infinite population inference,
    189–190
Interval estimate, 27

J

Jeffreys prior
  binomial success rate, 121–122
  Fisher information, 120–121
  multivariate parameters, 122
  probability matching, 123–124
Joint probability, definition, 10
Jolly-Seber model, *see also* CMSA
    model
  multistate models, 268–269
  overview, 254–255

K

Kleptoparasites, 23–24
Koala, mark-recapture model
    example, 218–223

L

Latent multinomial models
  model M_t, 226–227
  model $M_{t,\alpha}$
    extensions, 236–237
    ghost records, 226–227
    Gibbs sampling
        overview, 229–230, 232–233
    priors and full conditionals
        for p_t and α, 233

priors and full conditionals
    for X and N, 233–234
  implementation, 234–236
  latent histories, 227–229
  null space, 230–231
  null space basis vectors,
      231–232
  recorded frequency vector, 228
  overview, 225
Likelihood function, 26
Line-transect model, 182–186
Logit survival rate, autoregressive
    smoothing, 291–292
Lower tail probability, 32

M

Marginal distribution function, 17
Markov chain Monte Carlo
    (MCMC), Monte Carlo
    calculation of posterior
    distribution
  autocorrelation, 62–64, 70, 72–74
  BUGS code, 70
  diagnostics, 70–74
  Gibbs sampling algorithm, 67–70
  Metropolis within Gibbs
      algorithm, 69
  Metropolis–Hastings algorithm,
      65–66
  overview, 48, 54–55
  posterior distribution calculation
      in hidden data models
      using BUGS
    complete data likelihood,
        169–171
    Gibbs sampler coding,
        173–174
    observed data likelihood,
        171–172
    partially integrated complete
        data likelihood, 172–173
  reversible jump Markov chain
      Monte Carlo for multimodel
      inference
    binomial model comparison,
        147–148
    Gibbs sampling, 146–147
    overview, 144–146
  stationary distributions, 59–65
Mark-recapture models, *see also*
    Latent multinomial models;
    Open population models

closed-population time and
    behavior models of
    Otis *et al.*
  Gibbs sample for model M_t,
      204–206
  heterogeneity models
    deer mouse example,
        211–214
    loglinear representation,
        215–217
    model M_b, 211
    model M_h, 209–211, 213–214,
        223–224
    Rasch model and M_h, 217–218
  types, 204
  vole example
    goodness of fit, 206, 208–209
    overview, 206–207
  koala example, 218–222
  missing data and data
      completion, 202–203
Maximum likelihood estimator
    (MLE), 26–27, 41, 70,
    166–167
MCMC, *see* Markov chain Monte
    Carlo
Mean residual lifetime, Bayesian
    prediction
  derived parameter ψ
    frequentist analysis, 81–82
    posterior distribution
        calculation via change of
        variables theorem, 82
    posterior features based on
        posterior of p, 82–83
  interval estimate comparison, 84
  overview, 81
  prior sensitivity, 84–85
Mean square error (MSE),
    minimum, 44–45
Median, 19
Metropolis within Gibbs
    algorithm, Markov chain
    Monte Carlo, 69
Metropolis–Hastings algorithm,
    Markov chain Monte Carlo,
    65–66
Minimum mean square error,
    44–45
MLE, *see* Maximum likelihood
    estimator
MMI, *see* Multimodel inference

Moment
  overview, 19–20
  properties, 301–303
Monte Carlo calculations, posterior
    distributions
  cumulative distribution function
      inversion, 55
  Markov chain Monte Carlo
      calculations
    autocorrelation, 62–64, 70,
        72–74
    BUGS code, 70
    diagnostics, 70–74
    Gibbs sampling algorithm,
        67–70
    hidden data model
        calculations using BUGS
      complete data likelihood,
          169–171
      Gibbs sampler coding,
          173–174
      observed data likelihood,
          171–172
      partially integrated
          complete data likelihood,
          172–173
    Metropolis–Hastings
        algorithm, 65–66
    Metropolis within Gibbs
        algorithm, 69
    overview, 48, 54–55
    stationary distributions, 59–65
  overview, 48
  rejection sampling, 56–59
  reversible jump Markov chain
      Monte Carlo for multimodel
      inference
    binomial model comparison,
        147–148
    Gibbs sampling, 146–147
    overview, 144–146
Moth
  CMSA model and coloration
      study, 256–260
  Cormack–Jolly–Seber model and
      capture-release study,
      101–104
Mourning dove, autoregressive
    smoothing of banding data
  higher order differences, 294–296
  modeling differences in
      parameter values, 291–292
  posterior medians, 292–293

  preliminary analysis, 288–291
  shrinkage comparisons, 292–293
MSE, see Mean square error
Multimodel inference (MMI)
  acceptability indices
    Akaike's information
        criterion, 150–153
    deviance information
        criterion, 153–154
    overview, 150
    trout return rate example and
        comparison of information
        criterion, 154–157
  Bayes factor
    estimation, 156–157
    likelihood ratio statistics, 134
    multiplier of odds, 135
    relative support measures, 136
    updating, 135–136
    vague prior problems, 137–139
  Bayesian multimodel inference
    objections, 130–131
    principles, 129–130
    specified model example,
        131–132
    unknown parameter example,
        132–134
  computation
    Bayesian information
        criterion, 148–149
    BUGS
      nested models, 141–144
      nonnested models, 139–141
    reversible jump Markov chain
        Monte Carlo
      binomial model
          comparison, 147–148
      Gibbs sampling, 146–147
      overview, 144–146
  model averaging, 159
  model complexity and Akaike's
      information criterion
      prejudice, 159–160
  overview, 127–129
  posterior Bayes factor, 138,
      150–151, 155
  repeat analyses, 158–159
  use and abuse, 158
Multinomial probability
    distribution, 21, 313–315
Multivariate normal distribution,
    309–310

Muskrat, finite population
    sampling
  asymptotic normality, 191–192
  Bayesian analysis, 189
  complete data likelihood,
      188–189
  directed acyclic graph, 189
  infinite population inference
      comparison, 189–190
Mutual independence, variables,
    20

N

Negative binomial distribution,
    321–322
Newton–Raphson algorithm, 70–71
Normal distribution, 308–309
Normalizing constant, posterior
    distributions, 47–48
Normal probability distribution, 21
North American Breeding Bird
    Survey (BBS), summary
    analysis with derived
    parameters
  comments on usefulness of
      index data, 86–87
  derived parameters, 87–89
  grassland bird trend summaries,
      89–92
  overview, 85–86, 93
Nuisance parameters, 52

O

Observed data likelihood (ODL)
  complete data likelihood
      comparison, 164–165
  posterior distribution calculation
      in hidden data models with
      Markov chain Monte Carlo
      using BUGS, 171–172
Occupancy models
  complex models, 180–181
  Gibbs sampler, 177–178
  overview, 176–177
  toad data example, 178–180
ODL, see Observed data likelihood
Open population models
  CMSA model
    complete data likelihood,
        255–256
    moth coloration study,
        256–260

Open population models
    (*continued*)
    multiple groups, 260–261
    multistate models, 268–269
    robust design
        overview, 261–262
        temporary emigration,
            264–268
        vole examples, 262–264,
            266–268
    complete data likelihood in full
        models, 253, 269–270
    continuous-time survival models
        hazard function, 240–241
        interval censoring, staggered
            entry, and known rates,
            243–244
        observed data likelihood,
            242–243
        right censoring and complete
            data likelihood, 241–242
    Cormack–Jolly–Seber model
        applications, 248
        Gibbs sampling algorithm,
            250–252
        observed data likelihood,
            248–250
        parameters and statistics, 249
    Jolly-Seber model
        multistate models, 268–269
        overview, 254–255
    mark-recapture–band recovery
        models
        complete data array, 245
        constrained models, 247
        dead recovery array, 244
        full conditional distributions
            for latent variables, 246–247
        reporting rate
            reparameterization, 245–246
    overview, 239–240
Outcome, probability theory, 15

                    P

PBF, *see* Posterior Bayes factor
Percentile, 19
Point estimation
    frequentist estimation, 43
    likelihood intervals, 28
    maximum likelihood oddity, 44
    minimum mean square error,
        44–45

Poisson probability distribution,
    21, 51–52, 312–313, 322
Population fitness
    definition, 271
    estimation, 285–286
    matrix for change through time,
        272
    R code for calculation, 273
Positive recurrent Markov chain,
    60
Posterior Bayes factor (PBF), 138,
    150–151, 155
Posterior density, trout counting in
    Hautapu River, 50
Posterior distribution
    conjugacy
        multivariate posterior
            distribution, 52–54
        tick counting on sheep, 51–52
        trout counting in Hautapu
            River, 49–51
    hidden data model calculations
        with Markov chain Monte
        Carlo using BUGS
        complete data likelihood,
            169–171
        Gibbs sampler coding,
            173–174
        observed data likelihood,
            171–172
        partially integrated complete
            data likelihood, 172–173
    mean residual lifetime
        calculation via change of
        variables theorem, 82
    Monte Carlo calculations
        cumulative distribution
            function inversion, 55
        Markov chain Monte Carlo
            calculations
            autocorrelation, 62–64, 70,
                72–74
            BUGS code, 70
            diagnostics, 70–74
            Gibbs sampling algorithm,
                67–70
            Metropolis within Gibbs
                algorithm, 69
            Metropolis–Hastings
                algorithm, 65–66
            overview, 48, 54–55
            stationary distributions,
                59–65

    overview, 48
    rejection sampling, 56–59
    normalizing constant, 47–48
    overview, 37, 47
Posterior predictive distribution
    model checking, 104–107
    overview, 79–81
Price, Richard, 5
Prior distribution
    Bayes factor and vague prior
        problems, 137–139
    Dusky seaside sparrow example
        extinction, 110–111
    hypergeometric distribution,
        111–113
    posterior distribution, 113–115
    objective Bayesian inference
        improper priors
            binomial success rate,
                118–119
            definition, 117–118
            normal mean, 119–120
        Jeffreys prior
            binomial success rate,
                121–122
            Fisher information, 120–121
            multivariate parameters,
                122
            probability matching,
                123–124
        overview, 115–116
        uniform priors and
            transformation invariance,
            116–117
    overview, 36, 109
    selection, 110, 124
Probability, *see also* Conditional
        probability; Joint probability
    Bayesian versus frequentist
        perspective, 7–8, 13
    bent coin flip outcomes, 5–7,
        11–12
    definition, 15
    distribution functions
        bracket notation, 18
        conditional distribution
            function, 17–18
        marginal distribution
            function, 17
        overview, 17
    independence of variables, 20
    moment, 19–20
    percentile, 19

probability distribution types, 20–21, 307–321
transforming variables, 20–21
uncertainty modeling, 13–14
Probability space, 15–16

Q

Quartile, 19

R

Rabbit, finite population sampling, 191
Rasch model, 217–218
Rejection sampling, posterior distribution calculation, 56–59
Reversible jump Markov chain Monte Carlo (RJMCMC), multimodel inference
binomial model comparison, 147–148
Gibbs sampling, 146–147
overview, 144–146
RJMCMC, *see* Reversible jump Markov chain Monte Carlo
Roseate tern, kleptoparasism and gender, 23–24

S

Scaled likelihood, 27
Scaled likelihood interval, 27
Simplicity, Bayesian approach, 7
state space, 59
Stationary distributions, Markov chain Monte Carlo, 59–65
Stratification, finite population sampling, 192–194
Support, distribution functions, 17

T

t-distribution, 321–321
Tern, *see* Roseate tern
Tick, counting on sheep and posterior distribution calculation, 51–52, 105–106
Toad, occupancy models, 178–180
Trout
counting in Hautapu River and posterior distribution calculation, 49–51
return rate example of multimodel inference, 154–157

U

Uniform distribution, 308
Uniform priors, 116–117

Uniform probability distribution, 21
Unknown event, *see* Bernoulli trial
Upper tail probability, 32

V

Vanadium, Bayesian prediction in dose–response study
derived parameters, 95
laboratory exposure, 94–95
mallard drake exposure to sodium metavanadate, 93–94
out of sample prediction, 95–98
Vole
CMSA modeling, 262–264, 266–268
mark-recapture model example
goodness of fit, 206, 208–209
overview, 206–207

W

Weibull distribution, exponential probability distribution, 317
Wishart distribution, gamma probability distribution, 319